Engineering Mechanics: Dynamics

E. W. NELSON, B.S.M.E., M.Adm.E.

Engineering Supervisor, Retired
Western Electric Company

CHARLES L. BEST, B.S.M.E., M.S., Ph.D.

Former Emeritus Professor
Lafayette College

W. G. McLEAN, B.S.E.E., Sc.M., Eng.D.

Former Emeritus Director of Engineering
Lafayette College

MERLE C. POTTER, B.S.M.E., M.S., Ph.D.

Emeritus Professor
Michigan State University

Schaum's Outline Series

New York   Chicago   San Francisco   Lisbon   London
Madrid   Mexico City   Milan   New Delhi   San Juan
Seoul   Singapore   Sydney   Toronto

6 7 8 9 0   CUS CUS   1 9 8 7 6 5

ISBN 978-0-07-171360-3
MHID 0-07-171360-3

This publication is designed to provide accurate and authoritative information in regard to the subject matter covered. It is sold with the understanding that neither the author nor the publisher is engaged in rendering legal, accounting, securities trading, or other professional services. If legal advice or other expert assistance is required, the services of a competent professional person should be sought.

> —*From a Declaration of Principles Jointly Adopted by a Committee of the American Bar Association and a Committee of Publishers and Associations*

Library of Congress Cataloging-in-Publication Data

Nelson, E. W. (Eric William)
    Engineering mechanics dynamics / E.W. Nelson, C.L. Best, W.G. McLean.—6th ed.
        p.   cm.—(Schaum's outlines)
    Rev. ed. of: Schaum's outline of theory and problems of engineering mechanics : statics and dynamics. c1988.
    ISBN-10: 0-07-171360-3 (alk. paper)
    ISBN-13: 978-0-07-171360-3 (alk. paper)   1. Mechanics, Applied—Outlines, syllabi, etc.
2. Dynamics—Outlines, syllabi, etc. I. Best, Charles L. II. McLean, W. G. (William G.)
III. Nelson, E. W. (Eric William). Schaum's outline of theory and problems of engineering mechanics.
IV. Title.
    TA350.M387   2010
    620.1'04—dc22

    2010018805

McGraw-Hill books are available at special quantity discounts to use as premiums and sales promotions or for use in corporate training programs. To contact a representative, please e-mail us at bulksales@mcgraw-hill.com.

Preface

This book is designed to supplement standard texts, primarily to assist students of engineering and science in acquiring a more thorough knowledge and proficiency in dynamics, the course that follows statics in the mechanics sequence. It is based on the authors' conviction that numerous solved problems constitute one of the best means for clarifying and fixing in mind of basic principles. While this book will not mesh precisely with any one text, the authors feel that it can be a very valuable adjunct to all.

The previous editions of this book have been very favorably received. This edition incorporates SI units only. This eliminates the problems encountered when mixing units and allows students to focus on the subject being studied.

The authors attempt to use the best mathematical tools available to students at the sophomore level. Thus the vector approach is applied in those chapters where its techniques provide an elegance and simplicity in theory and problems. On the other hand, we have not hesitated to use scalar methods elsewhere, since they provide entirely adequate solutions to many of the problems. Chapter 1 is a complete review of the minimum number of vector definitions and operations necessary for the entire book, and applications of this introductory chapter are made throughout the book.

Chapter topics correspond to material usually covered in a standard introductory dynamics course. Each chapter begins with statements of pertinent definitions and principles. The text material is followed by graded sets of solved and supplementary problems. The solved problems serve to illustrate and amplify the theory, present methods of analysis, provide practical examples, and bring into sharp focus those fine points that enable the student to apply the basic principles correctly and confidently. Numerous derivations of formulas are included among the solved problems. The many supplementary problems serve as a review of the material covered in each chapter.

In the first edition the authors gratefully acknowledged their indebtedness to Paul B. Eaton and J. Warren Gillon. In the second edition the authors received helpful suggestions and criticism from Charles L. Best and John W. McNabb. Also in that edition Larry Freed and Paul Gary checked the solutions to the problems. In the third and fourth editions, computer solutions were added to numerous problems; these solutions have been eliminated in this sixth edition since several software packages have been developed that allow students to perform such solutions. For the fifth edition the authors thank William Best for checking the solutions to the new problems and reviewing the added new material. For typing the manuscripts of the third and fourth editions we are indebted to Elizabeth Bullock.

E. W. NELSON
C. L. BEST
W. G. MCLEAN
M. C. POTTER

About the Authors

E. W. NELSON graduated from New York University with a B.S.M.E. and an M.Adm.E. He taught mechanical engineering at Lafayette College and later joined the engineering organization of the Western Electric Company (now Lucent Technologies). Retired from Western Electric, he is currently a Fellow of the American Society of Mechanical Engineers. He is a registered Professional Engineer and a member of Tau Beta Pi and Pi Tau Sigma.

CHARLES L. BEST (deceased) was Emeritus Professor of Engineering at Lafayette College. He held a B.S. in mechanical engineering from Princeton, an M.S. in mathematics from Brooklyn Polytechnic Institute, and a Ph.D. in applied mechanics from Virginia Polytechnic Institute. He is coauthor of two books on engineering mechanics and coauthor of another book on FORTRAN programming for engineering students. He was a member of Tau Beta Pi.

W. G. McLEAN (deceased) was Emeritus Director of Engineering at Lafayette College. He held a B.S.E.E. from Lafayette College, an Sc.M. from Brown University, and an honorary Eng.D. from Lafayette College. Professor McLean is the coauthor of two books on engineering mechanics, was past president of the Pennsylvania Society of Professional Engineers, and was active in the codes and standards committees of the American Society of Mechanical Engineers. He was a registered Professional Engineer and a member of Phi Beta Kappa and Tau Beta Pi.

MERLE C. POTTER has engineering degrees from Michigan Technological University and the University of Michigan. He has coauthored *Statics, Strength of Materials, Fluid Mechanics, The Mechanics of Fluids, Thermodynamics for Engineers, Thermal Sciences, Differential Equations, Advanced Engineering Mathematics,* and *Jump Start the HP-48G* in addition to numerous exam review books. His research involved fluid flow stability and energy-related topics. He has received numerous awards, including the ASME's 2008 James Harry Potter Gold Medal. He is Professor Emeritus of Mechanical Engineering at Michigan State University.

Contents

CHAPTER 1

Vectors

1.1 Definitions

Scalar quantities possess only magnitude; examples are time, volume, energy, mass, density, and work. Scalars are added by ordinary algebraic methods, for example, 2 s + 7 s = 9 s and 14 kg − 5 kg = 9 kg.

Vector quantities possess both magnitude and direction;[*] examples are force, displacement, and velocity. A vector is represented by an arrow at the given angle. The head of the arrow indicates the sense, and the length represents the magnitude of the vector. The symbol for a vector is shown in print in boldface type, such as **P**. The magnitude is represented by $|\mathbf{P}|$ or *P*. Often, when writing on paper, we would use $\vec{P}$.

A *free vector* may be moved anywhere in space provided it maintains the same direction and magnitude.

A *sliding vector* may be applied at any point along its line of action. By the *principle of transmissibility*, the external effects of a sliding vector remain the same.

A *bound* or *fixed vector* must remain at the same point of application.

A *unit vector* is a vector one unit in length. It is represented by **i**, **n**, or in written form by $\hat{i}$, $\hat{n}$.

The *negative* of a vector **P** is the vector −**P** that has the same magnitude and angle but is of the opposite sense.

The *resultant* of a system of vectors is the least number of vectors that will replace the given system.

1.2 Addition of Two Vectors

(*a*) The *parallelogram law* states that the resultant **R** of two vectors **P** and **Q** is the diagonal of the parallelogram for which **P** and **Q** are adjacent sides. All three vectors **P**, **Q**, and **R** are concurrent as shown in Fig. 1-1(*a*). **P** and **Q** are also called the components of **R**.

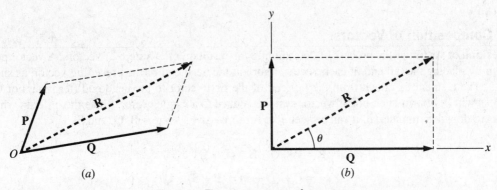

(*a*)   (*b*)

Fig. 1-1   The components of a vector.

---

[*]Direction is understood to include both the angle that the line of action makes with a given reference line and the sense of the vector along the line of action.

(b) If the sides of the parallelogram in Fig. 1-1(a) are perpendicular, the vectors **P** and **Q** are said to be *rectangular* components of the vector **R**. The rectangular components are illustrated in Fig. 1-1(b). The magnitudes of the rectangular components are given by

$$\mathbf{Q} = \mathbf{R}\cos\theta$$

$$\mathbf{P} = \mathbf{R}\cos(90° - \theta) = \mathbf{R}\sin\theta$$

(1)

(c) *Triangle law.* Place the tail end of either vector at the head end of the other. The resultant is drawn from the tail end of the first vector to the head end of the other. The triangle law follows from the parallelogram law because opposite sides of the parallelogram are free vectors, as shown in Fig. 1-2.

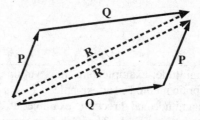

Fig. 1-2 The triangle law.

(d) Vector addition is commutative; that is, **P** + **Q** = **Q** + **P**.

1.3 Subtraction of a Vector

Subtraction of a vector is accomplished by adding the negative of the vector:

$$\mathbf{P} - \mathbf{Q} = \mathbf{P} + (-\mathbf{Q})$$

(2)

Note also that

$$-(\mathbf{P} + \mathbf{Q}) = -\mathbf{P} - \mathbf{Q}$$

1.4 Zero Vector

A *zero vector* is obtained when a vector is subtracted from itself; that is, **P** − **P** = **0**. This is also called a *null vector*.

1.5 Composition of Vectors

Composition of vectors is the process of determining the resultant of a system of vectors. A vector polygon is drawn by placing the tail end of each vector in turn at the head end of the preceding vector, as shown in Fig. 1-3. The resultant is drawn from the tail end of the first vector to the head end (terminus) of the last vector. As will be shown later, not all vector systems reduce to a single vector. Since the order in which the vectors are drawn is immaterial, it can be seen that for three given vectors **P**, **Q**, and **S**,

$$\mathbf{R} = \mathbf{P} + \mathbf{Q} + \mathbf{S} = (\mathbf{P} + \mathbf{Q}) + \mathbf{S}$$

$$= \mathbf{P} + (\mathbf{Q} + \mathbf{S}) = (\mathbf{P} + \mathbf{S}) + \mathbf{Q}$$

(3)

Equation (3) may be extended to any number of vectors.

1.6 Multiplication of Vectors by Scalars

(a) The product of vector **P** and scalar *m* is a vector *m***P** whose magnitude is |*m*| times as great as the magnitude of **P** and that is similarly or oppositely directed to **P**, depending on whether *m* is positive or negative.

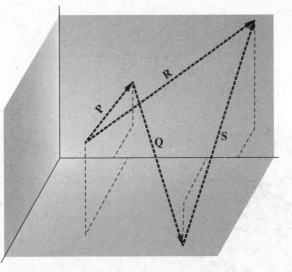

Fig. 1-3 Composition of a vector.

(*b*) Other operations with scalars *m* and *n* are

$$(m + n)\mathbf{P} = m\mathbf{P} + n\mathbf{P}$$

$$m(\mathbf{P} + \mathbf{Q}) = m\mathbf{P} + m\mathbf{Q} \tag{4}$$

$$m(n\mathbf{P}) = n(m\mathbf{P}) = (mn)\mathbf{P}$$

1.7 Orthogonal Triad of Unit Vectors

An *orthogonal triad* of unit vectors **i**, **j**, and **k** is formed by drawing unit vectors along the *x*, *y*, and *z* axes, respectively. A right-handed set of axes is shown in Fig. 1-4.

A vector **P** is written as

$$\mathbf{P} = P_x\mathbf{i} + P_y\mathbf{j} + P_z\mathbf{k} \tag{5}$$

where $P_x\mathbf{i}$, $P_y\mathbf{j}$, and $P_z\mathbf{k}$ are the vector components of **P** along the *x*, *y*, and *z* axes, respectively, as shown in Fig. 1-5.

Note that $P_x = P\cos\theta_x$, $P_y = P\cos\theta_y$, and $P_z = P\cos\theta_z$.

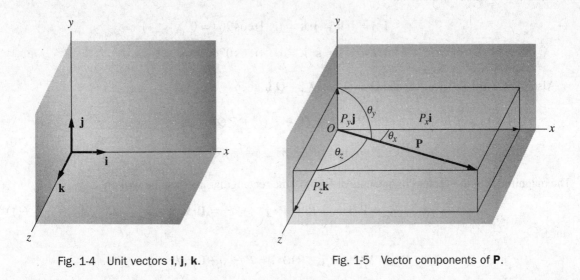

Fig. 1-4 Unit vectors **i**, **j**, **k**.

Fig. 1-5 Vector components of **P**.

1.8 Position Vector

The *position vector* **r** of a point (x, y, z) in space is written

$$\mathbf{r} = x\mathbf{i} + y\mathbf{j} + z\mathbf{k} \tag{6}$$

where $r = \sqrt{x^2 + y^2 + z^2}$ (see Fig. 1-6).

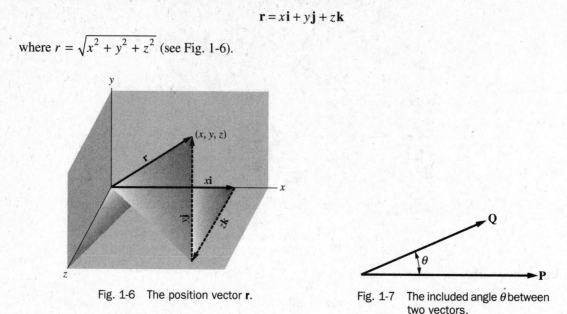

Fig. 1-6   The position vector **r**.

Fig. 1-7   The included angle θ between two vectors.

1.9 Dot or Scalar Product

The *dot* or *scalar product* of two vectors **P** and **Q**, written **P · Q**, is a scalar quantity and is defined as the product of the magnitudes of the two vectors and the cosine of their included angle θ (see Fig. 1-7). Thus,

$$\mathbf{P} \cdot \mathbf{Q} = PQ \cos \theta \tag{7}$$

The following laws hold for dot products, where m is a scalar:

$$\mathbf{P} \cdot \mathbf{Q} = \mathbf{Q} \cdot \mathbf{P}$$

$$\mathbf{P} \cdot (\mathbf{Q} + \mathbf{S}) = \mathbf{P} \cdot \mathbf{Q} + \mathbf{P} \cdot \mathbf{S}$$

$$(\mathbf{P} + \mathbf{Q}) \cdot (\mathbf{S} + \mathbf{T}) = \mathbf{P} \cdot (\mathbf{S} + \mathbf{T}) + \mathbf{Q} \cdot (\mathbf{S} + \mathbf{T}) = \mathbf{P} \cdot \mathbf{S} + \mathbf{P} \cdot \mathbf{T} + \mathbf{Q} \cdot \mathbf{S} + \mathbf{Q} \cdot \mathbf{T} \tag{8}$$

$$m(\mathbf{P} \cdot \mathbf{Q}) = (m\mathbf{P}) \cdot \mathbf{Q} = \mathbf{P} \cdot (m\mathbf{Q})$$

Since **i**, **j**, and **k** are orthogonal,

$$\mathbf{i} \cdot \mathbf{j} = \mathbf{i} \cdot \mathbf{k} = \mathbf{j} \cdot \mathbf{k} = (1)(1)\cos 90° = 0$$

$$\mathbf{i} \cdot \mathbf{i} = \mathbf{j} \cdot \mathbf{j} = \mathbf{k} \cdot \mathbf{k} = (1)(1)\cos 0° = 1 \tag{9}$$

Also, if $\mathbf{P} = P_x\mathbf{i} + P_y\mathbf{j} + P_z\mathbf{k}$ and $\mathbf{Q} = Q_x\mathbf{i} + Q_y\mathbf{j} + Q_z\mathbf{k}$, then

$$\mathbf{P} \cdot \mathbf{Q} = P_x Q_x + P_y Q_y + P_z Q_z$$

$$\mathbf{P} \cdot \mathbf{P} = P^2 = P_x^2 + P_y^2 + P_z^2 \tag{10}$$

The magnitudes of the vector components of **P** along the rectangular axes can be written

$$P_x = \mathbf{P} \cdot \mathbf{i} \qquad P_y = \mathbf{P} \cdot \mathbf{j} \qquad P_z = \mathbf{P} \cdot \mathbf{k} \tag{11}$$

since, e.g.,

$$\mathbf{P} \cdot \mathbf{i} = (P_x\mathbf{i} + P_y\mathbf{j} + P_z\mathbf{k}) \cdot \mathbf{i} = P_x + 0 + 0 = P_x$$

Similarly, the magnitude of the vector component of **P** along any line L can be written $\mathbf{P} \cdot \mathbf{e}_L$, where $\mathbf{e}_L$ is the unit vector along the line L. (Some authors use **u** as the unit vector.) Figure 1-8 shows a plane through the tail end A of vector **P** and a plane through the head B, both planes being perpendicular to line L. The planes intersect line L at points C and D. The vector **CD** is the component of **P** along L, and its magnitude equals $\mathbf{P} \cdot \mathbf{e}_L = P e_L \cos \theta$.

Applications of these principles can be found in Problems 1.15 and 1.16.

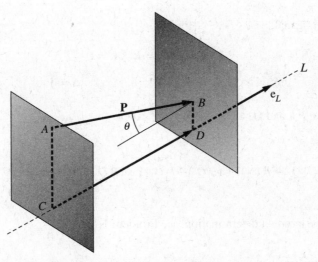

Fig. 1-8   The component of **P** along a line.

1.10   The Cross or Vector Product

The *cross* or *vector product* of two vectors **P** and **Q**, written $\mathbf{P} \times \mathbf{Q}$, is a vector **R** whose magnitude is the product of the magnitudes of the two vectors and the sine of their included angle. The vector $\mathbf{R} = \mathbf{P} \times \mathbf{Q}$ is normal to the plane of **P** and **Q** and points in the direction of advance of a right-handed screw when turned in the direction from **P** to **Q** through the smaller included angle θ. Thus if **e** is the unit vector that gives the direction of $\mathbf{R} = \mathbf{P} \times \mathbf{Q}$, the cross product can be written

$$\mathbf{R} = \mathbf{P} \times \mathbf{Q} = (PQ \sin\theta)\mathbf{e} \qquad 0 \le \theta \le 180° \qquad (12)$$

Figure 1-9 indicates that $\mathbf{P} \times \mathbf{Q} = -\mathbf{Q} \times \mathbf{P}$ (not commutative).

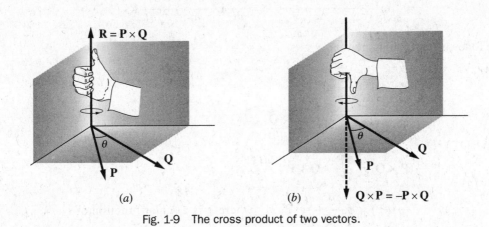

Fig. 1-9   The cross product of two vectors.

The following laws hold for cross products, where m is a scalar:

$$\mathbf{P} \times (\mathbf{Q} + \mathbf{S}) = \mathbf{P} \times \mathbf{Q} + \mathbf{P} \times \mathbf{S}$$

$$(\mathbf{P} + \mathbf{Q}) \times (\mathbf{S} + \mathbf{T}) = \mathbf{P} \times (\mathbf{S} + \mathbf{T}) + \mathbf{Q} \times (\mathbf{S} + \mathbf{T})$$

$$= \mathbf{P} \times \mathbf{S} + \mathbf{P} \times \mathbf{T} + \mathbf{Q} \times \mathbf{S} + \mathbf{Q} \times \mathbf{T} \qquad (13)$$

$$m(\mathbf{P} \times \mathbf{Q}) = (m\mathbf{P}) \times \mathbf{Q} = \mathbf{P} \times (m\mathbf{Q})$$

Since $\mathbf{i}$, $\mathbf{j}$, and $\mathbf{k}$ are orthogonal,

$$\mathbf{i} \times \mathbf{i} = \mathbf{j} \times \mathbf{j} = \mathbf{k} \times \mathbf{k} = 0$$

$$\mathbf{i} \times \mathbf{j} = \mathbf{k} \qquad \mathbf{j} \times \mathbf{k} = \mathbf{i} \qquad \mathbf{k} \times \mathbf{i} = \mathbf{j} \qquad (14)$$

Also, if $\mathbf{P} = P_x\mathbf{i} + P_y\mathbf{j} + P_z\mathbf{k}$ and $\mathbf{Q} = Q_x\mathbf{i} + Q_y\mathbf{j} + Q_z\mathbf{k}$, then

$$\mathbf{P} \times \mathbf{Q} = (P_yQ_z - P_zQ_y)\mathbf{i} + (P_zQ_x - P_xQ_z)\mathbf{j} + (P_xQ_y - P_yQ_x)\mathbf{k} = \begin{vmatrix} \mathbf{i} & \mathbf{j} & \mathbf{k} \\ P_x & P_y & P_z \\ Q_x & Q_y & Q_z \end{vmatrix} \qquad (15)$$

For proof of this cross-product determination, see Problem 1.12.

1.11 Vector Calculus

(*a*) *Differentiation* of a vector $\mathbf{P}$ that varies with respect to a scalar quantity such as time t is performed as follows.

Let $\mathbf{P} = \mathbf{P}(t)$; that is, $\mathbf{P}$ is a function of time t. A change $\Delta\mathbf{P}$ in $\mathbf{P}$ as time changes from t to $t + \Delta t$ is

$$\Delta\mathbf{P} = \mathbf{P}(t + \Delta t) - \mathbf{P}(t)$$

Then
$$\frac{d\mathbf{P}}{dt} = \lim_{\Delta t \to 0} \frac{\Delta\mathbf{P}}{\Delta t} = \lim_{\Delta t \to 0} \frac{\mathbf{P}(t + \Delta t) - \mathbf{P}(t)}{\Delta t} \qquad (16)$$

If $\mathbf{P}(t) = P_x\mathbf{i} + P_y\mathbf{j} + P_z\mathbf{k}$, where P_x, P_y, and P_z are functions of time t, we have

$$\frac{d\mathbf{P}}{dt} = \lim_{\Delta t \to 0} \frac{(P_x + \Delta P_x)\mathbf{i} + (P_y + \Delta P_y)\mathbf{j} + (P_z + \Delta P_z)\mathbf{k} - P_x\mathbf{i} - P_y\mathbf{j} - P_z\mathbf{k}}{\Delta t}$$

$$= \lim_{\Delta t \to 0} \frac{\Delta P_x\mathbf{i} + \Delta P_y\mathbf{j} + \Delta P_z\mathbf{k}}{\Delta t} = \frac{dP_x}{dt}\mathbf{i} + \frac{dP_y}{dt}\mathbf{j} + \frac{dP_z}{dt}\mathbf{k} \qquad (17)$$

The following operations are valid:

$$\frac{d}{dt}(\mathbf{P} + \mathbf{Q}) = \frac{d\mathbf{P}}{dt} + \frac{d\mathbf{Q}}{dt}$$

$$\frac{d}{dt}(\mathbf{P} \cdot \mathbf{Q}) = \frac{d\mathbf{P}}{dt} \cdot \mathbf{Q} + \mathbf{P} \cdot \frac{d\mathbf{Q}}{dt}$$

$$\frac{d}{dt}(\mathbf{P} \times \mathbf{Q}) = \frac{d\mathbf{P}}{dt} \times \mathbf{Q} + \mathbf{P} \times \frac{d\mathbf{Q}}{dt} \qquad (18)$$

$$\frac{d}{dt}(\phi\mathbf{P}) = \phi\frac{d\mathbf{P}}{dt} + \frac{d\phi}{dt}\mathbf{P} \qquad \text{where } \phi \text{ is a scalar function of } t$$

(b) *Integration* of a vector **P** that varies with respect to a scalar quantity, such as time t, is performed as follows. Let **P** = **P**(t); that is, **P** is a function of time t. Then

$$\int_{t_0}^{t_1} \mathbf{P}(t)\, dt = \int_{t_0}^{t_1} (P_x \mathbf{i} + P_y \mathbf{j} + P_z \mathbf{k})\, dt$$

$$= \mathbf{i} \int_{t_0}^{t_1} P_x\, dt + \mathbf{j} \int_{t_0}^{t_1} P_y\, dt + \mathbf{k} \int_{t_0}^{t_1} P_z\, dt \tag{19}$$

1.12　Dimensions and Units

In the study of mechanics, the characteristics of a body and its motion can be described in terms of a set of fundamental quantities called *dimensions*. In the United States, engineers have been accustomed to a gravitational system using the dimensions of force, length, and time. Most countries throughout the world use an absolute system in which the selected dimensions are mass, length, and time. There is a growing trend to use this second system in the United States.

Both systems derive from Newton's second law of motion, which is often written as

$$\mathbf{R} = m\mathbf{a} \tag{20}$$

where **R** is the resultant of all forces acting on a particle, **a** is the acceleration of the particle, and m is the constant of proportionality called the *mass*.

The International System (SI)

In the International System (SI),[*] the unit of mass is the kilogram (kg), the unit of length is the meter (m), and the unit of time is the second (s). The unit of force is the newton (N) and is defined as the force that will accelerate a mass of one kilogram one meter per second squared (m/s^2). Thus,

$$1\ \text{N} = (1\ \text{kg})(1\ \text{m/s}^2) = 1\ \text{kg·m/s}^2 \tag{21}$$

A mass of 1 kg falling freely near the surface of the earth has an acceleration of gravity g that varies from place to place. In this book we assume an average value of 9.80 m/s^2. Thus the force of gravity acting on a 1-kg mass becomes

$$W = mg = (1\ \text{kg})(9.80\ \text{m/s}^2) = 9.80\ \text{kg·m/s}^2 = 9.80\ \text{N} \tag{22}$$

Of course, problems in statics involve forces; but, in a problem, a mass given in kilograms is not a force. The gravitational force acting on the mass, referred to as the *weight*, must be used. In all work involving mass, the student must remember to multiply the mass in kilograms by 9.80 m/s^2 to obtain the gravitational force in newtons. A 5-kg mass has a gravitational force of $5 \times 9.8 = 49$ N acting on it.

In solving statics problems, the mass may not be mentioned. It is important to realize that the mass in kilograms is a constant for a given body. On the surface of the moon, this same given mass will have acting on it a force of gravity approximately one-sixth of that on the earth.

The student should also note that, in SI, the millimeter (mm) is the standard linear dimension unit for engineering drawings. Centimeters are tolerated in SI and can be used to avoid the many zeros required when using millimeters. Further, a space should be left between the number and unit symbol, for example, 2.85 mm, not 2.85mm. When using five or more figures, space them in groups of 3 starting at the decimal point as 12 830 000. Do not use commas in SI. A number with four figures can be written without the space unless it is in a column of quantities involving five or more figures.

Tables of SI units, SI prefixes, and conversion factors for the modern metric system (SI) are included in Appendix A. In this sixth edition, all the problems are in SI units.

---

[*]SI is the acronym for Système International d'Unités (modernized international metric system).

SOLVED PROBLEMS

1.1. In a plane, find the resultant of a 300-N force at 30° and a −250-N force at 90°, using the parallelogram method. Refer to Fig. 1-10(*a*). Also, find the angle α between the resultant and the *y* axis. (Angles are always measured counterclockwise from the positive *x* axis.)

SOLUTION

Draw a sketch of the problem, not necessarily to scale. The negative sign indicates that the 250-N force acts along the 90° line downward toward the origin. This is equivalent to a positive 250-N force along the 270° line, according to the principle of transmissibility.

As in Fig. 1-10(*b*), place the tail ends of the two vectors at a common point. Complete the parallelogram. Consider the triangle, one side of which is the *y* axis, in Fig. 1-10(*b*). The sides of this triangle are *R*, 250, and 300. The angle between the 250 and 300 sides is 60°. Applying the law of cosines gives

$$R^2 = 300^2 + 250^2 - 2(300)(250)\cos 60° \qquad \therefore R = 278.3$$

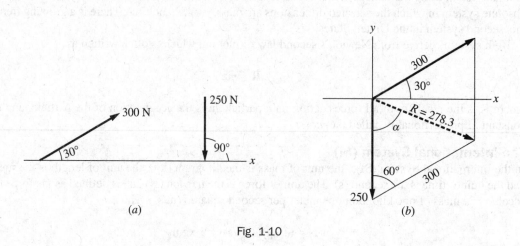

(*a*)   (*b*)

Fig. 1-10

Now applying the law of sines, we get

$$\frac{300}{\sin \alpha} = \frac{278.3}{\sin 60°} \qquad \therefore \alpha = 69°$$

Note: If the forces and angles are drawn to scale, the magnitude of *R* and the angle α could be measured from the drawing.

1.2. Use the triangle law and solve Problem 1.1 (see Fig. 1-11).

SOLUTION

It is immaterial which vector is chosen first. Take the 300-N force. To the head of this vector attach the tail end of the 250-N force. Sketch the resultant from the tail end of the 300-N force to the head end of the 250-N force. Using the triangle shown, the results are the same as in Problem 1.1

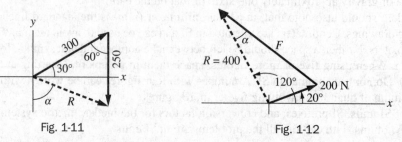

Fig. 1-11   Fig. 1-12

1.3. The resultant of two forces in a plane is 400 N at 120°, as shown in Fig. 1-12. One of the forces is 200 N at 20°. Determine the missing force *F* and the angle α.

SOLUTION

Select a point through which to draw the resultant and the given 200-N force. Draw the force connecting the head ends of the given force and the resultant. This represents the missing force **F**.

The result is obtained by the laws of trigonometry. The angle between **R** and the 200-N force is 100°, and hence, by the law of cosines, the unknown force F is

$$F^2 = 400^2 + 200^2 - 2(400)(200)\cos 100° \qquad \therefore F = 477 \text{ N}$$

Then, by the law of sines, the angle α is found:

$$\frac{477}{\sin 100°} = \frac{200}{\sin \alpha} \qquad \therefore \alpha = 24.4°$$

1.4. In a plane, subtract 130 N at 60° from 280 N at 320° (see Fig. 1-13).

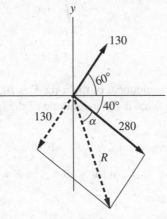

Fig. 1-13

SOLUTION

To the 280-N, 320° force add the negative of the 130-N, 60° force. The resultant is found as follows:

$$R^2 = 280^2 + 130^2 - 2(280)(130)\cos 100° \qquad \therefore R = 329 \text{ N}$$

The law of sines allows us to find α:

$$\frac{329}{\sin 100°} = \frac{130}{\sin \alpha} \qquad \therefore \alpha = 22.9°$$

Thus, R makes an angle of −62.9° with the x axis.

1.5. Determine the resultant of the following coplanar system of forces: 26 N at 10°; 39 N at 114°; 63 N at 183°; 57 N at 261° (see Fig. 1-14).

SOLUTION

This problem can be solved by using the idea of rectangular components. Resolve each force in Fig. 1-14 into x and y components. Since all the x components are collinear, they can be added algebraically, as can the y components. Now, if the x components and y components are added, the two sums form the x and y components of the resultant. Thus,

$$R_x = 26\cos 10° + 39\cos 114° + 63\cos 183° + 57\cos 261° = -62.1$$

$$R_y = 26\sin 10° + 39\sin 114° + 63\sin 183° + 57\sin 261° = -19.5$$

$$R = \sqrt{(-62.1)^2 + (-19.5)^2} \qquad \therefore R = 65 \text{ lb}$$

$$\tan \theta = \frac{-19.5}{-62.1} \qquad \therefore \theta = 17°$$

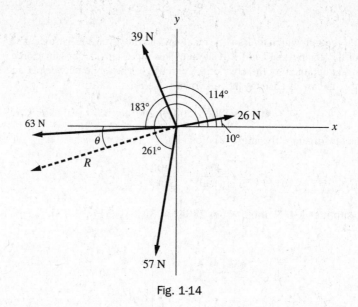

Fig. 1-14

1.6. In Fig. 1-15 the rectangular component of the force **F** is 10 N in the direction of *OH*. The force **F** acts at 60° to the positive *x* axis. What is the magnitude of the force?

SOLUTION

The component of **F** in the direction of *OH* is $F\cos\theta$. Hence,

$$F\cos 15° = 10 \qquad \therefore F = 10.35 \text{ N}$$

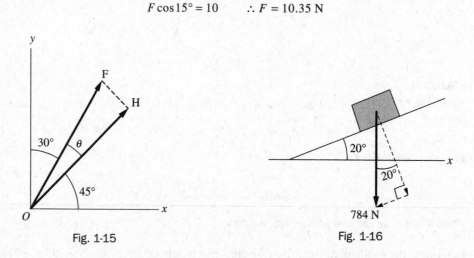

Fig. 1-15   Fig. 1-16

1.7. An 80-kg block is positioned on a board inclined 20° with the horizontal. What is the gravitational component (*a*) normal to the board and (*b*) parallel to the board? See Fig. 1-16.

SOLUTION

(*a*) The normal component is at an angle of 20° with the gravitational force vector (the weight), which has a magnitude of 80(9.8) = 784 N. The normal component is

$$F_\perp = 784\cos 20° = 737 \text{ N}$$

(*b*) The parallel component is

$$F_\parallel = 784\cos 70° = 268 \text{ N}$$

1.8. A force *P* of 235 N acts at an angle of 60° with the horizontal on a block resting on a 22° inclined plane. Determine (*a*) the horizontal and vertical components of *P* and (*b*) the components of *P* perpendicular to and along the plane. Refer to Fig. 1-17(*a*).

SOLUTION

(a) The horizontal component P_h acts to the left and is

$$P_h = 235\cos 60° = 118 \text{ N}$$

The vertical component P_v acts up and is

$$P_v = 235\sin 60° = 204 \text{ N}$$

as shown in Fig. 1-17(b).

(b) The component $P_{\parallel}$ parallel to the plane

$$P_{\parallel} = 235\cos(60° - 22°) = 185 \text{ N}$$

acting up the plane. The component $P_{\perp}$ normal to the plane

$$P_{\perp} = 235\sin 38° = 145 \text{ N}$$

as shown in Fig. 1-17(c).

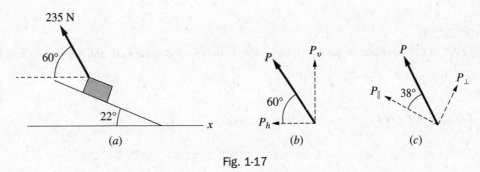

Fig. 1-17

1.9. The three forces shown in Fig. 1-18 produce a resultant force of 20 N acting upward along the y axis. Determine the magnitudes of **F** and **P**.

SOLUTION

For the resultant to be a force of 20 N upward along the y axis, $R_x = 0$ and $R_y = 20$ N. As the sum of the x components must be equal to the x component of the resultant

$$R_x = P\cos 30° - 90\cos 40° = 0 \qquad \therefore P = 79.6 \text{ N}$$

Similarly,

$$R_y = P\sin 30° + 90\sin 40° - F = 20 \qquad \therefore F = 77.7 \text{ N}$$

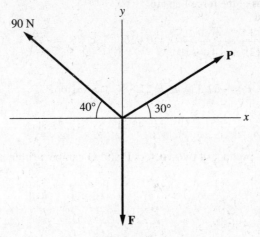

Fig. 1-18

1.10. Refer to Fig. 1-19. The x, y, and z edges of a rectangular parallelepiped are 4, 3, and 2 m, respectively. If the diagonal OP drawn from the origin represents a 50-N force, determine the x, y, and z components of the force. Express the force as a vector in terms of the unit vectors $\mathbf{i}$, $\mathbf{j}$, and $\mathbf{k}$.

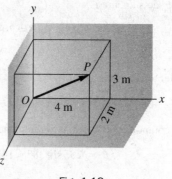

Fig. 1-19

SOLUTION

Let θ_x, θ_y, θ_z represent, respectively, the angles between the diagonal OP and the x, y, z axes. Then

$$P_x = P\cos\theta_x \qquad P_y = P\cos\theta_y \qquad P_z = P\cos\theta_z$$

Length of $OP = \sqrt{4^2 + 3^2 + 2^2} = 5.38$ m. Hence,

$$\cos\theta_x = \frac{4}{5.38} \qquad \cos\theta_y = \frac{3}{5.38} \qquad \cos\theta_z = \frac{2}{5.38}$$

Since each component in the sketch is in the positive direction of the axis along which it acts,

$$P_x = 50\cos\theta_x = 37.2 \text{ N} \qquad P_y = 50\cos\theta_y = 27.9 \text{ N} \qquad P_z = 50\cos\theta_z = 18.6 \text{ N}$$

The vector $\mathbf{P}$ is written as

$$\mathbf{P} = P_x\mathbf{i} + P_y\mathbf{j} + P_z\mathbf{k} = 37.2\mathbf{i} + 27.9\mathbf{j} + 18.6\mathbf{k} \text{ N}$$

1.11. Determine the x, y, and z components of a 100-N force passing from the origin through the point $(2, -4, 1)$. Express the vector in terms of the unit vectors $\mathbf{i}$, $\mathbf{j}$, and $\mathbf{k}$.

SOLUTION

The direction cosines of the force line are

$$\cos\theta_x = \frac{2}{\sqrt{(2)^2 + (-4)^2 + (1)^2}} = 0.436 \qquad \cos\theta_y = \frac{-4}{\sqrt{21}} = -0.873 \qquad \cos\theta_z = 0.218$$

Hence, $P_x = 43.7$ N, $P_y = -87.3$ N, $P_z = 21.8$ N. The vector $\mathbf{P}$ is

$$\mathbf{P} = 43.7\mathbf{i} - 87.3\mathbf{j} + 21.8\mathbf{k} \text{ N}$$

1.12. Show that the cross product of two vectors $\mathbf{P}$ and $\mathbf{Q}$ can be written as

$$\mathbf{P} \times \mathbf{Q} = \begin{vmatrix} \mathbf{i} & \mathbf{j} & \mathbf{k} \\ P_x & P_y & P_z \\ Q_x & Q_y & Q_z \end{vmatrix}$$

SOLUTION

Write the given vectors in component form and expand the cross product to obtain

$$\mathbf{P} \times \mathbf{Q} = (P_x\mathbf{i} + P_y\mathbf{j} + P_z\mathbf{k}) \times (Q_x\mathbf{i} + Q_y\mathbf{j} + Q_z\mathbf{k})$$

$$= (P_xQ_x)\mathbf{i} \times \mathbf{i} + (P_xQ_y)\mathbf{i} \times \mathbf{j} + (P_xQ_z)\mathbf{i} \times \mathbf{k}$$
$$+ (P_yQ_x)\mathbf{j} \times \mathbf{i} + (P_yQ_y)\mathbf{j} \times \mathbf{j} + (P_yQ_z)\mathbf{j} \times \mathbf{k}$$
$$+ (P_zQ_x)\mathbf{k} \times \mathbf{i} + (P_zQ_y)\mathbf{k} \times \mathbf{j} + (P_zQ_z)\mathbf{k} \times \mathbf{k}$$

But $\mathbf{i} \times \mathbf{i} = \mathbf{j} \times \mathbf{j} = \mathbf{k} \times \mathbf{k} = 0$; and $\mathbf{i} \times \mathbf{j} = \mathbf{k}$ and $\mathbf{j} \times \mathbf{i} = -\mathbf{k}$, etc. Hence,

$$\mathbf{P} \times \mathbf{Q} = (P_xQ_y)\mathbf{k} - (P_xQ_z)\mathbf{j} - (P_yQ_x)\mathbf{k} + (P_yQ_z)\mathbf{i} + (P_zQ_x)\mathbf{j} - (P_zQ_y)\mathbf{i}$$

These terms can be grouped as

$$\mathbf{P} \times \mathbf{Q} = (P_yQ_z - P_zQ_y)\mathbf{i} + (P_zQ_x - P_xQ_z)\mathbf{j} + (P_xQ_y - P_yQ_x)\mathbf{k}$$

or in determinant form as

$$\mathbf{P} \times \mathbf{Q} = \begin{vmatrix} \mathbf{i} & \mathbf{j} & \mathbf{k} \\ P_x & P_y & P_z \\ Q_x & Q_y & Q_z \end{vmatrix}$$

Be careful to observe that the scalar components of the first vector **P** in the cross product are written in the middle row of the determinant.

1.13. A force $\mathbf{F} = 2.63\mathbf{i} + 4.28\mathbf{j} - 5.92\mathbf{k}$ N acts through the origin. What is the magnitude of this force and what angles does it make with the x, y, and z axes?

SOLUTION

$$F = \sqrt{(2.63)^2 + (4.28)^2 + (-5.92)^2} = 7.75 \text{ N}$$

$$\cos\theta_x = +\frac{2.63}{7.75} \qquad \theta_x = 70.2°$$

$$\cos\theta_y = +\frac{4.28}{7.75} \qquad \theta_y = 56.3°$$

$$\cos\theta_z = -\frac{5.92}{7.75} \qquad \theta_z = 139.8°$$

1.14. Find the dot product of $\mathbf{P} = 4.82\mathbf{i} - 2.33\mathbf{j} + 5.47\mathbf{k}$ N and $\mathbf{Q} = -2.81\mathbf{i} - 6.09\mathbf{j} + 1.12\mathbf{k}$ m.

SOLUTION

$$\mathbf{P} \cdot \mathbf{Q} = P_xQ_x + P_yQ_y + P_zQ_z = (4.82)(-2.81) + (-2.33)(-6.09) + (5.47)(1.12) = 6.72 \text{ N·m}$$

1.15. Determine the unit vector $\mathbf{e}_L$ for a line L that originates at point $(2, 3, 0)$ and passes through point $(-2, 4, 6)$. Next determine the projection of the vector $\mathbf{P} = 2\mathbf{i} + 3\mathbf{j} - \mathbf{k}$ along the line L.

SOLUTION

The line L changes from $+2$ to -2 in the x direction, or a change of -4. The change in the y direction is $4 - 3 = 1$. The change in the z direction is $6 - 0 = 6$. The unit vector is

$$\mathbf{e}_L = \frac{-4\mathbf{i} + \mathbf{j} + 6\mathbf{k}}{\sqrt{(-4)^2 + 1^2 + 6^2}} = -0.549\mathbf{i} + 0.137\mathbf{j} + 0.823\mathbf{k}$$

The projection of **P** is then

$$\mathbf{P}\cdot\mathbf{e}_L = 2(-0.549) + 3(0.137) - 1(0.823) = -1.41$$

1.16. Determine the projection of the force $\mathbf{P} = 10\mathbf{i} - 8\mathbf{j} + 14\mathbf{k}$ N on the directed line L which originates at point $(2, -5, 3)$ and passes through point $(5, 2, -4)$.

SOLUTION

The unit vector along L is

$$\mathbf{e}_L = \frac{(5-2)\mathbf{i} + [2-(-5)]\mathbf{j} + (-4-3)\mathbf{k}}{\sqrt{3^2 + 7^2 + (-7)^2}}$$

$$= 0.290\mathbf{i} + 0.677\mathbf{j} - 0.677\mathbf{k}$$

The projection of **P** on L is

$$\mathbf{P}\cdot\mathbf{e}_L = (10\mathbf{i} - 8\mathbf{j} + 14\mathbf{k})\cdot(0.29\mathbf{i} + 0.677\mathbf{j} - 0.677\mathbf{k})$$

$$= 2.90 - 5.42 - 9.48 = -12.0 \text{ N}$$

The minus sign indicates that the projection is directed opposite to the direction of L.

1.17. Find the cross product of $\mathbf{P} = 2.85\mathbf{i} + 4.67\mathbf{j} - 8.09\mathbf{k}$ m and $\mathbf{Q} = 28.3\mathbf{i} + 44.6\mathbf{j} + 53.3\mathbf{k}$ N.

SOLUTION

$$\mathbf{P}\times\mathbf{Q} = \begin{vmatrix} \mathbf{i} & \mathbf{j} & \mathbf{k} \\ P_x & P_y & P_z \\ Q_x & Q_y & Q_z \end{vmatrix} = \begin{vmatrix} \mathbf{i} & \mathbf{j} & \mathbf{k} \\ 2.85 & 4.67 & -8.09 \\ 28.3 & 44.6 & 53.3 \end{vmatrix}$$

$$= \mathbf{i}[(4.67)(53.3) - (44.6)(-8.09)] - \mathbf{j}[(2.85)(53.3) - (28.3)(-8.09)]$$

$$+ \mathbf{k}[(2.85)(44.6) - (28.3)(4.67)]$$

$$= \mathbf{i}(249 + 361) - \mathbf{j}(152 + 229) + \mathbf{k}(127 - 132) = 610\mathbf{i} - 381\mathbf{j} - 5\mathbf{k} \text{ N·m}$$

1.18. Determine the time derivative of the position vector $\mathbf{r} = x\mathbf{i} + 6y^2\mathbf{j} - 3z\mathbf{k}$, where $\mathbf{i}, \mathbf{j}, \mathbf{k}$ are fixed vectors.

SOLUTION

The time derivative is

$$\frac{d\mathbf{r}}{dt} = \frac{dx}{dt}\mathbf{i} + 12y\frac{dy}{dt}\mathbf{j} - 3\frac{dz}{dt}\mathbf{k}$$

1.19. Determine the time integral from time $t_1 = 1$ s to time $t_2 = 3$ s of the velocity vector

$$\mathbf{v} = t^2\mathbf{i} + 2t\mathbf{j} - \mathbf{k} \text{ m/s}$$

where $\mathbf{i}, \mathbf{j}$, and $\mathbf{k}$ are fixed vectors.

SOLUTION

$$\int_1^3 (t^2\mathbf{i} + 2t\mathbf{j} - \mathbf{k})\,dt = \mathbf{i}\int_1^3 t^2\,dt + \mathbf{j}\int_1^3 2t\,dt - \mathbf{k}\int_1^3 dt = 8.67\mathbf{i} + 8.00\mathbf{j} - 2.00\mathbf{k}$$

SUPPLEMENTARY PROBLEMS

1.20. Determine the resultant of the coplanar forces 100 N at 0° and 200 N at 90°.

   Ans. 224 N, $\theta_x = 64°$

1.21. Determine the resultant of the coplanar forces 32 N at 20° and 64 N at 190°.

   Ans. 33.0 N, $\theta_x = 180°$

1.22. Find the resultant of the coplanar forces 80 N at −30° and 60 N at 60°.

   Ans. 100 N, $\theta_x = 6.87°$

1.23. Find the resultant of the concurrent coplanar forces 120 N at 78° and 70 N at 293°.

   Ans. 74.7 N, $\theta_x = 45.2°$

1.24. The resultant of two coplanar forces is 18 N at 30°. If one of the forces is 28 N at 0°, determine the other.

   Ans. 15.3 N, 144°

1.25. The resultant of two coplanar forces is 36 N at 45°. If one of the forces is 24 N at 0°, find the other force.

   Ans. 25.5 N, 87°

1.26. The resultant of two coplanar forces is 50 N at 143°. One of the forces is 120 N at 238°. Determine the missing force.

   Ans. 134 N, $\theta_x = 79.6°$

1.27. The resultant of two forces, one in the positive *x* direction and the other in the positive *y* direction, is 100 N at 50° counterclockwise from the positive *x* direction. What are the two forces?

   Ans. $R_x = 64.3$ N, $R_y = 76.6$ N

1.28. A force of 120 N has a rectangular component of 84 N acting along a line making an angle of 20° counterclockwise from the positive *x* axis. What angle does the 120-N force make with the positive *x* axis?

   Ans. 65.6°

1.29. Determine the resultant of the coplanar forces: 6 N at 38°; 12 N at 73°; 18 N at 67°; 24 N at 131°.

   Ans. 50.0 N, $\theta_x = 91°$

1.30. Determine the resultant of the coplanar forces: 20 N at 0°; 20 N at 30°; 20 N at 60°; 20 N at 90°; 20 N at 120°; 20 N at 150°.

   Ans. 77.2 N, $\theta_x = 75°$

1.31. Determine the single force that will replace the following coplanar forces: 120 N at 30°; 200 N at 110°; 340 N at 180°; 170 N at 240°; 80 N at 300°.

   Ans. 351 N, 175°

1.32. Find the single force to replace the following coplanar forces: 150 N at 78°; 320 N at 143°; 485 N at 249°; 98 N at 305°; 251 N at 84°.

   Ans. 321 N, 171°

1.33. A sled is being pulled by a force of 100 N exerted in a rope inclined 30° with the horizontal. What is the effective component of the force pulling the sled? What is the component tending to lift the sled vertically?

   Ans. $P_h = 86.6$ N, $P_v = 50$ N

1.34. Determine the resultant of the following coplanar forces: 15 N at 30°; 55 N at 80°; 90 N at 210°; 130 N at 260°.

Ans.   136 N, $\theta_x = 235°$

1.35. A car is traveling at a constant speed in a tunnel, up a 1 percent grade. If the car and passenger weigh 12.4 kN, what tractive force must the engine supply to just overcome the component of the gravitational force on the car along the bottom of the tunnel?

Ans.   124 N

1.36. A telephone pole is supported by a guy wire that exerts a pull of 800 N on the top of the pole. If the angle between the wire and the pole is 50°, what are the horizontal and vertical components of the pull on the pole?

Ans. $P_h = 613$ N, $P_v = 514$ N

1.37. A boat is being towed through a canal by a horizontal cable that makes an angle of 10° with the shore. If the pull on the cable is 200 N, find the force tending to move the boat along the canal.

Ans.   197 N

1.38. Express in terms of the unit vectors **i**, **j**, and **k** the force of 200 N that starts at the point (2, 5, −3) and passes through the point (−3, 2, 1).

Ans. $\mathbf{F} = -141\mathbf{i} - 84.9\mathbf{j} + 113\mathbf{k}$ N

1.39. Determine the resultant of the three forces $\mathbf{F}_1 = 2.0\mathbf{i} + 3.3\mathbf{j} - 2.6\mathbf{k}$ N, $\mathbf{F}_2 = -\mathbf{i} + 5.2\mathbf{j} - 2.9\mathbf{k}$ N, and $\mathbf{F}_3 = 8.3\mathbf{i} - 6.6\mathbf{j} + 5.8\mathbf{k}$ N, which are concurrent at the point (2, 2, −5).

Ans. $\mathbf{R} = 9.3\mathbf{i} + 1.9\mathbf{j} + 0.3\mathbf{k}$ N at (2, 2, −5)

1.40. The pulley shown in Fig. 1-20 is free to ride on the supporting guide wire. If the pulley supports a 160-N weight, what is the tension in the wire?

Ans. $T = 234$ N

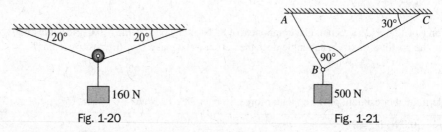

Fig. 1-20   Fig. 1-21

1.41. Two cables support a 500-N weight as shown in Fig. 1-21. Determine the tension in each cable.

Ans. $T_{AB} = 433$ N, $T_{BC} = 250$ N

1.42. What horizontal force P is required to hold the 10-N weight W in the position shown in Fig. 1-22?

Ans. $P = 3.25$ N

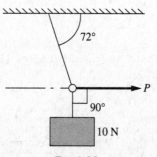

Fig. 1-22

1.43. A charged particle is at rest under the action of three other charged particles. The forces exerted by two of the particles are shown in Fig. 1-23. Determine the magnitude and direction of the third force.

Ans. $F = 0.147$ N, $\theta_x = 76.8°$

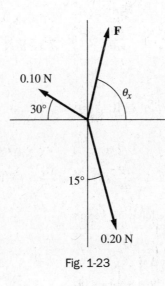

Fig. 1-23

1.44. Determine the resultant of the coplanar forces 200 N at 0° and 400 N at 90°.

Ans. 448 N, $\theta_x = 64°$

(Since each force in Problem 1.20 has been multiplied by the scalar 2, the magnitude of the resultant in this problem should be double that of Problem 1.20. The angle should be the same.)

1.45. What vector must be added to the vector $\mathbf{F} = 30$ N, 60° to yield the zero vector?

Ans. 30 N, $\theta_x = 240°$

1.46. At time $t = 2$ s, a point moving on a curve has coordinates $(3, -5, 2)$. At time $t = 3$ s, the coordinates of the point are $(1, -2, 0)$. What is the change in the position vector?

Ans. $\Delta \mathbf{r} = -2\mathbf{i} + 3\mathbf{j} - 2\mathbf{k}$

1.47. Determine the dot product of $\mathbf{P} = 4\mathbf{i} + 2\mathbf{j} - \mathbf{k}$ and $\mathbf{Q} = -3\mathbf{i} + 6\mathbf{j} - 2\mathbf{k}$.

Ans. $+2$

1.48. Find the dot product of $\mathbf{P} = 2.12\mathbf{i} + 8.15\mathbf{j} - 4.28\mathbf{k}$ N and $\mathbf{Q} = 6.29\mathbf{i} - 8.93\mathbf{j} - 10.5\mathbf{k}$ m.

Ans. -14.5 N·m

1.49. Determine the cross product of the vectors in Problem 1.47.

Ans. $\mathbf{P} \times \mathbf{Q} = 2\mathbf{i} + 11\mathbf{j} + 30\mathbf{k}$

1.50. Determine the cross product of $\mathbf{P} = 2.12\mathbf{i} + 8.15\mathbf{j} - 4.28\mathbf{k}$ and $\mathbf{Q} = 2.29\mathbf{i} - 8.93\mathbf{j} - 10.5\mathbf{k}$.

Ans. $-124\mathbf{i} + 12.5\mathbf{j} - 37.6\mathbf{k}$

1.51. Determine the derivative with respect to time of $\mathbf{P} = x\mathbf{i} + 2y\mathbf{i} - z^2\mathbf{k}$.

Ans. $\dfrac{d\mathbf{P}}{dt} = \dfrac{dx}{dt}\mathbf{i} + 2\dfrac{dy}{dt}\mathbf{j} - 2z\dfrac{dz}{dt}\mathbf{k}$

1.52. If $\mathbf{P} = 2t\mathbf{i} + 3t^2\mathbf{j} - t\mathbf{k}$ and $\mathbf{Q} = t\mathbf{i} + t^2\mathbf{j} + t^3\mathbf{k}$, show that

$$\frac{d}{dt}(\mathbf{P} \cdot \mathbf{Q}) = 4t + 8t^3$$

Check the result by using

$$\frac{d\mathbf{P}}{dt} \cdot \mathbf{Q} + \mathbf{P} \cdot \frac{d\mathbf{Q}}{dt} = \frac{d}{dt}(\mathbf{P} \cdot \mathbf{Q})$$

1.53. In Problem 1.52 show that

$$\frac{d}{dt}(\mathbf{P} \times \mathbf{Q}) = (15t^4 + 3t^2)\mathbf{i} - (8t^3 + 2t)\mathbf{j} - 3t^2\mathbf{k}$$

Check the result by using

$$\frac{d\mathbf{P}}{dt} \times \mathbf{Q} + \mathbf{P} \times \frac{d\mathbf{Q}}{dt} = \frac{d}{dt}(\mathbf{P} \times \mathbf{Q})$$

1.54. Determine the dot product for the following vectors.

P	**Q**
(*a*) $3\mathbf{i} - 2\mathbf{j} + 8\mathbf{k}$	$-\mathbf{i} - 2\mathbf{j} - 3\mathbf{k}$
(*b*) $0.86\mathbf{i} + 0.29\mathbf{j} - 0.37\mathbf{k}$	$1.29\mathbf{i} - 8.26\mathbf{j} + 4.0\mathbf{k}$
(*c*) $a\mathbf{i} + b\mathbf{j} - c\mathbf{k}$	$d\mathbf{i} - e\mathbf{j} + f\mathbf{k}$

Ans.

−23

−2.77

$ad - be - cf$

1.55. Determine the cross products for the following vectors.

P	**Q**
(*a*) $3\mathbf{i} - 2\mathbf{j} + 8\mathbf{k}$	$-\mathbf{i} - 2\mathbf{j} - 3\mathbf{k}$
(*b*) $0.86\mathbf{i} + 0.29\mathbf{j} - 0.37\mathbf{k}$	$1.29\mathbf{i} - 8.26\mathbf{j} + 4.0\mathbf{k}$
(*c*) $a\mathbf{i} + b\mathbf{j} - c\mathbf{k}$	$d\mathbf{i} - e\mathbf{j} + f\mathbf{k}$

Ans.

$22\mathbf{i} + \mathbf{j} - 8\mathbf{k}$

$-1.90\mathbf{i} - 3.92\mathbf{j} - 7.48\mathbf{k}$

$(bf - ec)\mathbf{i} - (af + cd)\mathbf{j} - (ae + bd)\mathbf{k}$

1.56. Determine the component of the vector $\mathbf{Q} = 10\mathbf{i} - 20\mathbf{j} - 20\mathbf{k}$ along a line drawn from point $(2, 3, -2)$ through the point $(1, 0, 5)$.

Ans.   −11.72

1.57. Determine the component of the vector $\mathbf{P} = 1.52\mathbf{i} - 2.63\mathbf{j} + 0.83\mathbf{k}$ on the line that originates at the point $(2, 3, -2)$ and passes through the point $(1, 0, 5)$.

Ans. $P_L = 1.59$

1.58. Given the vector $\mathbf{P} = \mathbf{i} + P_y\mathbf{j} - 3\mathbf{k}$ and $\mathbf{Q} = 4\mathbf{i} + 3\mathbf{j}$, determine the value of P_y so that the cross product of the two vectors will be $9\mathbf{i} - 12\mathbf{j}$.

Ans. $P_y = 0.75$

1.59. Given the vectors $\mathbf{P} = \mathbf{i} - 3\mathbf{j} + P_z\mathbf{k}$ and $\mathbf{Q} = 4\mathbf{i} - \mathbf{k}$, determine the value of P_z so that the dot product of the two vectors will be 14.

Ans. $P_z = -10$

1.60. Express the vectors shown in Fig. 1-24 in $\mathbf{i}, \mathbf{j}, \mathbf{k}$ notation.

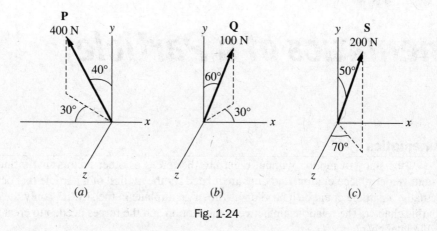

Fig. 1-24

Ans.   (a) $\mathbf{P} = -223\mathbf{i} + 306\mathbf{j} - 129\mathbf{k}$; (b) $\mathbf{Q} = 75\mathbf{i} + 50\mathbf{j} - 43.3\mathbf{k}$; (c) $\mathbf{S} = 144\mathbf{i} + 129\mathbf{j} + 52.4\mathbf{k}$

CHAPTER 2

Kinematics of a Particle

2.1 Kinematics

Kinematics is the study of motion without regard to the forces or other factors that influence the motion. The position, velocity, acceleration, and time are related for the motion of a particle that occupies a point in space. Actually, a particle could be a bead on a wire or an airplane in the sky. It is only the motion that is of interest in this chapter. The relationship between the motion and the forces needed to create the motion will be studied in later chapters.

2.2 Rectilinear Motion

Rectilinear motion is motion of a point P along a straight line, which for convenience here will be chosen as the x axis. Vector symbols are unnecessary in this part.

The *position* of point P at any time t is expressed in terms of its distance x from a fixed origin O on the x axis. This distance x is positive or negative according to the usual sign convention.

The *average velocity* v_{av} of point P during the time interval between t and $t + \Delta t$ during which its position changes from x to $x + \Delta x$ is the quotient $\Delta x / \Delta t$. Mathematically this is

$$v_{av} = \frac{\Delta x}{\Delta t} \tag{1}$$

The *instantaneous velocity* v of point P at time t is the limit of the average velocity as the increment of time approaches zero as a limit. Mathematically this is

$$v = \lim_{\Delta t \to 0} \frac{\Delta x}{\Delta t} = \frac{dx}{dt} \tag{2}$$

The *average acceleration* a_{av} of point P during the time interval between t and $t + \Delta t$ during which its velocity changes from v to $v + \Delta v$ is the quotient $\Delta v / \Delta t$. Mathematically this is

$$a_{av} = \frac{\Delta v}{\Delta t} \tag{3}$$

The *instantaneous acceleration* a of point P at time t is the limit of the average acceleration as the increment of time approaches zero as a limit. Mathematically this is

$$a = \lim_{\Delta t \to 0} \frac{\Delta v}{\Delta t} = \frac{dv}{dt} = \frac{d^2 x}{dt^2} \tag{4}$$

Or

$$a = \frac{dv}{dx}\frac{dx}{dt} = v\frac{dv}{dx} \tag{5}$$

For *constant acceleration* $a = a_0$, the following formulas are valid:

$$v = v_0 + a_0 t \tag{6}$$

$$v^2 = v_0^2 + 2a_0 s \tag{7}$$

$$s = v_0 t + \frac{1}{2} a_0 t^2 \tag{8}$$

$$s = \frac{1}{2}(v + v_0)t \tag{9}$$

where v_0 = initial velocity
v = final velocity
a_0 = constant acceleration
t = time
s = displacement

Simple harmonic motion is rectilinear motion in which the acceleration is negatively proportional to the displacement. Mathematically this is written as

$$a = -k^2 x \tag{10}$$

As an example, Eq. (10) is satisfied by a point vibrating so that its displacement x is given by the equation

$$x = b \sin \omega t \tag{11}$$

where b = amplitude in meters
ω = constant *circular frequency* in radians per second
t = time in seconds

Thus, since $x = b \sin \omega t$, then $v = dx/dt = \omega b \cos \omega t$ and $a = d^2 x/dt^2 = -\omega^2 b \sin \omega t = -\omega^2 x$. That is, $a = -k^2 x$, where $k = \omega$, a constant, and the motion is simple harmonic.

2.3 Curvilinear Motion

Curvilinear motion in a plane is motion along a plane curve (path). The velocity and acceleration of a point on such a curve will be expressed in rectangular components, tangential and normal components, and radial and transverse components.

Rectangular Components

The position vector $\mathbf{r}$ of a point P on such a curve in terms of the unit vectors $\mathbf{i}$ and $\mathbf{j}$ along the x and y axes, respectively, is written

$$\mathbf{r} = x\mathbf{i} + y\mathbf{j} \tag{12}$$

As P moves, $\mathbf{r}$ changes and the velocity $\mathbf{v}$ can be expressed as

$$\mathbf{v} = \frac{d\mathbf{r}}{dt} = \frac{dx}{dt}\mathbf{i} + \frac{dy}{dt}\mathbf{j} \tag{13}$$

Using $dx/dt = \dot{x}$ and $dy/dt = \dot{y}$ and $d\mathbf{r}/dt = \dot{\mathbf{r}}$ as convenient symbols, we have

$$\mathbf{v} = \dot{\mathbf{r}} = \dot{x}\mathbf{i} + \dot{y}\mathbf{j} \tag{14}$$

The speed of the point is the magnitude of the velocity $\mathbf{v}$; that is,

$$|\mathbf{v}| = \sqrt{\dot{x}^2 + \dot{y}^2} \tag{15}$$

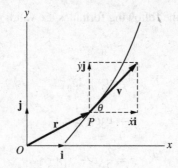

Fig. 2-1   The velocity of a particle.

If θ is the angle that the vector $\mathbf{v}$ makes with the x axis, we can write

$$\tan\theta = \frac{\dot{y}}{\dot{x}} = \frac{dy/dt}{dx/dt} = \frac{dy}{dx} \tag{16}$$

Thus, the velocity vector $\mathbf{v}$ is tangent to the path at point P (see Fig. 2-1). The acceleration vector $\mathbf{a}$ is the time rate of change of $\mathbf{v}$; that is,

$$\mathbf{a} = \frac{d\mathbf{v}}{dt} = \frac{d^2\mathbf{r}}{dt^2} = \frac{d^2x}{dt^2}\mathbf{i} + \frac{d^2y}{dt^2}\mathbf{j} \tag{17}$$

Using the symbolic notation $\mathbf{a} = \dot{\mathbf{v}} = \ddot{\mathbf{r}}$, $\ddot{x} = d^2x/dt^2$, and $\ddot{y} = d^2y/dt^2$, we can write

$$\mathbf{a} = \dot{\mathbf{v}} = \ddot{\mathbf{r}} = \ddot{x}\mathbf{i} + \ddot{y}\mathbf{j} \tag{18}$$

The magnitude of the acceleration vector $\mathbf{a}$ is

$$|\mathbf{a}| = \sqrt{\ddot{x}^2 + \ddot{y}^2} \tag{19}$$

In general, $\mathbf{a}$ is not tangent to the path at point P.

Tangential and Normal Components

In the preceding discussion the velocity vector $\mathbf{v}$ and acceleration vector $\mathbf{a}$ were expressed in terms of the orthogonal unit vectors $\mathbf{i}$ and $\mathbf{j}$ along the x and y axes, respectively. The following discussion shows how to express the same vector $\mathbf{v}$ and the same vector $\mathbf{a}$ in terms of the unit vector $\mathbf{e}_t$ tangent to the path at point P and the unit vector $\mathbf{e}_n$ at right angles to $\mathbf{e}_t$.

In Fig. 2-2, point P is shown on the curve at a distance s along the curve from a reference point P_0. The position vector $\mathbf{r}$ of point P is a function of the scalar quantity s. To study this relationship, let Q be a point on the curve near P.

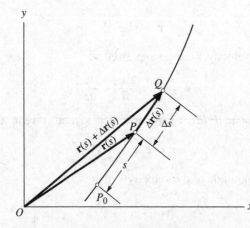

Fig. 2-2   The position vector $\mathbf{r}$.

The position vectors $\mathbf{r}(s)$ and $\mathbf{r}(s) + \Delta\mathbf{r}(s)$ for points P and Q, respectively, are shown as well as the change $\Delta\mathbf{r}(s)$, which is the directed straight line PQ. The distance along the curve from P to Q is Δs. The derivative of $\mathbf{r}(s)$ with respect to s is

$$\frac{d\mathbf{r}(s)}{ds} = \lim_{\Delta s \to 0} \frac{\mathbf{r}(s) + \Delta\mathbf{r}(s) - \mathbf{r}(s)}{\Delta s} = \lim_{\Delta s \to 0} \frac{\Delta\mathbf{r}(s)}{\Delta s} \tag{20}$$

As Q approaches P, the ratio of the magnitude of the straight line $\Delta\mathbf{r}(s)$ to the arc length Δs approaches unity. Also, the straight line $\Delta\mathbf{r}(s)$ approaches the tangent to the path at P. Thus, in the limit, a unit vector $\mathbf{e}_t$ is defined as

$$\frac{d\mathbf{r}(s)}{ds} = \mathbf{e}_t \tag{21}$$

Next consider how $\mathbf{e}_t$ changes with s. As shown in Fig. 2-3(a), the center of curvature C is a distance ρ, the *radius of curvature*, from P. If we assume point Q is close to P, the unit tangent vectors at P and Q are $\mathbf{e}_t$ and $\mathbf{e}_t + \Delta\mathbf{e}_t$, respectively. Since the tangents at P and Q are perpendicular to the radii drawn to C, the angle between $\mathbf{e}_t$ and $\mathbf{e}_t + \Delta\mathbf{e}_t$ as shown in Fig. 2-3(b) is also $\Delta\theta$. Because $\mathbf{e}_t$ and $\mathbf{e}_t + \Delta\mathbf{e}_t$ are unit vectors, $\Delta\mathbf{e}_t$ represents only a change in direction (but not magnitude). Thus the triangle in Fig. 2-3(b) is isosceles and is shown drawn to a larger scale in Fig. 2-3(c). From Fig. 2-3(c) it should be evident that

$$\frac{\left|\frac{1}{2}\Delta\mathbf{e}_t\right|}{1} = \sin\left(\frac{1}{2}\Delta\theta\right) \approx \frac{1}{2}\Delta\theta \qquad \text{from which} \qquad \left|\Delta\mathbf{e}_t\right| \approx \Delta\theta \tag{22}$$

But from Fig. 2-3(a), $\Delta s = \rho\Delta\theta$; hence, we can write $\Delta s \approx \rho\left|\Delta\mathbf{e}_t\right|$. Thus,

$$\lim_{\Delta s \to 0} \frac{\left|\Delta\mathbf{e}_t\right|}{\Delta s} = \frac{1}{\rho} \tag{23}$$

Also, in the limit $\Delta\mathbf{e}_t$ is perpendicular to $\mathbf{e}_t$ and is directed toward the center of curvature C. Let $\mathbf{e}_n$ be the unit vector that is perpendicular to $\mathbf{e}_t$ and directed toward the center of curvature C. Then

$$\frac{d\mathbf{e}_t}{ds} = \lim_{\Delta s \to 0} \frac{\left|\Delta\mathbf{e}_t\right|}{\Delta s}\mathbf{e}_n = \frac{1}{\rho}\mathbf{e}_n \tag{24}$$

The velocity vector $\mathbf{v}$ may now be given in terms of the unit vectors $\mathbf{e}_t$ and $\mathbf{e}_n$. Using Eq. (21) and noting $ds/dt = \dot{s}$ is the speed of P along the path, we can write

$$\mathbf{v} = \frac{d\mathbf{r}}{dt} = \frac{d\mathbf{r}}{ds}\frac{ds}{dt} = \dot{s}\mathbf{e}_t \tag{25}$$

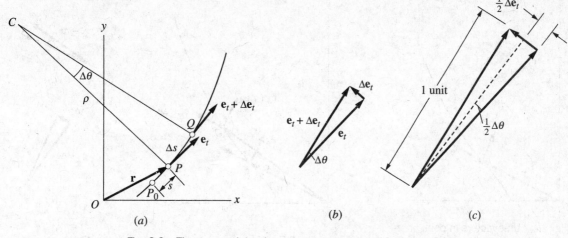

(a)   (b)   (c)

Fig. 2-3   The tangential unit vector $\mathbf{e}_t$ and radius of curvature ρ.

The acceleration vector **a** is the time derivative of the velocity vector **v** defined in Eq. (25):

$$\mathbf{a} = \frac{d\mathbf{v}}{dt} = \ddot{s}\mathbf{e}_t + \dot{s}\frac{d\mathbf{e}_t}{dt} \tag{26}$$

But

$$\frac{d\mathbf{e}_t}{dt} = \frac{d\mathbf{e}_t}{ds}\frac{ds}{dt}$$

and from Eq. (24) this may be written as

$$\frac{d\mathbf{e}_t}{dt} = \frac{\dot{s}}{\rho}\mathbf{e}_n \tag{27}$$

Then

$$\mathbf{a} = \ddot{s}\mathbf{e}_t + \frac{\dot{s}^2}{\rho}\mathbf{e}_n \tag{28}$$

Note that $\ddot{s}$ along the tangent is the time rate of change of the speed of the point.

Radial and Transverse Components

The point P on the curve may be located with polar coordinates in terms of any point chosen as a pole. Figure 2-4 shows the origin O as the pole. Polar coordinates are useful in studying the motion of planets and other central force problems. The velocity vector **v** and the acceleration vector **a** are now derived in terms of unit vectors along and perpendicular to the radius vector. Note that there is an infinite set of unit vectors because any point may be chosen as a pole.

The radius vector **r** makes an angle ϕ with the x axis. The unit vector $\mathbf{e}_r$ is chosen outward along **r**. The unit vector $\mathbf{e}_\phi$ is perpendicular to **r** and in the direction of increasing ϕ.

Since the vector **r** is r units long in the $\mathbf{e}_r$ direction, we can write

$$\mathbf{r} = r\mathbf{e}_r \tag{29}$$

The velocity vector **v** is the time derivative of the product in Eq. (29):

$$\mathbf{v} = \dot{\mathbf{r}} = \dot{r}\mathbf{e}_r + r\dot{\mathbf{e}}_r \tag{30}$$

where $\dot{\mathbf{e}}_r = d\mathbf{e}_r/dt$.

To evaluate $\dot{\mathbf{e}}_r$ and $\dot{\mathbf{e}}_\phi$, allow P to move to a nearby point Q with a corresponding set of unit vectors $\mathbf{e}_r + \Delta\mathbf{e}_r$ and $\mathbf{e}_\phi + \Delta\mathbf{e}_\phi$ as shown in Fig. 2-5(a).

Figure 2-5(b) and (c) illustrate these unit vectors. Since the triangles are isosceles, we can deduce the following conclusions by reasoning similar to that used in the explanation of the $\mathbf{e}_t$ and $\mathbf{e}_n$ vectors: $d\mathbf{e}_r$ in

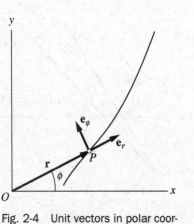

Fig. 2-4   Unit vectors in polar coordinates.

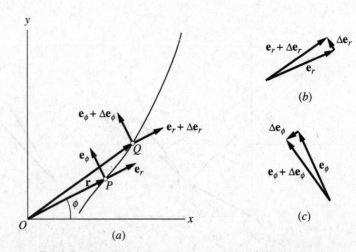

Fig. 2-5   Changes in $\mathbf{e}_r$ and $\mathbf{e}_\phi$ over a small displacement.

the limit has a magnitude $d\phi$ in the $\mathbf{e}_\phi$ direction, and $d\mathbf{e}_\phi$ in the limit has a magnitude $d\phi$ in the negative $\mathbf{e}_r$ direction. Hence,

$$\dot{\mathbf{e}}_r = \frac{d\mathbf{e}_r}{d\phi}\frac{d\phi}{dt} = \dot{\phi}\mathbf{e}_\phi \qquad \text{and} \qquad \dot{\mathbf{e}}_\phi = \frac{d\mathbf{e}_\phi}{d\phi}\frac{d\phi}{dt} = -\dot{\phi}\mathbf{e}_r \tag{31}$$

where $\dot{\phi}$ is the angular velocity ω, the time derivative of the angle ϕ that the radius vector $\mathbf{r}$ makes with the x axis.

The velocity vector $\mathbf{v}$ may now be written

$$\mathbf{v} = \dot{r}\mathbf{e}_r + r\dot{\phi}\mathbf{e}_\phi \tag{32}$$

The acceleration vector $\mathbf{a}$ is the time derivative of the terms in Eq. (32):

$$\mathbf{a} = \ddot{r}\mathbf{e}_r + \dot{r}\dot{\mathbf{e}}_r + \dot{r}\dot{\phi}\mathbf{e}_\phi + r\ddot{\phi}\mathbf{e}_\phi + r\dot{\phi}\dot{\mathbf{e}}_\phi$$

$$= \ddot{r}\mathbf{e}_r + \dot{r}\dot{\phi}\mathbf{e}_\phi + \dot{r}\dot{\phi}\mathbf{e}_\phi + r\ddot{\phi}\mathbf{e}_\phi + r\dot{\phi}^2\mathbf{e}_r \tag{33}$$

where $\ddot{\phi}$ is the angular acceleration α (time derivative of the angular velocity $\dot{\phi}$). Collecting terms, this becomes

$$\mathbf{a} = (\ddot{r} - r\dot{\phi}^2)\mathbf{e}_r + (2\dot{r}\dot{\phi} + r\ddot{\phi})\mathbf{e}_\phi \tag{34}$$

As a special case of curvilinear motion, consider a point moving in a circular path of radius R. Substituting R for r in Eqs. (32) and (33), noting $\dot{R} = \ddot{R} = 0$, we obtain

$$\mathbf{v} = R\dot{\phi}\mathbf{e}_\phi \qquad \text{(tangent to the path)} \tag{35}$$

$$\mathbf{a} = -R\dot{\phi}^2\mathbf{e}_r + R\ddot{\phi}\mathbf{e}_\phi \tag{36}$$

Thus, the acceleration has a tangential component of magnitude $R\ddot{\phi}$ and a normal component directed toward the center of magnitude $R\dot{\phi}^2$.

2.4 Units

Units have been omitted purposely in the foregoing discussion. The following table lists the units used in the SI system and in the U.S. Customary (so-called engineering) System. We will use only SI units in the problems, but because engineering units are not obsolete, they are included in this table. A number of conversions are listed in Appendix A, along with information about the SI system of units.

Symbol	SI Units	Engineering Units
s, ρ, R, x, y	m	ft
$v, \dot{x}, \dot{y}, \dot{s}$	m/s	ft/s or fps
$a, \ddot{x}, \ddot{y}, \ddot{s}$	m/s^2	ft/s^2
θ, ϕ	radians (rad)	radians (rad)
$\omega, \dot{\theta}, \dot{\phi}$	rad/s	rad/s
$\alpha, \ddot{\theta}, \ddot{\phi}$	rad/s^2	rad/s^2

Table 2-1 lists the prefixes that are often used in the SI system of units.

We finish this section with comments on significant figures. In most calculations, a material property is involved. Material properties are seldom known to four significant figures and often only to three. Thus, it is not appropriate to express answers to five or six significant figures. Our calculations are only as accurate as the least accurate number in our equation. For example, we use gravity as 9.80 m/s^2, only three significant figures. A dimension is stated as 10 mm; it is assumed accurate to three and at most four significant figures. It is usually acceptable to express answers using four significant figures, but not five

Table 2-1   Prefixes for SI Units

Multiplication Factor	Prefix	Symbol
10^{12}	tera	T
10^{9}	giga	G
10^{6}	mega	M
10^{3}	kilo	k
10^{-2}	centi*	c
10^{-3}	milli	m
10^{-6}	micro	μ
10^{-9}	nano	n
10^{-12}	pico	p

*Discouraged except in cm, cm^2, or cm^3.

or six. The use of calculators may even provide eight. The engineer does not, in general, work with five or six significant figures. Note that if the leading digit in an answer is 1, it does not count as a significant figure; for example, 12.48 has three significant figures.

SOLVED PROBLEMS

2.1. A rocket car moves along a straight track according to the equation $x = 3t^3 + t + 2$, where x is in meters and t is in seconds. Determine the displacement, velocity, and acceleration when $t = 4$ s.

SOLUTION

$$x = 3t^3 + t + 2 = 3(4)^3 + 4 + 2 = 198 \text{ m}$$

$$v = \frac{dx}{dt} = 9t^2 + 1 = 9(4)^2 + 1 = 145 \text{ m/s}$$

$$a = \frac{dv}{dt} = 18t = 18(4) = 72 \text{ m/s}^2$$

2.2. In Problem 2.1, what is the average acceleration during the fifth second?

SOLUTION

The velocity at the end of the fifth second is $v = 9(5)^2 + 1 = 226$ m/s. Hence, the change in velocity during the fifth second is 226 m/s − 145 m/s = 81 m/s.
   The average acceleration is

$$a_{av} = \frac{\Delta v}{\Delta t} = \frac{81 \text{ m/s}}{1 \text{ s}} = 81 \text{ m/s}^2$$

Or the acceleration after 5 s is $18(5) = 90$ m/s^2. So the average acceleration during the fifth second is

$$a_{av} = \frac{90 + 72}{2} = 81 \text{ m/s}^2$$

2.3. A point moves along a straight line such that its displacement is $s = 8t^2 + 2t$, where s is in meters and t is in seconds. Plot the displacement, velocity, and acceleration against time. These are called $s - t, v - t, a - t$ diagrams.

SOLUTION

Differentiating $s = 8t^2 + 2t$ yields $v = ds/dt = 16t + 2$ and $a = dv/dt = d^2s/dt^2 = 16$.
   This shows that the acceleration is constant, 16 m/s^2.

To determine values for plotting, use the following tabular form, where t is in seconds, s is in meters, and v is in meters per second.

t	t^2	$8t^2$	$2t$	$s = 8t^2 + 2t$	$16t$	$v = 16t + 2$	$a = 16$
0	0	0	0	0	0	2	16
1	1	8	2	10	16	18	16
2	4	32	4	36	32	34	16
3	9	72	6	78	48	50	16
4	16	128	8	136	64	66	16
5	25	200	10	210	80	82	16
10	100	800	20	820	160	162	16

These data are plotted in the s, v, and a diagrams below. Some valuable relationships may be deduced from these diagrams. The slope of the s–t curve at any time t is the height or ordinate of the v–t curve at time t. This follows since $v = ds/dt$.

Again, the slope of the v–t curve (in this particular case the slope is the same at any point of the straight line, that is, 16 m/s^2) at any time t is the ordinate of the a–t curve at any time t. This follows since $a = dv/dt$.

The two equations just given may also be written as

$$a\,dt = dv \qquad \text{and} \qquad v\,dt = ds$$

Integration between proper limits yields

$$\int_{t_0}^{t} a\,dt = \int_{v_0}^{v} dv = v - v_0 \qquad \text{and} \qquad \int_{t_0}^{t} v\,dt = \int_{s_0}^{s} ds = s - s_0 \qquad (1)$$

where $\displaystyle\int_{t_0}^{t} a\,dt$ = area under a–t diagram for time interval from t_0 to t

$\displaystyle\int_{t_0}^{t} v\,dt$ = area under v–t diagram for time interval from t_0 to t

$v - v_0$ = change in velocity in same time interval t_0 to t

$s - s_0$ = change in displacement in same time interval t_0 to t

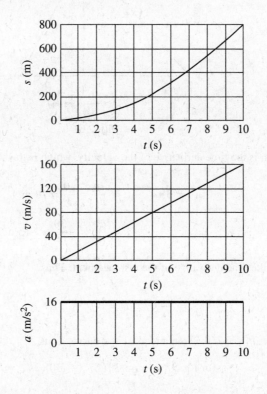

⋅ The first equation in (1) states that the change in the ordinate of the v–t diagram for any time interval is equal to the area under the a–t diagram within that time interval. A similar statement may be made for the change in the ordinate of the s–t diagram in the second equation in (1).

2.4. An automobile accelerates uniformly from rest to 90 km/h in 25 s. Find its constant acceleration and its displacement during this time.

SOLUTION

The following data are given: $v_0 = 0$, $v = 90$ km/h $= 25$ m/s, $t = 25$ s.

To determine the acceleration, which is a constant a_0, apply the formula $v = v_0 + a_0$:

$$a_0 = \frac{v - v_0}{t} = \frac{(25 - 0)\,\text{m/s}}{25\,\text{s}} = 1\ \text{m/s}^2$$

To determine the displacement using only the original data,

$$s = \frac{v + v_0}{2} t = \frac{(25 + 0)\,\text{m/s}}{2} \times 25\ \text{s} = 312.5\ \text{m}$$

2.5. A particle moves with rectilinear motion. The speed increases from 0 to 30 m/s in 3 s and then decreases to 0 in 2 s.

(a) Sketch the v–t curve.
(b) What is the acceleration during the first 3 s and during the next 2 s?
(c) What is the distance traveled in 5 s?
(d) How long does it take the particle to go 50 m?

SOLUTION

(a) The sketch of the v–t curve is shown in Fig. 2-6.

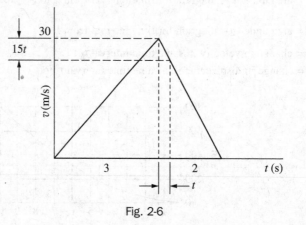

Fig. 2-6

(b) The acceleration is the time derivative of the velocity, which is the slope of the v–t curve. Thus,

$$\text{at } t = 3\text{ s} \qquad a = \frac{dv}{dt} = \frac{30}{3} = 10\ \text{m/s}^2$$

$$\text{at } t = 5\text{ s} \qquad a = \frac{dv}{dt} = -\frac{30}{2} = -15\ \text{m/s}^2$$

(c) The velocity is the time derivative of the displacement. Thus,

$$v = \frac{ds}{dt} \qquad ds = v\,dt \qquad \Delta s = \int v\,dt$$

The integral of $v\,dt$ is also the area under the v–t curve. Thus,

$$\text{for } t = 5 \qquad s = (30)(3)/2 + (30)(2)/2 = 75\ \text{m}$$

(d) The distance traveled in the first 3 s is 45 m as calculated from the area under the v–t curve. The velocity is given by the equation of the v–t curve for the region $t = 3$ to $t = 5$ s. Thus, for the added 5 m traveled, the area under the v–t curve is the sum of the rectangle and triangle between the dashed vertical lines. Or

$$(30 - 15t)t + (15t)\frac{t}{2} = 5$$

Solving the resulting quadratic equation, $t = 0.175$ s. Note that the other solution of the quadratic equation is 3.826 s, which is greater than 2 s, the maximum value that t can have. Hence the total time is $T = 3 + 0.175 = 3.175$ s.

2.6. A balloon is rising with a velocity of 2 m/s when a bag of sand is released. If the height at the time of release is 120 m, how long does it take the bag of sand to reach the ground?

SOLUTION

The sand is rising at the same rate as the balloon at the instant of release. Hence,

$$v_0 = 2 \text{ m/s} \qquad y = 120 \text{ m} \qquad g = a = 9.8 \text{ m/s}^2$$

First, solve using the ground as the datum ($y = 0$), with up being positive. (*Note:* $y = 0$ as the sand reaches the ground.)

$$y = y_0 + v_0 t + \frac{1}{2} a t^2$$

$$0 = 120 + 2t + \frac{1}{2}(-9.8)t^2$$

$$\therefore t = 5.16 \text{ s}$$

Next, solve using the balloon as the datum. Use up as positive. (*Note:* $y = -120$ m as the sand reaches the ground.)

$$y = y_0 + v_0 t + \frac{1}{2} a t^2$$

$$-120 = 0 + 2t + \frac{1}{2}(-9.8)t^2$$

This, of course, yields

$$t = 5.16 \text{ s}$$

2.7. A ball is projected vertically upward with a velocity of 40 m/s. Three seconds later a second ball is projected vertically upward with a velocity of 30 m/s. At what point above the surface of the earth will they meet?

SOLUTION

Let t be the time after the first ball is projected that the two meet. The second ball will then have been traveling for $t - 3$ s. The displacements for both balls will be the same at time t.

Let s_1 and s_2 be the displacements of the first and second balls, respectively. Then

$$s_1 = (v_0)_1 t - \frac{1}{2} g t^2 \qquad \text{and} \qquad s_2 = (v_0)_2 (t - 3) - \frac{1}{2} g (t - 3)^2$$

Equating s_1 and s_2 and substituting the given values of $(v_0)_1$ and $(v_0)_2$, we obtain

$$40t - 4.9t^2 = 30(t - 3) - 4.9(t - 3)^2 \qquad \therefore t = 6.91 \text{ s}$$

Substituting this value of t in the equation for s_1 (or s_2), the displacement is

$$s_1 = 40 \text{ m/s} \times 6.91 \text{ s} - \frac{1}{2}(9.8 \text{ m/s}^2)(6.91 \text{ s})^2 = 42.4 \text{ m}$$

2.8. A ball is thrown at an angle of 40° to the horizontal. With what initial speed should the ball be thrown in order to land 100 m away? Neglect air resistance.

SOLUTION

Choose the xy axes with the origin at the point where the ball is thrown. By neglecting air resistance the x component of the acceleration is zero. The y component of the acceleration is $-g$.
From Eq. (8) with $a_x = 0$ and $a_y = -9.8$ m/s^2,

$$x = v_{0x}t \quad \text{and} \quad y = v_{0y}t - \frac{1}{2}(9.8)t^2$$

Given that when $x = 100$, $y = 0$ and $v_{0x} = v_0 \cos 40°$, $v_{0y} = v_0 \sin 40°$, the above equations become

$$100 = v_0 \cos 40°(t)$$

$$0 = v_0 \sin 40°(t) - \frac{1}{2}(9.8)t^2$$

Solve the first equation for t, substitute in the second equation, and solve for v_0:

$$v_0 = 31.5 \text{ m/s}$$

2.9. A particle moves along a horizontal straight line with an acceleration $a = 6\sqrt[3]{s}$ m/s^2. When $t = 2$ s, its displacement $s = 27$ m and its velocity $v = 27$ m/s. Calculate the velocity and acceleration of the point when $t = 4$ s.

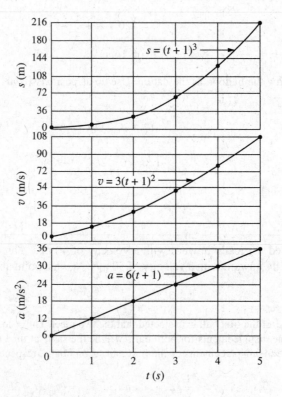

SOLUTION

Since the acceleration is given as a function of the displacement, use the differential equation $a\,ds = v\,dv$. Then

$$\int 6s^{1/3}ds = \int v\,dv \quad \text{or} \quad \frac{9}{2}s^{4/3} = \frac{1}{2}v^2 + C_1$$

Since $v = 27$ when $s = 27$, $C_1 = 0$ and $v = 3s^{2/3}$.

Next use $v = ds/dt$ to obtain $ds/s^{2/3} = 3\,dt$; from this,

$$3s^{1/3} = 3t + C_2$$

Substitute the condition $s = 27$ when $t = 2$ to obtain $C_2 = 3$ and $s = (t + 1)^3$.
   The equations are therefore

$$s = (t + 1)^3 \qquad v = 3(t + 1)^2 \qquad a = 6(t + 1)$$

When $t = 4$ s,

$$s = 125 \text{ m} \qquad v = 75 \text{ m/s} \qquad a = 30 \text{ m/s}^2$$

A plot of these quantities against time is shown above. Note that the ordinate of the v–t curve at any time t is the slope of the s–t curve at the same time. Also, the ordinate of the a–t curve at any time t is the slope of the v–t curve at that time.

2.10. A particle moves on a vertical line with an acceleration $a = 2\sqrt{v}$ m/s^2. When $t = 2$ s, its displacement $s = 64/3$ m and its velocity $v = 16$ m/s. Determine the displacement, velocity, and acceleration of the particle when $t = 3$ s.

SOLUTION

Since $a = dv/dt$, then $2\sqrt{v} = dv/dt$. Separating the variables, $2\,dt = dv/v^{1/2}$. Integrating,

$$2t + C_1 = 2v^{1/2}$$

But $v = 16$ m/s when $t = 2$ s; hence, $C_1 = 4$.
   The equation becomes

$$t + 2 = v^{1/2} \qquad \text{or} \qquad v = (t + 2)^2 = \frac{ds}{dt}$$

Then $ds = (t + 2)^2\,dt$. Integrating,

$$s = \frac{1}{3}(t + 2)^3 + C_2$$

But $s = 64/3$ m when $t = 2$ s; hence, $C_2 = 0$.
   The equations are therefore

$$s = \frac{1}{2}(t + 2)^3 \qquad v = (t + 2)^2 \qquad a = 2(t + 2)$$

When $t = 3$ s,

$$s = 41.7 \text{ m} \qquad v = 25 \text{ m/s} \qquad a = 10 \text{ m/s}^2$$

2.11. The acceleration of a point moving on a vertical line is given by the equation $a = 12t - 20$. It is known that its displacement $s = -10$ m at time $t = 0$ and that its displacement $s = 10$ m at time $t = 5$ s. Derive the equation of its motion $s(t)$.

SOLUTION

Integrate $a = dv/dt = 12t - 20$ to obtain

$$v = 6t^2 - 20t + C_1$$

Integrate this once more to obtain

$$s = 2t^3 - 10t^2 + C_1 t + C_2$$

The constants of integration may now be evaluated. Substitute the known values of s and t:

$$-10 = 2(0)^3 - 10(0)^2 + C_1(0) + C_2 \qquad \therefore C_2 = -10$$

$$10 = 2(5)^3 - 10(5)^2 + C_1(5) - 10 \qquad \therefore C_1 = 4$$

The equation of motion is

$$s(t) = 2t^3 - 10t^2 + 4t - 10$$

2.12. In the system shown in Fig. 2-7(*a*), determine the velocity and acceleration of block 2 at the instant.

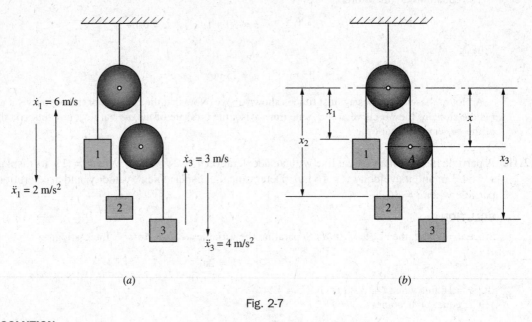

(*a*)   (*b*)

Fig. 2-7

SOLUTION

Figure 2-7(*b*) is drawn to show the position of each weight relative to the fixed support. The length of the cord between weight 1 and point *A* is a constant and equals one-half the circumference of the top pulley plus $x_1 + x$. The length of the cord between weights 2 and 3 is a constant and equals one-half the circumference of pulley *A* plus $x_2 - x + x_3 - x$.

Thus, $x_1 + x = $ constant; $x_2 + x_3 - 2x = $ constant. Time derivatives then show

$$\dot{x}_1 + \dot{x} = 0 \qquad (1) \qquad\qquad \dot{x}_2 + \dot{x}_3 - 2\dot{x} = 0 \qquad (3)$$

$$\ddot{x}_1 + \ddot{x} = 0 \qquad (2) \qquad\qquad \ddot{x}_2 + \ddot{x}_3 - 2\ddot{x} = 0 \qquad (4)$$

Calling the upward direction positive and substituting $\dot{x}_1 = 6$ m/s into Eq. (1), we find $\dot{x} = -6$ m/s. Substituting this value together with $\dot{x}_3 = 3$ m/s into Eq. (3), we find

$$\dot{x}_2 = 2\dot{x} - \dot{x}_3 = 2(-6) - (3) = -15 \text{ m/s (down)}$$

Similar reasoning for the accelerations shows

$$\ddot{x} = 2 \qquad \text{and} \qquad \ddot{x}_2 = 2\ddot{x} - \ddot{x}_3 = 2(+2) - (-4) = 8 \text{ m/s}^2$$

2.13. Show that the curvature of a plane curve at point *P* may be expressed as

$$\frac{1}{\rho} = \frac{\dot{x}\ddot{y} - \ddot{x}\dot{y}}{(\dot{x}^2 + \dot{y}^2)^{3/2}}$$

where ρ is the radius of curvature, $\dot{x}$ and $\dot{y}$ are the *x* and *y* components of the speed of *P*, and $\ddot{x}$ and $\ddot{y}$ are the *x* and *y* components of the magnitude of the acceleration of *P*.

SOLUTION

From calculus, the curvature of any curve $y = f(x)$ at a point *P* is

$$\frac{1}{\rho} = \frac{d^2y/dx^2}{[1 + (dy/dx)^2]^{3/2}} \qquad\qquad (1)$$

But $\quad \dfrac{dy}{dx} = \dfrac{dy}{dt}\dfrac{dt}{dx} = \dfrac{\dot{y}}{\dot{x}}\quad$ and $\quad \dfrac{d^2y}{dx^2} = \dfrac{d}{dx}\left(\dfrac{dy}{dx}\right) = \dfrac{d}{dt}\left(\dfrac{dy}{dx}\right)\dfrac{dt}{dx} = \dfrac{d}{dt}\left(\dfrac{\dot{y}}{\dot{x}}\right)\dfrac{1}{\dot{x}} = \dfrac{\dot{x}\ddot{y}-\dot{y}\ddot{x}}{\dot{x}^2\ddot{x}}$

Substituting in (1), we obtain the equation given in the problem statement.

2.14. A particle describes a path $y = 3.6x^2$, where x and y are in meters. The velocity has a constant x component of 2 m/s. Assume that the particle is at the origin at the start of the motion, and solve for the components of displacement, velocity, and acceleration in terms of time.

SOLUTION

Since $dx/dt = 2$ m/s, we can integrate to obtain $x = 2t + C_1$. But $x = 0$ at $t = 0$; hence, $C_1 = 0$.
Thus,

$$x = 2t \text{ m}$$

Also, $y = 3.6x^2 = 3.6(2t)^2 = 14.4t^2$ m. Thus,

$$\frac{dy}{dt} = 28.8t \text{ m/s}$$

Finally, $d^2x/dt^2 = 0$ so

$$\frac{d^2y}{dt^2} = \frac{d^2}{dt^2}(14.4t^2) = 28.8 \text{ m/s}^2$$

2.15. A particle describes the path $y = 4x^2$ with constant speed $v = 10$ m/s, where x and y are in meters. What is the normal component of the acceleration at the point where $y = 1$ m?

SOLUTION

From calculus, the curvature is

$$\frac{1}{\rho} = \frac{d^2y/dx^2}{[1+(dy/dx)^2]^{3/2}} = \frac{8}{[1+(8x)^2]^{3/2}}$$

The normal component of acceleration is then

$$a_n = \frac{v^2}{\rho} = \frac{8v^2}{[1+64x^2]^{3/2}} = \frac{8\times10^2}{[1+64(1/4)]^{3/2}} = 0.194 \text{ m/s}^2$$

2.16. A particle moves on a path with a velocity vector of $\mathbf{v} = 3t^2\mathbf{i} - 4t\mathbf{j} + 2\mathbf{k}$ m/s. The particle is at the origin when $t = 0$.

(a) Determine the coordinates of its position after 4 s.
(b) Determine the equation of its path.
(c) Determine the projection of the velocity vector in the direction of the vector $\mathbf{n} = 4\mathbf{i} + \mathbf{j} - 3\mathbf{k}$ when $t = 4$ s.

SOLUTION

(a) The position vector is the integral of the velocity vector. This can be seen from the definition of velocity $\mathbf{v} = d\mathbf{r}/dt$ so that

$$\mathbf{r} = \int \mathbf{v}\, dt = t^3\mathbf{i} - 2t^2\mathbf{j} + 2t\mathbf{k}$$

At $t = 4$ s, $\mathbf{r} = 64\mathbf{i} - 32\mathbf{j} - 8\mathbf{k}$ m. The coordinates of the position at $t = 4$ s are then

$$x = 64 \text{ m} \qquad y = -32 \text{ m} \qquad \text{and} \qquad z = 8 \text{ m}$$

(b) At any time t the so-called parametric equations of position are given by the coefficients of $\mathbf{i}$, $\mathbf{j}$, and $\mathbf{k}$ in $\mathbf{r}$. Thus, $x = t^3$, $y = -2t^2$, and $z = 2t$. Eliminating t from these parametric equations yields

$$t = x^{1/3} \qquad y = -2x^{2/3} \qquad z^2 = 4x^{2/3}$$

Combining equations gives

$$y + \left(\frac{z}{2}\right)^2 = -x^{2/3}$$

or

$$4x^{2/3} + 4y + z^2 = 0$$

the equation of the path. Let the reader show that at $t = 4$ s this equation is satisfied.

(c) The unit vector in the desired direction is

$$\mathbf{e}_L = \frac{4\mathbf{i} + \mathbf{j} - 3\mathbf{k}}{\sqrt{4^2 + 1^2 + (-3)^2}}$$

Hence, the projection of $\mathbf{v}$ on $\mathbf{n}$ at $t = 4$ s is

$$\mathbf{v} \cdot \mathbf{e}_L = \left[\frac{4\mathbf{i} + \mathbf{j} - 3\mathbf{k}}{\sqrt{26}}\right] \cdot [3(4)^2\mathbf{i} - 4(4)\mathbf{j} + 2\mathbf{k}] = 33.3 \text{ m/s}$$

2.17. A particle moves along the path whose equation is $r = 2\theta$ m. If the angle $\theta = t^2$ rad, determine the velocity of the particle when θ is 60°. Use two methods.

SOLUTION

A plot of the path is shown in Fig. 2-8(a) with unit vectors $\mathbf{e}_r$ along $\mathbf{r}$ and $\mathbf{e}_\theta$ perpendicular to $\mathbf{r}$ and in the direction of increasing θ.

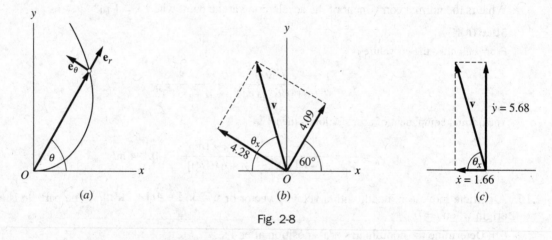

Fig. 2-8

(a) *Polar coordinates.* See Fig. 2-8(b).

Since $\theta = t^2$, $\dot\theta = 2t$. Then $r = 2\theta = 2t^2$ and $\dot r = 4t$.

The velocity vector $\mathbf{v}$ at $\theta = \pi/3$ rad is found as follows:

$$\theta = \frac{\pi}{3} = t^2 \qquad \text{or} \qquad t = 1.023 \text{ s}$$

Using Eq. (32), the velocity is

$$\mathbf{v} = \dot r \mathbf{e}_r + r\dot\theta \mathbf{e}_\theta = 4(1.023)\mathbf{e}_r + [2(1.023)^2][2(1.023)]\mathbf{e}_\theta = 4.09\mathbf{e}_r + 4.28\mathbf{e}_\theta$$

Then

$$v = \sqrt{(4.09)^2 + (4.28)^2} = 5.92 \text{ m/s} \qquad \text{with} \qquad \theta_x = 30° + \tan^{-1}(4.09/4.28) = 73.7°$$

(b) *Cartesian coordinates.* See Fig. 2-8(c).

$$x = r\cos\theta = 2\theta\cos\theta \qquad\qquad y = r\sin\theta = 2\theta\sin\theta$$

$$= 2t^2\cos t^2 \qquad\qquad\qquad = 2t^2\sin t^2$$

Then, at $t = 1.023$ s ($\cos t^2 = \cos \pi/3$, $\sin t^2 = \sin \pi/3$)

$$\dot{x} = 4t \cos t^2 + 2t^2(-\sin t^2)(2t) = -1.66 \text{ m/s}$$

$$\dot{y} = 4t \sin t^2 + 2t^2(\cos t^2)(2t) = 5.68 \text{ m/s}$$

Hence,

$$v = \sqrt{(-1.66)^2 + (5.68)^2} = 5.92 \text{ m/s} \qquad \text{with} \qquad \theta_x = \tan^{-1}\frac{5.68}{1.66} = 73.7°$$

2.18. In the preceding problem determine the magnitude and acceleration of the particle using the same two methods.

SOLUTION

(*a*) *Polar coordinates.* See Fig. 2-9. Using Eq. (34),

$$\mathbf{a} = (\ddot{r} - r\dot{\theta}^2)\mathbf{e}_r + (2\dot{r}\dot{\theta} + r\ddot{\theta})\mathbf{e}_\theta = -4.77\mathbf{e}_r + 20.94\mathbf{e}_\theta$$

since $\theta = t^2$, $\dot{\theta} = 2t$, $\ddot{\theta} = 2$; $r = 2\theta = 2t^2$, $\dot{r} = 4t$, $\ddot{r} = 4$; $t = 1.023$ s at $\theta = \pi/3$. Thus,

$$a = \sqrt{(-4.77)^2 + (20.94)^2} = 21.5 \text{ m/s}^2 \qquad \text{with} \qquad \theta_x = 30° - 12.8° = 17.2°$$

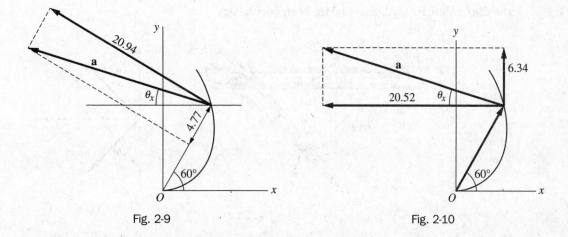

Fig. 2-9   Fig. 2-10

(*b*) *Cartesian coordinates.* See Fig. 2-10.

Continue the time derivatives, using x and y from Problem 2.17(*b*), and evaluate at $t = 1.023$ ($t^2 = \pi/3$):

$$\ddot{x} = 4\cos t^2 + 4t(-\sin t^2)(2t) + 12t^2(-\sin t^2) - 4t^3(\cos t^2)(2t) = -20.52 \text{ m/s}^2$$

$$\ddot{y} = 4\sin t^2 + 4t(\cos t^2)(2t) + 12t^2(\cos t^2) + 4t^3(-\sin t^2)(2t) = 6.34 \text{ m/s}^2$$

Hence, $\quad a = \sqrt{(-20.52)^2 + (6.34)^2} = 21.5 \text{ m/s}^2 \qquad \text{with} \qquad \theta_x = \tan^{-1}\frac{6.34}{20.52} = 17.2°$

2.19. In the Stotch yoke shown in Fig. 2-11 the crank *OA* is turning with a constant angular velocity ω rad/s. Derive the expressions for the displacement, velocity, and acceleration of the sliding member.

SOLUTION

Let *B* represent the position of the left end of the slider when $\theta = 0°$. The displacement x is written $x = OB - \ell - OA \cos \theta$. When the crank is horizontal, $OB = \ell + OA$, hence

$$x = \ell + OA - \ell - OA \cos \theta = OA(1 - \cos \theta)$$

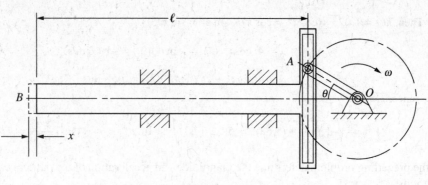

Fig. 2-11

Let $OA = R$. Also, since the crank is turning with constant angular velocity ω, the expression ωt may be substituted for θ. Differentiating $x = R(1 - \cos \omega t)$ yields

$$v = \frac{dx}{dt} = R\omega \sin \omega t \qquad \text{and} \qquad a = \frac{dv}{dt} = R\omega^2 \cos \omega t$$

2.20. In Fig. 2-12, the oscillating arm OD is rotating clockwise with a constant angular velocity 10 rad/s. Block A slides freely in the slot in the arm OD and is pinned to block B, which slides freely in the horizontal slot in the framework. Determine, for $\theta = 45°$, the total velocity of pin P as a point in block B using (*a*) cartesian coordinates and (*b*) polar coordinates.

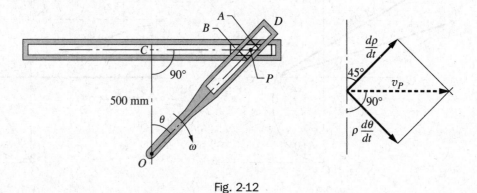

Fig. 2-12

SOLUTION

Note that the total or absolute velocity of P as a point in block B can only be horizontal, since it is a point in B and all points in B move horizontally. But as a point in block A it has this same velocity also.

(*a*) Let $x =$ distance of P from C. Then $x = 0.5 \tan \theta$ (θ is a function of t) and

$$v_P = \frac{dx}{dt} = 0.5 (\sec^2 \theta) \frac{d\theta}{dt}$$

When $\theta = 45°$,

$$v_P = 0.5(\sec^2 45°)10 = 10 \text{ m/s}$$

(*b*) Let $\rho =$ distance of P from O, which will be used as a pole in studying the motion. Then $\rho = 0.5 \sec \theta$. The radial component of the velocity along OP is

$$\frac{d\rho}{dt} = 0.5 \sec \theta \tan \theta \frac{d\theta}{dt}$$

For $\theta = 45°$, this becomes

$$v_r = 0.5 \sec 45° \tan 45° \times 10 = 7.07 \text{ m/s}$$

This component is directed outward along *OP*.

The transverse component of the velocity is

$$\rho \frac{d\theta}{dt} = 0.5 \sec\theta \frac{d\theta}{dt}$$

For $\theta = 45°$, this becomes

$$v_\theta = 0.5 \sec 45° \times 10 = 7.07 \text{ m/s}$$

This component is perpendicular to the arm *OP* and acts down to the right because ω is clockwise. The two components are shown to the right of the figure. Hence,

$$v_P = \sqrt{\left(\frac{d\rho}{dt}\right)^2 + \left(\rho \frac{d\theta}{dt}\right)^2} = \sqrt{7.07^2 + 7.07^2} = 10 \text{ m/s}$$

and is horizontal.

2.21. The bar *AB* shown in Fig. 2-13 moves so that its lowest point *A* travels horizontally to the right with constant velocity $v_A = 5$ m/s. What is the velocity of point *B* when $\theta = 70°$? The length of the bar is 6.24 m.

SOLUTION

Let *x* and *y* be the distances of *A* and *B* from point *O* at any time during the motion. Since $x^2 + y^2 = \ell^2$, then

$$2x \frac{dx}{dt} + 2y \frac{dy}{dt} = 0$$

and

$$v_B = \frac{dy}{dt} = -\frac{x}{y}\frac{dx}{dt} = -(\cot\theta)(v_A) = \frac{1}{\tan 70°} \times 5 = -1.82 \text{ m/s}$$

The minus sign indicates that *B* is traveling down. Note that v_B is independent of ℓ.

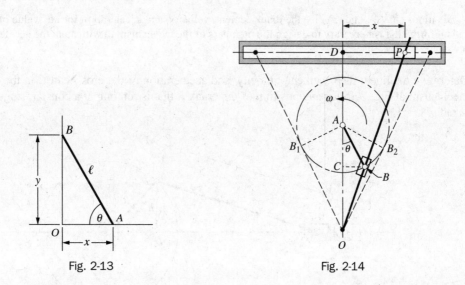

Fig. 2-13   Fig. 2-14

2-22. In the *quick-return mechanism* shown in Fig. 2-14, the crank *AB* is driven at a constant angular velocity ω rad/s. The slider at *B* slides along the rod *OP*, causing *OP* to oscillate about the pin *O*. In turn, *OP* slides in and out of the slider at *P* thereby moving the second slider pinned at *P* along the horizontal slot. A cutting tool attached to this second slider will be subjected to a reciprocating motion. The cutting tool reaches the extremes of its horizontal travel when *OP* is tangent to the crank circle at B_1 and B_2. Cutting occurs while the crank pin moves from B_2 counterclockwise to B_1. The return stroke occurs in the remaining arc from B_1 to B_2. Since the speed is constant, the times of the working and return strokes are proportional to the angles traversed. The cutting stroke occurs during the larger angle and hence takes the longer time. The return stroke is faster; hence, the name *quick-return mechanism* is appropriate. Determine the expressions for displacement, velocity, and acceleration of the cutting tool at P.

SOLUTION

From the figure,

$$\frac{x}{BC} = \frac{OD}{OC}$$

Let $OD = \ell$, $AB = R$, $OA = d$. Then $BC = R\sin\theta$, $OC = OA - AC = d - R\cos\theta$. Substituting in the original expression in x,

$$\frac{x}{R\sin\theta} = \frac{\ell}{d - R\cos\theta} \qquad \text{or} \qquad x = \frac{R\ell\sin\theta}{d - R\cos\theta} \tag{1}$$

and

$$v = \frac{dx}{dt} = \frac{R\ell[(d - R\cos\theta)\cos\theta\, d\theta/dt - \sin\theta(R\sin\theta\, d\theta/dt)]}{(d - R\cos\theta)^2}$$

Since $d\theta/dt$ is the angular velocity ω, the equations for the velocity and acceleration become

$$v = \frac{R\ell\omega(d\cos\theta - R\cos^2\theta - R\sin^2\theta)}{(d - R\cos\theta)^2} \qquad \text{or} \qquad v = R\ell\omega\frac{d\cos\theta - R}{(d - R\cos\theta)^2} \tag{2}$$

$$a = \frac{dv}{dt} = R\ell\omega\frac{(d - R\cos\theta)^2(-d\sin\theta\, d\theta/dt) - (d\cos\theta - R)[2(d - R\cos\theta)R\sin\theta\, d\theta/dt]}{(d - R\cos\theta)^4}$$

$$= \frac{-R\ell\omega^2\sin\theta(d^2 - 2R^2 + Rd\cos\theta)}{(d - R\cos\theta)^3} \tag{3}$$

Equations (1), (2), and (3) give the displacement, velocity, and acceleration for any value of the angle θ. This information is necessary to design the members of the mechanism to withstand the accelerating forces involved.

2.23. Determine the linear displacement, velocity, and acceleration of the crosshead C in the slider crank mechanism of Fig. 2-15 for any position of the crank R that is rotating at a constant angular velocity ω rad/s.

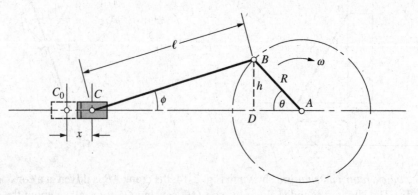

Fig. 2-15

SOLUTION

Let C_0 be the extreme left position of the crosshead, which travels horizontally along the centerline as shown in Fig. 2-15. It is evident from the figure that $x = C_0A - CA$ and $CA = CD + DA$.

When C is at C_0, B is on the centerline; hence, $C_0A = \ell + R$. Also, $CA = \ell \cos\phi + R\cos\theta$. Then

$$x = \ell + R - \ell\cos\phi - R\cos\theta$$

The relation between ϕ and θ is derived from the right triangles ADB and DCB:

$$h = \ell\sin\phi = R\sin\theta$$

Then

$$\sin\phi = \frac{R}{\ell}\sin\theta \qquad \text{and} \qquad \cos\phi = \sqrt{1 - \sin^2\phi} = \sqrt{1 - \frac{R^2}{\ell^2}\sin^2\theta}$$

and the displacement is

$$x = \ell + R - \ell\sqrt{1 - \frac{R^2}{\ell^2}\sin^2\theta} - R\cos\theta$$

Differentiation of this expression with respect to time is somewhat involved because of the radical. However, an approximation of the radical that is sufficiently accurate when $R/\ell < \frac{1}{4}$ is obtained by using the first two terms of the power series expansion of the square root term:

$$\sqrt{1 - \frac{R^2}{\ell^2}\sin^2\theta} \approx 1 - \frac{1}{2}\left(\frac{R^2}{\ell^2}\right)\sin^2\theta$$

Making this substitution, the displacement becomes

$$x = \ell + R - \ell + \frac{R^2}{2\ell}\sin^2\theta - R\cos\theta = R(1 - \cos\theta) + \frac{R^2}{2\ell}\sin^2\theta$$

Differentiation yields the velocity and acceleration of C:

$$v = \frac{dx}{dt} = R\sin\theta\frac{d\theta}{dt} + \frac{R^2}{2\ell}2\sin\theta\cos\theta\frac{d\theta}{dt} = R\omega\left(\sin\theta + \frac{R}{2\ell}\sin 2\theta\right)$$

$$a = \frac{dv}{dt} = R\omega\left(\cos\theta\frac{d\theta}{dt} + \frac{R}{2\ell}2\cos 2\theta\frac{d\theta}{dt}\right) = R\omega^2\left(\cos\theta + \frac{R}{\ell}\cos 2\theta\right)$$

2.24. A point P moves on a circular path with a constant speed (magnitude of its linear velocity) of 12 m/s. If the radius of the path is 2 m, study the motion of the projection of the point on a horizontal diameter. Refer to Fig. 2-16.

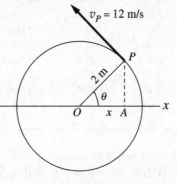

Fig. 2-16

SOLUTION

Point A is the projection on the horizontal diameter of point P. Assume the origin is at the center of the circle. The displacement x of the point A is the projection of the radius vector OP on the x axis (along the horizontal diameter).

Since the line OP sweeps out equal angles in equal times (the angular velocity is constant), the expression for θ may be written $\theta = \omega t$. The x coordinate of P is therefore,

$$x = OP \cos \theta = 2 \cos \omega t$$

The angular velocity of the radius is $\omega = v/r = 12/2 = 6$ rad/s. Then $x = 2 \cos 6t$ and

$$v = \frac{dx}{dt} = -12 \sin 6t \qquad \text{and} \qquad a = \frac{dv}{dt} = -72 \cos 6t$$

The equation for a may be rewritten $a = -(36)(2 \cos 6t) = -36x$. But this means that point A moves so that its acceleration a is negatively proportional to its displacement x. This is the requirement of simple harmonic motion, so if point P moves on a circular path with constant speed, the projection of point P on a diameter moves with simple harmonic motion (see the plots in the next problem).

2.25. Plot the displacement, velocity, and acceleration of point A in Problem 2.24 against time.

SOLUTION

The amplitude of the displacement x is the maximum value, which occurs when $\cos 6t$ assumes its maximum value of unity (either plus or minus). The amplitude is therefore 2. The amplitude of velocity v is 12, and the amplitude of acceleration a is 72.

The period is the time T to complete one cycle. It is evident that the motion will repeat itself after the radius vector OP has gone through a complete revolution $\theta = 2\pi$ rad. Equating the function $6t$ to the value of $\theta = 2\pi$ for a cycle yields

$$6 \text{ rad/s} \times T = 2\pi \text{ rad} \qquad \text{or} \qquad T = \frac{2\pi}{6} = 1.05 \text{ s}$$

It is now possible to correlate the angle θ with time t. For example, when $\theta = \pi/2$ or $90°$, t is one-fourth of a revolution, that is $\frac{1}{4} \times 1.05$ s $= 0.263$ s. Proceeding in this fashion, the following table displays the quantities as a function of θ.

θ	t	$\cos 6t$ or $\cos \theta$	$\sin 6t$ or $\sin \theta$	$x = 2 \cos 6t$	$v = -12 \sin 6t$	$a = -72 \cos 6t$
0°	0	1.000	0	2.00	0	−72.0
30°	0.088	0.886	0.500	1.73	−6.00	−62.3
60°	0.175	0.500	0.866	1.00	−10.4	−36.3
90°	0.263	0	1.000	0	−12.0	0
120°	0.350	−0.500	0.866	−1.00	−10.4	36.0
150°	0.438	−0.866	0.500	−1.73	−6.00	62.3
180°	0.525	−1.000	0	−2.00	0	72.0
210°	0.612	−0.866	−0.500	−1.73	6.00	62.3
240°	0.700	−0.500	−0.866	−1.00	10.4	36.0
270°	0.788	0	−1.000	0	12.0	0
300°	0.875	0.500	−0.866	1.00	10.4	−36.0
330°	0.962	0.866	−0.500	1.73	6.00	−62.3
360°	1.05	1.000	0	2.00	0	−72.0

The plotting of these points provides a visual picture of the motion during one cycle. Naturally, as time increases, these curves duplicate in succeeding cycles. The following graphs indicate the motion during one cycle.

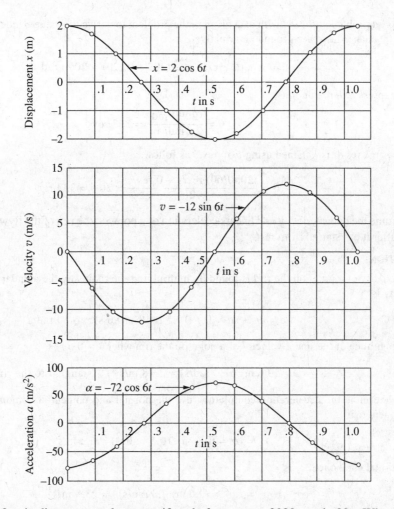

2.26. A flywheel 1.2 m in diameter accelerates uniformly from rest to 2000 rpm in 20 s. What is its angular acceleration?

SOLUTION

In the analysis of this problem, first note that a constant acceleration is involved. This means that the formulas for constant acceleration may be used. These are similar in angular motion to those in rectilinear motion; that is, ω replaces v, α replaces a_0, and θ replaces s.

The wheel starts from rest; hence, $\omega_0 = 0$. The three known quantities are ω_0, ω, t. The quantity sought is angular acceleration α. The formula involving these four quantities is

$$\omega = \omega_0 + \alpha t$$

A word of warning is in order regarding units: ω in rad/s and α in rad/s^2 when t is in seconds. The quantities of interest are

$$\omega_0 = 0 \qquad \omega = 2000 \text{ rpm} = \frac{2000}{60} \frac{\text{rev}}{\text{s}} \times 2\pi \frac{\text{rad}}{\text{rev}} = 209 \text{ rad/s}$$

Hence,

$$\alpha = \frac{\omega - \omega_0}{t} = \frac{209 \text{ rad/s} - 0 \text{ rad/s}}{20 \text{ s}} = 10.45 \text{ rad/s}^2$$

2.27. In Problem 2.26, how many revolutions does the flywheel make in attaining its speed of 2000 rpm?

SOLUTION

To determine the number of revolutions θ, select the equation expressing the relation between θ and the three given quantities ω_0, ω, t. Of course, a formula may be used involving the angular acceleration α just determined, but it is advisable to proceed with data given in the problem to derive the value θ independently of α, which could by chance have been found incorrectly:

$$\theta = \frac{1}{2}(\omega + \omega_0)t = \frac{1}{2}(209 + 0)(20) = 2090 \text{ rad}$$

To express θ in revolutions,

$$\theta = \frac{2090 \text{ rad}}{2\pi \text{ rad/rev}} = 333 \text{ rev}$$

The same result is obtained using ω in rev/s as follows:

$$\theta = \frac{(2000/60) \text{ rev/s} + 0 \text{ rev/s}}{2} \times 20 \text{ s} = 333 \text{ rev}$$

2.28. Determine the linear velocity and linear acceleration of a point on the rim of the flywheel in Problem 2.26, 0.6 s after it has started from rest.

SOLUTION

The velocity of a point on the rim is found by multiplying the radius by the angular velocity. The angular velocity is

$$\omega = \omega_0 + \alpha t = 0 + (10.5 \text{ rad/s}^2)(0.6 \text{ s}) = 6.30 \text{ rad/s}$$

The magnitude of the linear velocity of a point on the rim when $t = 0.6$ s is

$$v = r\omega = (0.6 \text{ m})(6.30 \text{ rad/s}) = 3.78 \text{ m/s} \qquad \text{(tangent to the rim)}$$

To determine the acceleration completely, use the normal and tangential components. The tangential component a_t is

$$a_t = r\alpha = (0.6 \text{ m})(10.5 \text{ rad/s}^2) = 6.3 \text{ m/s}^2$$

The normal component a_n is

$$a_n = r\omega^2 = (0.6 \text{ m})(6.3 \text{ rad/s})^2 = 23.8 \text{ m/s}^2$$

Figure 2-17 illustrates these components for any point P on the rim.

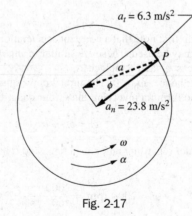

Fig. 2-17

The total acceleration a is the vector sum of the two components a_t and a_n. Let ϕ be the angle between the total acceleration and the radius. The normal acceleration a_n is directed toward the center of the circle:

$$a = \sqrt{a_t^2 + a_n^2} = \sqrt{(6.3)^2 + (23.8)^2} = 24.6 \text{ m/s}^2$$

$$\phi = \tan^{-1}\frac{a_t}{a_n} = \tan^{-1}\frac{6.3}{23.8} = 0.259 \text{ rad} = 14.8°$$

where the angle ϕ is shown in Fig. 2-17.

2.29. A uniform slender rod is 2 m long and rotates on a horizontal plane about a vertical axis through one end. If the rod accelerates uniformly from 40 to 60 rpm in a 5-s interval, what is the linear speed of the center of the rod at the beginning and end of that time interval?

SOLUTION

The speed of the center is $v = r\omega$. Thus, the speeds at the beginning and end of the time interval are, respectively,

$$v_B = r\omega_B = (1 \text{ m})\left(\frac{40 \times 2\pi}{60}\text{ rad/s}\right) = 4.19 \text{ m/s}$$

$$v_E = r\omega_E = (1 \text{ m})\left(\frac{60 \times 2\pi}{60}\text{ rad/s}\right) = 6.28 \text{ m/s}$$

2.30. In Problem 2.29, determine the normal and tangential components of the acceleration of the center of the rod 2 s after acceleration begins.

SOLUTION

The uniform angular acceleration α at any time during the 5-s interval is

$$\alpha = \frac{\omega_E - \omega_B}{t} = \frac{\frac{120}{60}\pi - \frac{80}{60}\pi}{5} = 0.419 \text{ rad/s}^2$$

The angular velocity ω after 2 s is

$$\omega = \omega_B + \alpha t = \frac{80}{60}\pi + 0.419(2) = 5.03 \text{ rad/s}$$

The components of the desired acceleration are

$$a_t = r\alpha = 1(0.419) = 0.419 \text{ m/s}^2$$

$$a_n = r\omega^2 = 1(5.03)^2 = 25.3 \text{ m/s}^2$$

2.31. A wheel 200 mm in diameter coasts to rest from a speed of 800 rpm in 600 s. Determine the angular acceleration.

SOLUTION

Given $\omega_0 = 800$ rpm $= 83.8$ rad/s and $\omega = 0$ at $t = 600$ s, then

$$\alpha = \frac{\omega - \omega_0}{t} = \frac{-83.8 \text{ rad/s}}{600 \text{ s}} = -0.14 \text{ rad/s}^2 \qquad \text{(deceleration)}$$

The acceleration is negative. This means that the angular velocity is in one direction, while the angular acceleration is oppositely directed, thereby indicating a slowing down of the wheel.

2.32. A wheel accelerates uniformly from rest to a speed of 200 rpm in $\frac{1}{2}$ s. It then rotates at that speed for 2 s before decelerating to rest in $\frac{1}{3}$ s. How many revolutions does it make during the entire time interval?

SOLUTION

From $t = 0$ to $t = 0.5$: $\qquad \theta_1 = \frac{1}{2}(\omega_0 + \omega)t = \frac{1}{2}(0 + 200/60 \text{ rev/s})(0.5 \text{ s}) = 0.83 \text{ rev}$

From $t = 0.5$ to $t = 2.5$: $\qquad \theta_2 = \omega t = (200/60 \text{ rev/s})(2 \text{ s}) = 6.67 \text{ rev}$

From $t = 2.5$ to rest: $\qquad \theta_3 = \frac{1}{2}(\omega_0 + \omega)t = \frac{1}{2}(200/60 \text{ rev/s} + 0)(0.333 \text{ s}) = 0.56 \text{ rev}$

Total number of revolutions

$$\theta = \theta_1 + \theta_2 + \theta_3 = 8.06 \text{ rev}$$

2.33. Two friction disks are shown in Fig. 2-18. Derive the expression for the angular velocity ratio in terms of the radii.

SOLUTION

The linear velocities of the mating points A and B on the two wheels are equal. If this were not true, the wheels would slip relative to each other. The linear velocities of points A and B are, respectively,

$$v_A = R_1\omega_1 \qquad v_B = R_2\omega_2$$

But $v_A = v_B$ if the drive is positive, i.e., without slip. Then

$$R_1\omega_1 = R_2\omega_2 \qquad \text{or} \qquad \frac{\omega_1}{\omega_2} = \frac{R_2}{R_1}$$

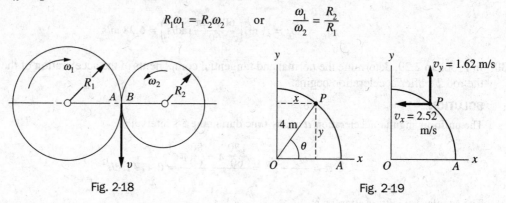

Fig. 2-18   Fig. 2-19

2.34. A bead P moves on a circular path in a counterclockwise direction so that the length of arc it sweeps out is $s = t^3 + 3$. The radius of the path is 4 m. The units of s and t are m and s, respectively. Determine the components of velocity (v_x, v_y) of the bead when $t = 1$ s. Refer to Fig. 2-19.

SOLUTION

The distance AP is traversed in 1 s, or $AP = s = 1^3 + 3 = 4$ m. By inspection, $x = 4 \cos \theta$ and $y = 4 \sin \theta$. Differentiating,

$$v_x = (-4 \sin \theta)\frac{d\theta}{dt} \qquad \text{and} \qquad v_P = (4 \cos \theta)\frac{d\theta}{dt}$$

These may be evaluated provided that θ is found as a function of time. The relation $s = r\theta$ yields

$$\theta = \frac{s}{r} = \frac{(t^3 + 3)}{4}$$

where θ must be in radians. Differentiate to obtain

$$\frac{d\theta}{dt} = \frac{3}{4}t^2$$

When $t = 1$ s, $\theta = 1$ rad and $d\theta/dt = 0.75$ rad/s.
  Substitution yields

$$v_x = -2.52 \text{ m/s} \qquad v_y = 1.62 \text{ m/s}$$

The negative sign indicates that the x component of the velocity is directed to the left. The y component of the velocity is directed up.

The total velocity $v = \sqrt{(v_x)^2 + (v_y)^2} = 3.0$ m/s. This could be obtained directly from $v = rd\theta/dt = 4(0.75t^2)$ when $t = 1$ s, or from $s = t^3 + 3$ and hence $v = ds/dt = 3t^2$.

2.35. In Problem 2.34, determine the axial components of the acceleration a_x and a_y when $t = 1$ s.

SOLUTION

Differentiate the expression for v_x to obtain

$$a_x = -4 \cos \theta \left(\frac{d\theta}{dt}\right)^2 - 4 \sin \theta \frac{d^2\theta}{dt^2}$$

Since $d\theta/dt = 0.75t^2$, then $d^2\theta/dt^2 = 1.5t$.
  At $t = 1$ s,

$$a_x = -4(\cos 1)(0.75)^2 - 4(\sin 1)(1.5) = -6.27 \text{ m/s}^2 \qquad \text{(to the left)}$$

Similarly,

$$a_y = -4 \sin\theta \left(\frac{d\theta}{dt}\right)^2 + 4\cos\theta \frac{d^2\theta}{dt^2} = 1.35 \text{ m/s}^2 \qquad \text{(up)}$$

The total acceleration

$$a = \sqrt{(a_x)^2 + (a_y)^2} = 6.41 \text{ m/s}^2$$

This could be obtained also by combining the tangential component a_t and the normal component a_n of the acceleration. These are

$$a_t = r\alpha = r\frac{d^2\theta}{dt^2} = 4(1.5t) \text{ or 6 m/s}^2$$

$$a_n = r\omega^2 = r\left(\frac{d\theta}{dt}\right)^2 = 4(0.75)^2 \text{ or 2.25 m/s}^2$$

Hence,

$$a = \sqrt{(a_t)^2 + (a_n)^2} = 6.41 \text{ m/s}^2$$

Note that $a_t = d^2s/dt^2 = 6t$ and $a_n = v^2/r = 9t^4/4$ give the same results with $t = 1$.

2.36. The x and y components of the displacement in meters of a point are given by the equations

$$x = 4t^2 - 3t \qquad y = t^3 - 10$$

Determine the velocity and acceleration of the point when $t = 2$ s.

SOLUTION

The velocity components obtained by differentiation are

$$v_x = \frac{dx}{dt} = 8t - 3 \qquad v_y = 3t^2$$

At $t = 2$ s, $v_x = 13$ m/s and $v_y = 12$ m/s. Hence,

$$v = \sqrt{(v_x)^2 + (v_y)^2} = 17.7 \text{ m/s} \qquad \text{and} \qquad \theta_x = \tan^{-1}\frac{v_y}{v_x} = \tan^{-1}\frac{12}{13} = 42.7°$$

where θ_x is the angle between the total velocity and the x axis.
  A second differentiation yields the acceleration components:

$$a_x = \frac{dv_x}{dt} = 8 \qquad a_y = \frac{dv_y}{dt} = 6t$$

At $t = 2$ s, $a_x = 8$ m/s^2 and $a_y = 12$ m/s^2. Hence,

$$a = \sqrt{(a_x)^2 + (a_y)^2} = 14.4 \text{ m/s}^2 \qquad \text{and} \qquad \phi_x = \tan^{-1}\frac{a_y}{a_x} = \tan^{-1}\frac{12}{8} = 56.3°$$

2.37. An automobile is moving south with an absolute velocity of 40 km/h. An observer at O is stationed 50 m to the east of the line of travel. When the automobile is directly west of the observer, what is the angular velocity relative to the observer? After the automobile moves 50 m south, what is its angular velocity relative to the observer at O?

SOLUTION

As indicated in Fig. 2-20, the velocity $v_{A/O}$ of the automobile at A relative to O is 40 km/h or 11.11 m/s. However, the linear velocity of A relative to O (in this case it is perpendicular to OA) is the product of the distance OA and the angular velocity of A relative to O. Then

$$v_{A/O} = OA \times \omega_{A/O}$$

$$11.11 \text{ m/s} = 50 \text{ m} \times \omega_{A/O} \qquad \therefore \omega_{A/O} = 0.222 \text{ rad/s}$$

For the next part of the problem note that the absolute velocity v_B of the vehicle at B is still south 40 km/h (11.11 m/s). The component $v_{B/O}$ (velocity of B relative to O) is perpendicular to the arm BO; hence, $v_{B/O} = 11.11 \cos 45° = 7.85$ m/s. Then

$$v_{B/O} = OB \times \omega_{B/O} \qquad 7.85 = \left(\frac{50}{\cos 45°} \right)(\omega_{B/O}) \qquad \therefore \omega_{B/O} = 0.111 \text{ rad/s}$$

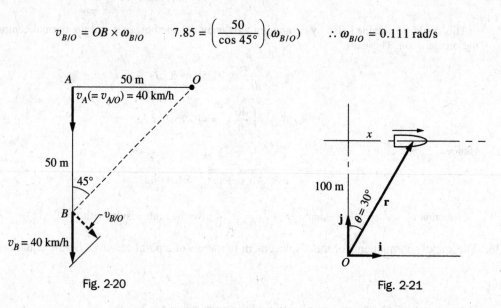

Fig. 2-20　　　　　　　　　　　　　　Fig. 2-21

2.38. A boat is traveling 20 km/h due east. An observer is stationed 100 m south of the line of travel. Determine the angular velocity of the boat relative to the observer when in the position shown in Fig. 2-21.

SOLUTION

Select unit vectors $\mathbf{i}$ and $\mathbf{j}$ in the east and north directions, respectively. Let $\mathbf{r}$ be the position vector of the boat relative to the observer O. Then

$$\mathbf{r} = x\mathbf{i} + 100\mathbf{j} = 100 \tan \theta \mathbf{i} + 100\mathbf{j}$$

The velocity $\mathbf{v}$ of the boat is

$$\mathbf{v} = \dot{\mathbf{r}} = \frac{100\dot{\theta}}{\cos^2 \theta} \mathbf{i} + 0\mathbf{j}$$

Since the speed $v = 40$ km/h = 11.11 m/s and $\theta = 30°$, we have

$$11.11 = \frac{100\dot{\theta}}{\cos^2 30°} \qquad \text{or} \qquad \omega = \dot{\theta} = 0.0833 \text{ rad/s} \qquad \text{clockwise}$$

2.39. The motion of a point is described by the following equations:

$$v_x = 20t + 5 \qquad v_y = t^2 - 20$$

In addition it is known that $x = 5$ m and $y = -15$ m when $t = 0$. Determine the displacement, velocity, and acceleration when $t = 2$ s.

SOLUTION

Rewriting the given equations as

$$v_x = \frac{dx}{dt} = 20t + 5 \qquad v_y = \frac{dy}{dt} = t^2 - 20$$

and integrating, the expressions for x and y are

$$x = 10t^2 + 5t + C_1 \qquad y = \frac{1}{3}t^3 - 20t + C_2$$

To evaluate C_1, substitute $x = 5$ and $t = 0$ in the x equation. Then $C_1 = 5$. To evaluate C_2, substitute $y = -15$ and $t = 0$ in the y equation. Then $C_2 = -15$. Substituting the values of C_1 and C_2, the equations for displacement become

$$x = 10t^2 + 5t + 5 \qquad y = \frac{1}{2}t^2 - 20t - 15$$

Differentiate v_x and v_y to obtain the equations for acceleration:

$$a_x = \frac{dv_x}{dt} = 20 \qquad a_y = \frac{dv_y}{dt} = 2t$$

Substituting $t = 2$ s in the expressions for displacement, velocity, and acceleration, the following values are obtained:

$$x = 55 \text{ m} \qquad y = -53 \text{ m} \qquad v_x = 45 \text{ m/s} \qquad v_y = -16 \text{ m/s} \qquad a_x = 20 \text{ m/s}^2 \qquad a_y = 4 \text{ m/s}^2$$

The magnitudes and directions of the total displacement, velocity, and acceleration can be found by combining their components as before.

2.40. The 100-mm-diameter pulley on a generator is being turned by a belt moving at 20 m/s and accelerating at 6 m/s². A fan with an outside diameter of 150 mm is attached to the pulley shaft. What are the linear velocity and acceleration of the tip of the fan?

SOLUTION

In Fig. 2-22 point A on the pulley has the same velocity as the belt with which it coincides at the instant. Hence, the angular velocity ω of the pulley (and also of the fan keyed to the same shaft) is

$$\omega = \frac{v}{r} = \frac{20}{0.05} = 400 \text{ rad/s}$$

The linear velocity of the fan tip is

$$v_B = (0.075)(400) = 30 \text{ m/s}$$

The tangential component of the linear acceleration of point A is equal to the acceleration of the belt; that is,

$$a_t = r\alpha \qquad \text{or} \qquad 6 = 0.05\alpha$$

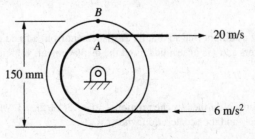

Fig. 2-22

From this, the angular acceleration α of the system is 120 rad/s^2. Then the tangential component of the acceleration of point B is

$$a_t = 0.075 \times 120 = 9 \text{ m/s}^2$$

It has a normal component that equals

$$a_n = 0.075 \times 400^2 = 12\,000 \text{ m/s}^2$$

Hence, the magnitude of the linear acceleration is

$$a = \sqrt{(12\,000)^2 + (9)^2} = 12\,000 \text{ m/s}^2$$

2.41. A ball is thrown at an angle of 40° to the horizontal. What height will the ball reach if it lands 100 m away? Neglect air resistance.

SOLUTION

Choose the x and y axes with the origin at the point where the ball is thrown. Neglecting air resistance, the x component of the acceleration is zero. The y component of the acceleration is $-g$. With $a_x = 0$ and $a_y = -9.8$ m/s^2, Eq. (8) provides

$$x = v_{0x}t \qquad \text{and} \qquad y = v_{0y}t - \frac{1}{2}(9.8)t^2$$

Given that when $x = 100$, $y = 0$, $v_{0x} = v_0 \cos 40°$ and $v_{0y} = v_0 \sin 40°$, the above equations become

$$100 = v_0 t \cos 40°$$

$$0 = v_0 t \sin 40° - \frac{1}{2}(9.8)t^2$$

Solving the first equation for v_0, substituting this in the second equation, and solving for t give $t = 4.138$ s. Substituting this value in the first equation yields $v_0 = 31.5$ m/s. The maximum height occurs at one-half the distance. Therefore, for $t = 2.069$ s

$$y_{max} = 31.5 \times 2.069 \sin 40° - \frac{1}{2} \times 9.8 \times 2.069^2 = 20.9 \text{ m}$$

SUPPLEMENTARY PROBLEMS

2.42. If a car moves at the rate of 50 km/h for 6 min, then 80 km/h for 10 min, and finally 10 km/h for 3 min, what is the average velocity in the total interval?

Ans.   16.52 m/s

2.43. A jet-propelled object has straight-line motion according to the equation $x = 2t^3 - t^2 - 2$, where x is in meters and t is in seconds. What is the change in displacement while the speed changes from 4 to 48 m/s?

Ans. $\Delta x = 44$ m

2.44. A body moves along a straight line so that its displacement from a fixed point on the line is given by $s = 3t^2 + 2t$, where s is in meters and t is in seconds. Find the displacement, velocity, and acceleration at the end of 3 s.

Ans.   33 m, 20 m/s, 6 m/s^2

2.45. The motion of a particle is defined by the relation $s = t^4 - 3t^2 + 2t - 8$, where s is in meters and t is in seconds. Determine the velocity and the acceleration when $t = 2$ s.

Ans. $\dot{s} = 22$ m/s, $\ddot{s} = 42$ m/s^2

2.46. A motorbike travels along a straight road between two points at a mean speed of 30 m/s. It returns at a mean speed of 15 m/s. What is the mean speed for the round trip?

   Ans. 20 m/s

2.47. An automobile accelerates uniformly from rest to 72 km/h, and then the brakes are applied so that it decelerates uniformly to a stop. If the total time is 15 s, what distance was traveled?

   Ans. $s = 150$ m

2.48. A bullet is fired with a muzzle velocity of 600 m/s. If the length of the barrel is 750 mm, what is the average acceleration?

   Ans. 2.4×10^5 m/s^2

2.49. An automobile is accelerating from rest at a uniform rate of 3 m/s^2. How long will it take to reach 50 km/h and in what distance?

   Ans. 4.63 s, 32.2 m

2.50. A stone is dropped from a balloon that is ascending at a uniform rate of 10 m/s. If it takes the stone 10 s to reach the ground, how high was the balloon at the instant the stone was dropped?

   Ans. 485 m

2.51. A person in a balloon rising with a constant velocity of 4 m/s propels a ball upward with velocity of 1.2 m/s relative to the balloon. After what time interval will the ball return to the balloon?

   Ans. $t = 0.245$ s

2.52. A stone is dropped with zero initial velocity into a well. The sound of the splash is heard 3.63 s later. How far below the ground surface is the surface of the water? Assume that the velocity of sound is 330 m/s.

   Ans. $s = 58.4$ m

2.53. Child *A* throws a ball vertically up with a speed of 10 m/s from the top of a shed 3 m high. Child *B* on the ground at the same instant throws a ball vertically up with a speed of 13.5 m/s. Determine the time at which the two balls will be at the same height above the ground. What is the height?

   Ans. $t = 0.857$ s, $h = 15.17$ m

2.54. A truck, traveling at constant speed, passes a parked police car. The policeman gives chase immediately, accelerating at a constant rate to 100 km/h in 10 s, after which he maintains a constant speed. If the police car overtakes the truck in 800 m, what is the constant speed of the truck?

   Ans. $v = 85$ km/h

2.55. A radar-equipped police car notes a car traveling 95 km/h. The police car starts pursuit 30 s after the observation and accelerates to 135 km/h in 20 s. Assuming the speeds are maintained on a straight road, how far from the observation post will the pursuit end?

   Ans. $s = 3560$ m

2.56. An automobile accelerates uniformly from rest on a straight level road. A second automobile starting from the same point 6 s later with initial velocity zero accelerates at 6 m/s^2 to overtake the first automobile 400 m from the starting point. What is the acceleration of the first automobile?

   Ans. $a = 2.62$ m/s^2

2.57. Plane *A* leaves an airport and flies north at 160 km/h. Plane *B* leaves the same airport 20 min later and flies north at 190 km/h. How long will it take *B* to overtake *A*?

   Ans. $t = 2.1$ h

2.58. A particle moves on a straight line with the acceleration shown in the graph in Fig. 2-23. Determine the velocity and displacement at time $t = 1, 2, 3$, and 4 s. Assume that the initial velocity is 3 m/s and the initial displacement is zero.

Ans. $v_1 = 1$ m/s, $s_1 = 2$ m; $v_2 = 3$ m/s, $s_2 = 4$ m; $v_3 = -1$ m/s, $s_3 = 5$ m; $v_4 = -3$ m/s, $s_4 = 3$ m

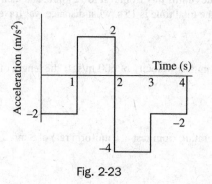

Fig. 2-23

2.59. A particle of dirt falls from an elevator that is moving up with a velocity of 3 m/s. If the particle reaches the bottom in 2 s, how high above the bottom was the elevator when the particle started falling?

Ans. $s = 13.6$ m

2.60. A bullet is fired vertically upward with a speed of 600 m/s. Neglecting drag, to what height would the bullet ascend?

Ans.   18.4 km

2.61. A ball is thrown vertically upward with a speed of 10 m/s from the edge of a cliff 20 m above sea level. What is the highest point above sea level reached? How long does it take the ball to hit the water? With what velocity does it hit the water?

Ans. $h = 25.1$ m, $t = 2.80$ s, $v = 17.4$ m/s

2.62. A particle moves with an acceleration $a = -6v$, where a is in m/s^2 and v is in m/s. When $t = 0$ s, the displacement $s = 0$ and the velocity $v = 9$ m/s. Find the displacement, velocity, and acceleration when $t = 0.5$ s.

Ans. $s = 1.43$ m, $\dot{s} = v = 0.448$ m/s, $\ddot{s} = a = 2.69$ m/s^2

2.63. A body moving with velocity v_0 enters a medium in which the resistance is proportional to the velocity squared. This means $a = -kv^2$. Determine the expression for the velocity in terms of time t.

Ans. $v = 1/(kt + 1/v_0)$

2.64. The speed of a particle is given by $v = 2t + 5t^2$, where t is in seconds. What distance does it travel while its speed increases from 7 to 99 m/s?

Ans. $s = 143$ m

2.65. A particle moves to the right from rest with an acceleration of 6 m/s^2 until its velocity is 12 m/s to the right. It is then subjected to an acceleration of 12 m/s^2 to the left until its total distance traveled is 36 m. Determine the total elapsed time.

Ans. $t = 4.73$ s

2.66. Water drips from a faucet at the rate of 8 drops per second. The faucet is 200 mm above the sink. When one drop strikes the sink, how far is the next drop above the sink?

Ans. $h = 29.1$ mm

2.67. A particle moving with a velocity of 6 m/s upward is subjected to an acceleration of 3 m/s^2 downward until its displacement is 2 m below its position when the acceleration began. The acceleration then ceases for 3 s.

The particle is then subjected to an acceleration of 4 m/s^2 upward for 5 s. Determine the displacement and the distance traveled.

Ans. $s = -7.4$ m relative to start, $d = 62.2$ m

2.68. The velocity–time relationship for a point moving on a straight line is $v(t) = 4 \cos \pi t/4$. How far does the point move in 2 s?

Ans. $x = 5.09$ m

2.69. An object moves on a straight line with constant acceleration 2 m/s^2. How long will it take to change its speed from 5 to 8 m/s? What change in displacement takes place during this time interval?

Ans. $t = 1.5$ s, $d = 9.75$ m

2.70. The motion of a particle along a straight path is given by the acceleration $a = t^3 - 2t^2 + 7$, where a is in m/s^2 and t is in seconds. The velocity is 3.58 m/s when $t = 1$ s, and the displacement is 9.39 m when $t = 1$ s. Calculate the displacement, velocity, and acceleration when $t = 2$ s.

Ans. $s = 15.9$ m, $v = 9.6$ m/s, $a = 7$ m/s^2

2.71. A particle moves in rectilinear motion along the x axis. Given $v = x^{1/2}$ m/s, determine the position, velocity, and acceleration at $t = 4$ s. At $t = 0$, $v = 1$ m/s.

Ans. $x = 9$ m, $v = 3$ m/s, $a = \frac{1}{2}$ m/s^2

2.72. In the system shown in Fig. 2-24, determine the velocity and acceleration of block 3 at the instant considered.

Ans. $v_3 = 10.5$ m/s up, $a_3 = 5.0$ m/s^2 up

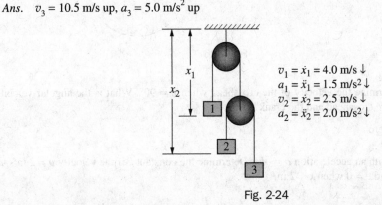

$$v_1 = \dot{x}_1 = 4.0 \text{ m/s} \downarrow$$
$$a_1 = \ddot{x}_1 = 1.5 \text{ m/s}^2 \downarrow$$
$$v_2 = \dot{x}_2 = 2.5 \text{ m/s} \downarrow$$
$$a_2 = \ddot{x}_2 = 2.0 \text{ m/s}^2 \downarrow$$

Fig. 2-24

2.73. A particle moves along the path $y = \frac{1}{3}x^2$ with a constant speed of 8 m/s. What are the x and y components of the velocity when $x = 3$ m? What is the acceleration of the particle when $x = 3$ m?

Ans. $\dot{x} = 3.58$ m/s, $\dot{y} = 7.16$ m/s, $a = 3.82$ m/s^2

2.74. A particle moves along the curve $y = \frac{1}{3}x^3 + 2$ where x and y are in meters. When $x = 0.8$ m, the x component of the velocity is 10 m/s, what is the total velocity?

Ans. $v = 11.87$ m/s, $\angle\ 32.6°$

2.75. A particle has a normal acceleration of 120 m/s^2 when traveling on a circular path with a constant peripheral speed of 80 m/s. What is the radius of the circle?

Ans. $r = 53.3$ m

2.76. An object P moves at a constant speed v in a counterclockwise direction along a circle of radius a, as shown In Fig. 2-25. Selecting a pole O at the left end of the horizontal diameter, derive expressions for the radial and transverse components of the acceleration. (*Hint:* $r = 2a \cos \theta$.)

Ans. $a_r = -(v^2/a) \cos \theta$, $a_\theta = -(v^2/a) \sin \theta$

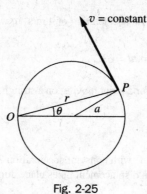

$v = \text{constant}$

Fig. 2-25

2.77. In the preceding problem show that total acceleration is v^2/a, which is the normal component for circular motion of a point with constant speed (there is no tangential component).

2.78. In the slider crank mechanism shown in Fig. 2-26, the crank is rotating at 200 rpm. Determine the velocity and acceleration of the crosshead when $\theta = 30°$.

Ans. $v = 7.6$ m/s, $a = 260$ m/s^2

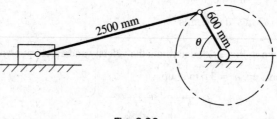

2500 mm

600 mm

θ

Fig. 2-26

2.79. In Problem 2.78, determine the velocity of the crosshead when $\theta = 90°$. What is the angular velocity of the member connecting the crosshead and the crank?

Ans. $v = 12.6$ m/s, zero

2.80. A particle oscillates with an acceleration $a = -kx$. Determine the constant k if the velocity $v = 2$ m/s when the displacement $x = 0$, and $v = 0$ when $x = 2$ m.

Ans. $k = 1\ s^{-2}$

2.81. A particle moves with simple harmonic motion with a frequency of 30 cycles/min and an amplitude of 6 mm. Determine the maximum velocity and acceleration.

Ans. $v_{max} = 18.8$ mm/s, $a_{max} = 59.2$ mm/s^2

2.82. A body having simple harmonic motion has a period of 6 s and an amplitude of 4 m. Determine the maximum velocity and acceleration of the body.

Ans. $\frac{4}{3}\pi$ m/s, $\frac{4}{9}\pi^2$ m/s^2

2.83. A particle having a constant speed of 15 m/s moves around a circle 10 m in diameter. What is the normal acceleration?

Ans. 45 m/s^2

2.84. The normal acceleration of a point on the rim of a 3-m-diameter flywheel is a constant 15 m/s^2. What is the angular speed of the flywheel?

Ans. 3.16 rad/s

2.85. A point moves on a path 3 m in diameter so that the distance traversed is $s = 3t^2$. What is the tangential acceleration at the end of 2 s?

Ans. 6 m/s^2

2.86. A particle moves on a 100-mm-radius circular path. The distance, measured along the path, is given by $s = 200t^3$ mm. What is the magnitude of the total acceleration after the particle has traveled around the circular path once?

Ans. $a = 16.65$ m/s^2

2.87. The flywheel of an automobile acquires a speed of 2000 rpm in 45 s. Find its angular acceleration. Assume uniform motion.

Ans. 4.65 rad/s^2

2.88. A rotor with a 15-mm diameter is spinning at 2 000 000 rpm in a high-vacuum chamber. What is the normal component of the acceleration of a point on the rim?

Ans. $a_n = 3.29 \times 10^8$ m/s^2

2.89. A point moves on a circular path with its position from rest defined by $s = t^3 + 5t$, where s is measured in meters. The magnitude of the acceleration is 8.39 m/s^2 when $t = 0.66$ s. What is the diameter of the path?

Ans. $d = 10.8$ m

2.90. A horizontal bar 1.2 m long rotates about a vertical axis through its midpoint. Its angular velocity changes uniformly from 0.5 to 2 rad/s in 20 s. What is the linear acceleration of a point on the end of the bar 5 s after the speedup occurs?

Ans. $a_t = 0.045$ m/s^2, $a_n = 0.46$ m/s^2

2.91. Particle P travels on a circular path with radius 2.5 m as shown in Fig. 2-27. The speed of P is decreasing (the tangential component of the acceleration is thus directed opposite to the velocity vector) at the instant considered. If the total acceleration vector is as shown, determine the velocity of P and the angular acceleration of the line OP at that instant.

Ans. $v = 6.07$ m/s, $\theta_x = 135°$; $\alpha = 3.4$ rad/s^2

2.92. Disk A drives disk B without slip occurring. Determine the velocity and acceleration of weight D, which is connected by a cord of drum C, which is keyed to disk B as shown in Fig. 2-28.

Ans. $v_D = 0.6$ m/s up, $a_D = 0.9$ m/s^2 up

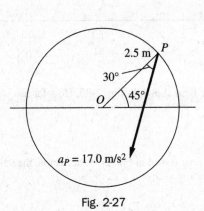

Fig. 2-27

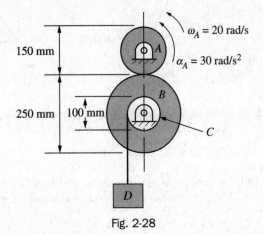

Fig. 2-28

2.93. The angular acceleration of a rotor is given by $\alpha = kt^{-1/2}$, where α is in rad/s^2 and t is in seconds. When $t = 1$ s, the angular velocity $v = 10$ rad/s and the angular displacement $\theta = 3.33$ rad. When $t = 0$ s, the angular displacement $\theta = -4$ rad. Determine θ, ω, and α when $t = 4$ s.

Ans. $\theta = 46.7$ rad, $\omega = 18$ rad/s, $\alpha = 2$ rad/s^2

2.94. The displacement in meters of a point is described in terms of x and y components as follows:

$$x = 2t^2 + 5t \qquad y = 4.9t^2$$

Determine the velocity and acceleration at the end of 4 s.

Ans. $v_x = 21$ m/s, $v_y = 39.2$ m/s; $a_x = 4$ m/s^2, $a_y = 9.8$ m/s^2

2.95. A bicycle rider travels 250 m north and then 160 m northwest. What is the rider's displacement? What distance does the rider cover?

Ans. $s = 381$ m, 72.7° north of west; $d = 410$ m

2.96. Car A is moving northwest with a speed of 100 km/h. Car B is moving east with a speed of 60 km/h. Determine the velocity of A relative to B. Determine the velocity of B relative to A.

Ans. $v_{A/B} = -131\mathbf{i} + 70.7\mathbf{j}$ km/h, $v_{B/A} = 131\mathbf{i} - 70.7\mathbf{j}$ km/h

2.97. Body A has a velocity of 15 km/h from west to east relative to body B, which in turn has a velocity of 50 km/h from northeast to southwest relative to body C. Determine the velocity of A relative to C.

Ans. $v_{A/C} = 40.8$ km/h, at 60° southwest

2.98. Refer to Fig. 2-29. A rotating spotlight is at perpendicular distance ℓ from a horizontal floor. The light revolves a constant N rpm about a horizontal axis perpendicular to the paper. Derive expressions for the velocity and acceleration of the light spot traveling along the floor. Let θ be the angle between the vertical line ℓ and the light beam at time t.

Ans. $\dot{x} = 0.105\ell N/\cos^2\theta$, $\ddot{x} = 0.022\ell N^2 \tan\theta/\cos^2\theta$

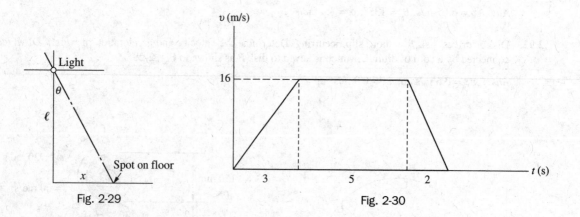

Fig. 2-29   Fig. 2-30

2.99. A particle moves in a straight line such that its v–t curve is as shown in Fig. 2-30. How far has it gone after 10 s? What is the acceleration at 9 s?

Ans. $s = 120$ m, $a = -8$ m/s^2

2.100. A particle moves with rectilinear motion. Given the a–s curve shown in Fig. 2-31, determine the velocity after the particle has traveled 30 m if the initial velocity is 10 m/s.

Ans. $v = 20$ m/s

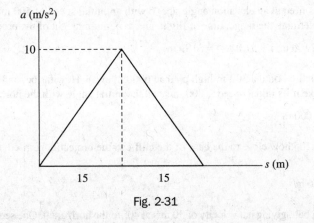

Fig. 2-31

2.101. A particle moves in a straight line with an *a–t* curve shown in Fig. 2-32. The initial displacement and velocity are zero. At what time and with what displacement will the particle come to rest again?

Ans. $t = 10$ s, $x = 29.3$ m

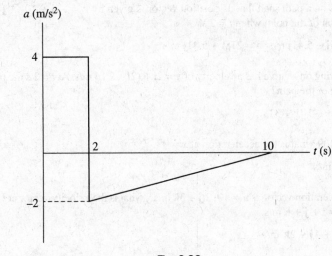

Fig. 2-32

2.102. A particle moves in rectilinear motion such that its *a–s* curve is as shown in Fig. 2-33. If, initially, $s = 0$ and $v = 4$ m/s, what is the velocity when the position is 8 m? 12 m?

Ans. $v = 10.6$ m/s, $v = 12$ m/s

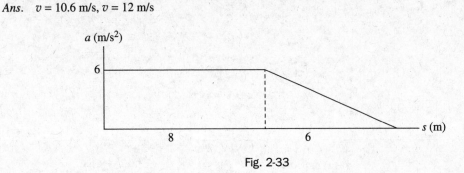

Fig. 2-33

2.103. A particle at rest starts to move on a circle of 0.8-m radius. Its tangential component of acceleration, as a function of arc distance, increases linearly from 0 to 3.2 m/s² in one trip around the circle, after which the tangential acceleration remains constant. What is the velocity and what is the normal component of the acceleration after the second trip around the circle?

Ans. $v = 6.95$ m/s, $a_n = 60.3$ m/s²

2.104. A projectile is fired at an elevation angle of 30° with an initial speed of 500 m/s on flat terrain. Neglecting air resistance, determine the range, time of flight, and maximum height of the projectile.

     Ans. $R = 22\,100$ m, $t = 51.0$ s, $h = 3189$ m

2.105. A projectile is fired onto a 300-m-high plateau 6000 m away. How far beyond the edge of the plateau will the projectile strike if its initial speed is 800 m/s and the initial angle with the horizontal of the projectile is 60°?

     Ans. $x = 11\,000$ m

2.106. In Problem 2.105 how close to the base of the cliff can the projectile be fired so that it will just clear the edge of the cliff?

     Ans. $x = 176$ m

2.107. A batter hits a ball, giving it a velocity of 30 m/s at 40° to the horizontal. One second after the ball is hit, a fielder standing in the direct path of the ball at a distance of 80 m starts running after the ball at a constant speed. How fast must he run to catch the ball?

     Ans. $v = 3.56$ m/s

2.108. A point moves on a path such that its position vector is given by $\mathbf{r} = e^{2t}\mathbf{i} + 40e^{-2t}\mathbf{j}$ m. Determine the velocity and acceleration of the point when $t = 2$ s.

     Ans. $\mathbf{v} = 109\mathbf{i} - 5.41\mathbf{j}$ m/s, $\mathbf{a} = 218\mathbf{i} + 5.41\mathbf{j}$ m/s^2

2.109. A particle moving on a curve has a velocity of $\mathbf{v} = 2\mathbf{i} + (2t + 20)\mathbf{j}$ m/s. At $t = 2$ s the position is $4\mathbf{i} + 75\mathbf{j}$ m. What is the equation of the path?

     Ans. $y = \frac{1}{4}x^2 + 10x + 31$

2.110. In Problem 2.109 what is the acceleration at $x = 0$?

     Ans. $\mathbf{a} = 2\mathbf{j}$ m/s^2

2.111. Given the acceleration vector as $\mathbf{a} = t\mathbf{i} + 2t\mathbf{j} - 3\mathbf{k}$ m/s^2, what is the velocity vector at $t = 2$ s^2? The velocity vector at $t = 1$ s is $\mathbf{v} = \mathbf{i} + \mathbf{j} + \mathbf{k}$ m/s.

     Ans. $\mathbf{v} = \frac{5}{2}\mathbf{i} + 4\mathbf{j} - 2\mathbf{k}$ m/s

2.112. In Problem 2.111 what is the component of the acceleration vector in the direction of the velocity vector at $t = 2$s? (*Note:* This is the tangential component of the acceleration.)

     Ans. $a_t = 5.27$ m/s^2

CHAPTER 3

Dynamics of a Particle

3.1 Newton's Laws of Motion

The dynamics of a particle involve the forces acting on the particle that result in an acceleration of the particle. The study of the dynamics of a particle is also referred to as *kinetics*. It requires the application of Newton's laws of motion.

Newton's first law: *A particle will maintain its state of rest or of uniform motion (at constant speed) along a straight line unless compelled by some force to change that state.* In other words, a particle accelerates only if an unbalanced force acts on it.

Newton's second law: *The time rate of change of the product of the mass and velocity of a particle is proportional to the force acting on the particle.* The product of the mass m and the velocity $\mathbf{v}$ is the linear momentum $\mathbf{L}$. Thus, the second law states

$$\mathbf{F} = \frac{d(m\mathbf{v})}{dt} = \frac{d\mathbf{L}}{dt} \tag{1}$$

If m is constant, the above equation becomes

$$\mathbf{F} = m\frac{d\mathbf{v}}{dt} = m\mathbf{a} \tag{2}$$

The units on the quantities in Newton's second law are as follows:

m = mass, kilograms (kg)

$\mathbf{a}$ = acceleration, m/s^2

$\mathbf{F}$ = force, newtons (N)

$\mathbf{L}$ = linear momentum, kg·m/s

Newton's third law: *To every action, or force, there is an equal and opposite reaction, or force.* In other words, if particle A exerts a force on particle B then particle B exerts a numerically equal and oppositely directed force on particle A.

3.2 Acceleration

The acceleration of the particle shown in Fig. 3-1 may be determined by the vector equation according to Newton's second law

$$\sum \mathbf{F} = m\mathbf{a} = m\ddot{\mathbf{r}} \tag{3}$$

where $\sum \mathbf{F}$ = vector sum of all the forces acting on the particle
$\quad m$ = mass of particle
$\quad \mathbf{a} = \ddot{\mathbf{r}}$ = acceleration

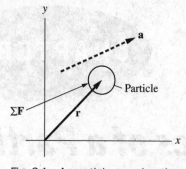

Fig. 3-1   A particle accelerating
under the action of the
resultant force $\Sigma\mathbf{F}$.

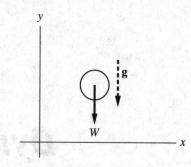

Fig. 3-2   A body in free fall.

A body in free fall with no drag force, shown in Fig. 3-2, has an acceleration equal to the acceleration of gravity g. The force acting on the body is the weight W. It follows from Eq. (3) that

$$W = mg \qquad (4)$$

where the weight W is measured in newtons, never kilograms.

3.3   D'Alembert's Principle

Jean D'Alembert suggested in 1743 that Newton's second law of motion, as given in Section 3.2, could be written

$$\sum \mathbf{F} - m\mathbf{a} = 0$$

Thus, an imaginary force (called an *inertia force*) that is collinear with $\Sigma\mathbf{F}$ but oppositely sensed and of magnitude $m\mathbf{a}$ would, if applied to the particle, cause it to be in equilibrium. The equations of equilibrium would then apply. Some authors state that the particle is in dynamic equilibrium. Actually the particle is *not* in equilibrium, but the equations of equilibrium can be applied.

3.4   Problems in Dynamics

The solutions of problems in dynamics vary with the type of force system. Many problems involve forces that are constant; Problems 3.1 through 3.16 are examples of this type. In other problems, the forces vary with position (rectilinear or angular); Problems 3.17 through 3.22 are examples of this type. The subject of vibrations is built on force systems that vary not only with distance but also with velocity. Problems 3.23 and 3.24 deal with forces that vary with the first and second powers of the velocity.

The subject of ballistics is introduced in an elementary manner in Problem 3.25, which deals with the motion of a projectile under the action of the constant force of gravity. To this solution may be added retarding forces that vary with the velocity of the projectile.

An object is said to be moving with central force motion if the force acting on the object is always directed through a central point. Satellites and planets are examples of central force motion. These are discussed in Problems 3.26 through 3.31.

SOLVED PROBLEMS

In solving problems, the vector equation $\mathbf{F} = m\mathbf{a}$ is replaced by scalar equations using components. In the diagrams, vectors are designated by their magnitudes when the directions are apparent.

3.1. A particle weighing 4 N is pulled up a smooth plane by a force F as shown in Fig. 3-3(a). Determine the force of the plane on the particle and the acceleration along the plane.

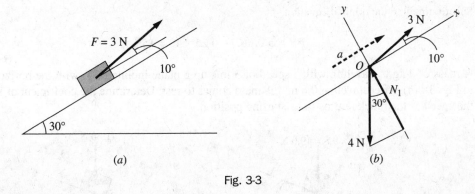

(a)   (b)

Fig. 3-3

SOLUTION

The free-body diagram is shown in Fig. 3-3(*b*). The acceleration *a* is shown as a dashed vector acting parallel to the plane and upward. If the value obtained is negative, this indicates that the acceleration acts parallel to the plane but downward.

It is important to keep in mind that the force system shown acting on the particle is not in equilibrium. If it were in equilibrium, the particle would not accelerate.

Applying Newton's laws, there result two equations along the *x* and *y* axes chosen, respectively, parallel and perpendicular to the plane:

$$\sum F_x = \frac{W}{g}a_x \qquad \text{or} \qquad 3\cos 10° - 4\sin 30° = \frac{4}{9.8}a_x$$

$$\sum F_y = \frac{W}{g}a_y \qquad \text{or} \qquad 3\sin 10° - 4\cos 30° + N_1 = 0$$

Assuming that the particle does not leave the plane, its velocity in the *y* direction is zero. Therefore a_y must also be zero.

The second equation yields the force of the plane on the particle $N_1 = 2.94$ N. From the first equation, $a_x = 2.34 \text{ m/s}^2$.

3.2. A particle having a mass of 5 kg starts from rest and attains a speed of 4 m/s in a horizontal distance of 12 m. Assuming a coefficient of friction of 0.25 and uniformly accelerated motion, what horizontal force *P* is required to accomplish this? Refer to Fig. 3-4.

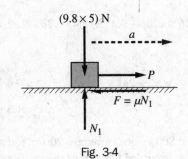

Fig. 3-4

SOLUTION

The equation of motion in the horizontal direction is

$$\sum F = P - 0.25N_1 = ma$$

By inspection, $N_1 = 9.8 \times 5 = 49$ N.

To determine the acceleration *a*, apply Eq. (7) in Chapter 2:

$$v^2 = v_0^2 + 2as$$

$$4^2 = 0 + 2a(12) \qquad \therefore a = \frac{(4 \text{ m/s})^2}{2(12 \text{ m})} = 0.667 \text{ m/s}^2$$

Substituting into the original equation,

$$P = 5(0.667) + 0.25 \times 49 = 15.6 \text{ N}$$

3.3. A mass of 2 kg is traveling with a speed of 3 m/s up a plane inclined 20° with the horizontal. Refer to Fig. 3-5(*a*). After traveling 0.8 m, the mass comes to rest. Determine the coefficient of friction and the speed as the block returns to its starting position.

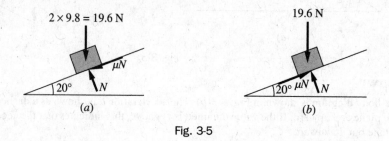

Fig. 3-5

SOLUTION

In the free-body diagram of Fig. 3-5(*a*), friction is shown acting down the plane. By inspection, the normal force $N = 19.6 \cos 20° = 18.42$ N. To determine the acceleration a, apply Eq. (7) in Chapter 2, $v^2 = v_0^2 + 2as$. Hence,

$$v^2 = v_0^2 + 2as$$

$$0 = 3^2 + 2a(0.8) \qquad \therefore a = \frac{-(3)^2}{2(0.8)} = -5.62 \text{ m/s}^2$$

Summing forces parallel to the plane (up being positive) yields

$$-19.6 \sin 20° - \mu(18.4) = 2(-5.62)$$

Thus $\mu = 0.247$.

To solve for the return speed, refer to Fig. 3-5(*b*), which shows the frictional force acting up the plane. Using the down direction as positive, the equation of motion becomes

$$19.6 \sin 20° - 0.247(18.4) = 2a \qquad \therefore a = 1.08 \text{ m/s}^2$$

Finally,

$$v^2 = v_0^2 + 2as \qquad \text{or} \qquad v^2 = 0 + 2(1.05)(0.8)$$

$$\therefore v = 1.3 \text{ m/s}$$

3.4. An automobile with a mass of 800 kg goes around a 700-m-radius curve at a constant speed of 60 km/h. If the road is not banked, what frictional force must the road exert on the tires so that they will maintain motion along the curve?

SOLUTION

In Fig. 3-6, O is the center of the curve 700 m from the automobile. The forces acting on the automobile are its weight W; the normal force N, which is equal to W; and the friction force F. Assuming to the left is positive, the equation of motion becomes

$$\sum F = \frac{W}{g} a_n = \frac{W}{g} \frac{v^2}{r} \qquad \text{or} \qquad F = \frac{800 \times 9.8}{9.8} \frac{(16.67)^2}{700} = 318 \text{ N}$$

where $v = 60 \times 1000/3600 = 16.67$ m/s.

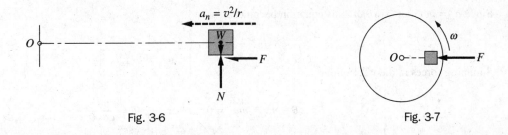

Fig. 3-6   Fig. 3-7

3.5. A small block of mass m is on a horizontal rotating turntable at a distance r from the center as shown in Fig. 3-7. Assuming a coefficient of friction μ between the mass and the turntable, what is the maximum angular velocity the turntable may have without the mass slipping?

SOLUTION

The only force acting horizontally is the friction F, which equals μN. Sum forces along the radius: $\sum F = ma_n$ or $F = ma_n$. Since the normal force N between the block and table is equal to mg and $a_n = v^2/r$, Newton's second law provides

$$\mu mg = \frac{mv^2}{r} = \frac{m(r\omega)^2}{r} \quad \text{or} \quad \omega = \sqrt{\frac{\mu g}{r}}$$

The units on ω would be rad/s.

3.6. Figure 3-8 indicates a particle of mass m that can move in a circular path about the y axis. The plane of the circular path is horizontal and perpendicular to the y axis. As the angular velocity ω increases, however, the particle rises, which means that the radius r of its circular path also increases. The mass and cord are called a *spherical* (sometimes *conical*) *pendulum*. Derive the relationship between θ and ω for constant angular velocity, and find the frequency in terms of θ.

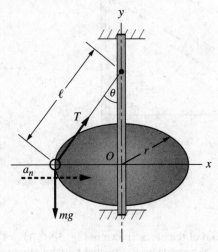

Fig. 3-8

SOLUTION

Assume that the constant angular velocity of the particle (or of its cord ℓ) is ω rad/s. The angle θ is the angle between the cord and the y axis. The forces acting on the particle are the force of gravity and the tension T in the cord.

    Since the particle is moving with constant angular velocity, its only linear acceleration is the normal component a_n directed toward the center of the path (i.e., toward the intersection of the horizontal plane of travel and the y axis). Summing forces along this normal,

$$\sum F_n = T \sin \theta = ma_n$$

Since $a_n = r\omega^2 = (\ell \sin \theta)\omega^2$ this equation becomes

$$T \sin \theta = m(\ell \sin \theta)\omega^2 \tag{1}$$

Summing forces in the y direction,

$$\sum F_y = T \cos \theta - mg = ma_y = 0 \qquad \text{or} \qquad T = \frac{mg}{\cos \theta}$$

Substitute this value T into equation (1):

$$\frac{mg}{\cos \theta} \sin \theta = m(\ell \sin \theta)\omega^2 \qquad \text{or} \qquad \omega = \sqrt{\frac{g}{\ell \cos \theta}}$$

If θ is known, this equation can be solved for the necessary angular velocity ω to maintain θ constant. Or, if ω is known, θ may be found. Since ω is constant for a given angle θ,

$$\text{Frequency } f = \frac{\omega \, \text{rad/s}}{2\pi \, \text{rad/rev}} = \frac{1}{2\pi} \sqrt{\frac{g}{\ell \cos \theta}} \text{ Hz}$$

This is the frequency about the y axis.

3.7. A block, assumed to be a particle and weighing 40 N, rests on a plane which can turn about the y axis [see Fig. 3-9(a)]. The length of the cord ℓ is 1.2 m. What is the tension in the cord when the angular velocity of the plane and block is 10 rev/min?

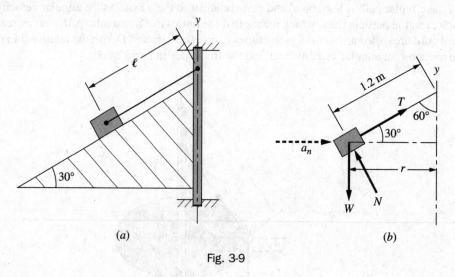

(a) (b)

Fig. 3-9

SOLUTION

From the free-body diagram of the block shown in Fig. 3-9(b), $r = 1.2 \cos 30° = 1.039$ m. The only acceleration present is the normal component a_n directed horizontally toward the y axis:

$$a_n = r\omega^2 = 1.039 \left(\frac{10 \times 2\pi}{60}\right)^2 = 1.139 \text{ m/s}^2$$

Sum forces horizontally along the radius r and along the y axis to obtain the following equations:

$$\sum F_n = T \cos 30° - N \sin 30° = ma_n = \frac{40}{9.8} \times 1.139 \tag{1}$$

$$\sum F_y = N \cos 30° + T \sin 30° - 40 = ma_y = 0 \tag{2}$$

Solve equation (2) for N:

$$N = \frac{40}{\cos 30°} - T \frac{\sin 30°}{\cos 30°}$$

Substituting into equation (1), there results

$$T \cos 30° - \left(\frac{40}{\cos 30°} - T \frac{\sin 30°}{\cos 30°} \right) \sin 30° = \frac{40}{9.8} \times 1.139 \qquad \text{or} \qquad T = 35.23 \text{ N}$$

3.8. The 20-N object at B in Fig. 3-10(a) moves in a circular, horizontal path under the action of a cord AB and a rigid bar BC, which can be considered weightless. At the instant B has a speed of 2 m/s, what are the forces in the cord AB and the bar BC?

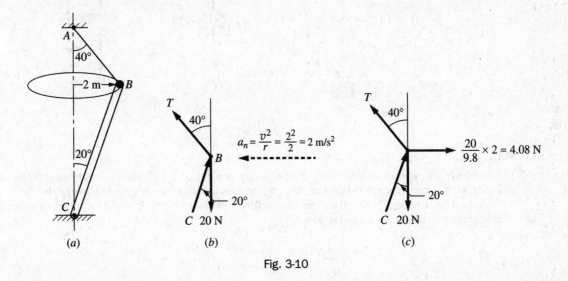

Fig. 3-10

SOLUTION

The free-body diagram in Fig. 3-10(b) shows the object under the action of its weight, tension T in the cord, and compression C in the bar. The normal acceleration of the object is $v^2/r = 2^2/2 = 2$ m/s^2 directed to the left. Summation of forces yields the following equations:

$$\sum F_v = T \cos 40° + C \cos 20° - 20 = 0$$

$$\sum F_h = T \sin 40° - C \sin 20° = m \frac{v^2}{r} = \frac{20}{9.8} \times 2 = 4.08$$

The solution of these equations yields

$$T = 12.3 \text{ N} \qquad \text{and} \qquad C = 11.2 \text{ N}$$

Another method to find the solution, using D'Alembert's principle and the "inertia force" is shown in Fig. 3-10(c). The figure is the free-body diagram with the inertia force $ma_n = 4.08$ N acting on the particle outwardly from the center of rotation. Thus,

$$\sum F_v = T \cos 40° + C \cos 20° - 20 = 0$$

$$\sum F_h = -T \sin 40° + C \sin 20° + 4.08 = 0$$

The solution of these equations yields

$$T = 12.3 \text{ N} \quad \text{and} \quad C = 11.2 \text{ N}$$

In circular motion, the inertia force is called the *centrifugal force*. The centrifugal force is often erroneously thought of as an actual force. It is not!

3.9. In a device as Atwood's machine, two equal masses M are connected, by a very light (negligible mass) tape passing over a frictionless cylinder, as shown in Fig. 3-11(a). A mass m whose magnitude is much less than M is added to one side, causing that mass to fall and the other to rise. The time is recorded by an inked stylus resting on the moving tape. Find an expression for the acceleration a.

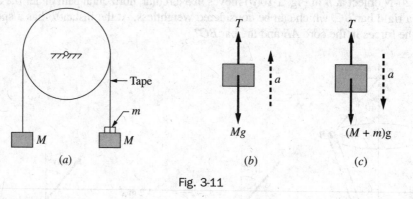

(a) (b) (c)

Fig. 3-11

SOLUTION

The free-body diagrams of the two mass systems are shown in Fig. 3-11(b) and (c). The same tension T is acting on each system through the tape because the friction of the pulley is assumed negligible.

The equations of motion, using the same acceleration (otherwise the tape would have broken or become slack) for the two free-body diagrams, are

$$\sum F = T - Mg = Ma \tag{1}$$

$$\sum F = Mg + mg - T = (M + m)a \tag{2}$$

Add equations (1) and (2) to eliminate the tension T and obtain

$$mg = 2Ma + ma \quad \text{or} \quad a = \frac{m}{2M + m}g$$

This expresses the relation between the acceleration of gravity g at the locality where the experiment is performed and the acceleration a of the masses as determined by measurement of distance and time on the tape.

3.10. Figure 3-12 shows a 2-kg mass resting on a frictionless plane inclined 20° to the horizontal. A cord that is parallel to the plane passes over a massless, frictionless pulley to a 4-kg mass which will drop vertically when released. What will be the speed of the 4-kg mass 4 s after it is released from rest?

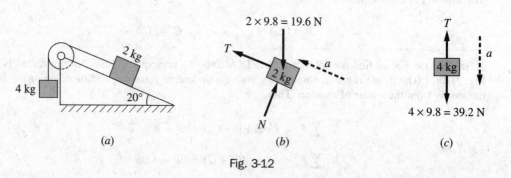

(a) (b) (c)

Fig. 3-12

SOLUTION

The free-body diagrams are shown in Fig. 3-12(*b*) and (*c*). Both bodies have the same magnitude of acceleration. The tension *T* acting on each free body is the same. The equations of motion are

$$\sum F_b = T - 19.6 \sin 20° = 2a$$

$$\sum F_c = 39.2 - T = 4a$$

Adding, the acceleration is

$$a = 5.42 \text{ m/s}^2$$

The velocity of the 4-kg mass after 4 s is

$$v = v_0 + at = 0 + 5.42(4) = 21.7 \text{ m/s}$$

3.11. Blocks *A* and *B*, weighing 100 and 300 N, respectively, are connected by a weightless rope passing over a frictionless pulley, as shown in Fig. 3-13(*a*). Assume a coefficient of friction of 0.30, and determine the velocity of the system 4 s after starting from rest.

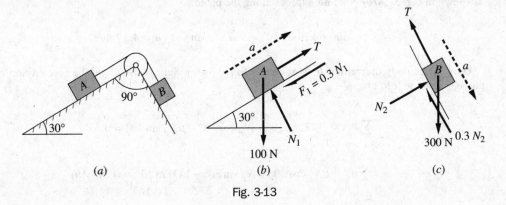

Fig. 3-13

SOLUTION

Free-body diagrams are drawn for bodies *A* and *B* [see Fig. 3-13(*b*) and (*c*)]. Summing forces perpendicular and parallel to the planes, the equations of motion are

$$\sum F_\perp = N_1 - 100 \cos 30° = \frac{100}{9.8}(0) = 0 \tag{1}$$

$$\sum F_\parallel = T - 100 \sin 30° - 0.30 N_1 = \frac{100}{9.8} a \tag{2}$$

$$\sum F_\perp = N_2 - 300 \cos 60° - \frac{300}{9.8}(0) = 0 \tag{3}$$

$$\sum F_\parallel = 300 \sin 60 - T - 0.30 N_2 = \frac{300}{9.8} a \tag{4}$$

Solve Eqs. (1) and (3) for N_1 and N_2. Substitute these values in Eqs. (2) and (4) and add the two equations to eliminate *T*. This yields an acceleration $a = 3.41 \text{ m/s}^2$.

Applying the kinematics equation

$$v = v_0 + at = 0 + 3.41(4) = 13.6 \text{ m/s}$$

3.12. Refer to Fig. 3-14(*a*). Determine the least coefficient of friction between *A* and *B* so that slip will not occur. Here *A* is a 40-kg mass, *B* is a 15-kg mass, and *F* is 500 N, parallel to the plane, which is frictionless.

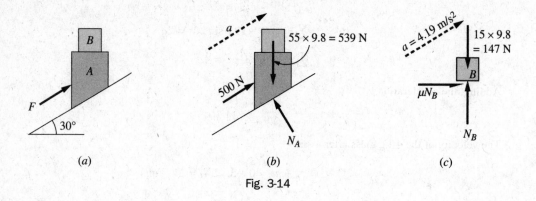

Fig. 3-14

SOLUTION

To determine the acceleration *a* of the system, draw a free-body diagram of the two masses taken as a unit, as shown in Fig. 3-14(*b*). Summing forces along the plane,

$$500 - 539\sin 30° = 55a \qquad \text{or} \qquad a = 4.19 \text{ m/s}^2$$

Draw a free-body diagram of *B*, as shown in Fig. 3-14(*c*). Sum forces along the acceleration vector and perpendicular to it to obtain

$$\sum F_\perp = -147\cos 30° + N_B \cos 30° - \mu N_B \sin 30° = 0 \qquad (1)$$

$$\sum F_\| = \mu N_B \cos 30° + N_B \sin 30° - 147\sin 30° = (15)(4.19) \qquad (2)$$

Multiply the first equation by cos 30° and the second by sin 30°. Then add to obtain $N_B = 178$ N. Substituting into either equation (1) or (2), $\mu = 0.30$.

3.13. A horizontal force $P = 70$ N is exerted on mass $A = 16$ kg as shown in Fig. 3-15(*a*). The coefficient of friction between *A* and the horizontal plane is 0.25. Block *B* has a mass of 4 kg, and the coefficient of friction between it and the plane is 0.50. The cord between the two masses makes an angle of 10° with the horizontal. What is the tension in the cord?

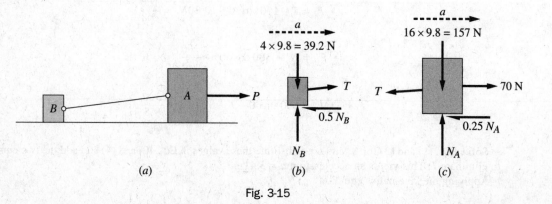

Fig. 3-15

SOLUTION

Newton's second law applied to the two free-body diagrams gives [see Fig. 3-15(b) and (c)]

$$\sum F_h = 70 - T\cos 10° - 0.25 N_A = 16a \tag{1}$$

Body B

$$\sum F_v = N_A - 157 - T\sin 10° = 0 \tag{2}$$

$$\sum F_h = T\cos 10° - 0.50 N_B = 4a \tag{3}$$

Body A

$$\sum F_v = N_B - 39.2 + T\sin 10° = 0 \tag{4}$$

Substitute N_A in terms of T from Eq. (2) into Eq. (1). Substitute N_B in terms of T from Eq. (4) into Eq. (3). Eliminate a between these two new equations to obtain $T = 20.5$ N.

3.14. A box is dropped onto a conveyor belt inclined 10° with the horizontal and traveling at 5 m/s. If the box is initially at rest and the coefficient of friction between the box and the belt is $\frac{1}{3}$, how long will it take before the box ceases to slip on the belt?

SOLUTION

Refer to Fig. 3-16. The equations of motion for the box are

$$\sum F_\perp = 0 = N - W\cos 10°$$

$$\sum F_\parallel = \frac{1}{3}N - W\sin 10° = \frac{W}{g}a$$

Solving these equations, we get $a = 1.515$ m/s^2.
   To find the time that will elapse before slipping ceases, use

$$v = v_0 + at \qquad 5 = 0 + 1.515t \qquad \therefore t = 3.3\text{ s}$$

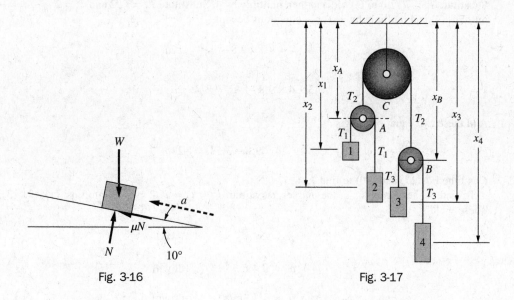

Fig. 3-16   Fig. 3-17

3.15. In the system shown in Fig. 3-17, the pulleys and cords may be considered massless and frictionless, as usual. The numbers indicate the masses in kilograms. Determine the acceleration of each mass and the tension T in the fixed cord holding pulley C.

SOLUTION

With the assumptions of no mass for the pulleys and cords, we can state $2T_1 = T_2$ and $2T_3 = T_2$. (A free-body diagram of each pulley shows these to be true.) Also the cords over pulleys A, B, and C are constant in length. Hence,

$$(x_1 - x_A) + (x_2 - x_A) = K_1 \qquad (x_3 - x_B) + (x_4 - x_B) = K_2 \qquad x_A + x_B = K_3$$

where K_1, K_2, and K_3 are constants. Second derivatives yield

$$a_1 + a_2 = 2a_A \qquad a_3 + a_4 = 2a_B \qquad a_A + a_B = 0$$

The equations of motion applied to free-body diagrams of each mass, assuming downward as the positive direction, are

$$1 \times 9.8 - T_1 = a_1 \tag{1}$$

$$2 \times 9.8 - T_1 = 2a_2 \tag{2}$$

$$3 \times 9.8 - T_3 = 3a_3 \tag{3}$$

$$4 \times 9.8 - T_3 = 4a_4 \tag{4}$$

Substitute $T_1 = T_2/2$ into Eq. (1) and then multiply by 2. Substitute $T_1 = T_2/2$ and $a_2 = 2a_A - a_1$ into Eq. (2). These equations become

$$2 \times 9.8 - \frac{2T_2}{2} = 2a_1 \tag{5}$$

$$2 \times 9.8 - \frac{T_2}{2} = 4a_A - 2a_1 \tag{6}$$

Add these two equations to get

$$4 \times 9.8 - 1.5T_2 = 4a_A \tag{7}$$

Substitute $T_3 = T_2/2$ into Eq. (3) and then multiply by 4. Substitute $T_3 = T_2/2$ and $a_4 = 2a_B - a_3 = -2a_A - a_3$ into Eq. (4) and multiply by 3. These equations become

$$4 \times 3 \times 9.8 - \frac{4T_2}{2} = 12a_3 \tag{8}$$

$$3 \times 4 \times 9.8 - \frac{3T_2}{2} = 3[4(-2a_A - a_3)] \tag{9}$$

Add these two equations to get

$$24 \times 9.8 - 3.5T_2 = -24a_A \tag{10}$$

Combine Eqs. (7) and (10) to find $T_2 = 37.6$ N.

To find the accelerations of the masses, substitute $T_1 = T_3 = T_2/2 = 18.8$ into Eqs. (1), (2), (3), and (4). These yield

$$a_1 = -9.0 \text{ m/s}^2 \qquad \text{(up)}$$

$$a_2 = 0.4 \text{ m/s}^2 \qquad \text{(down)}$$

$$a_3 = 3.53 \text{ m/s}^2 \qquad \text{(down)}$$

$$a_4 = 5.1 \text{ m/s}^2 \qquad \text{(down)}$$

The tension T in the fixed cord is $T = 2T_2 = 75.2$ N.

3.16. Two masses of 14 and 7 kg connected by a cord rest on a smooth plane inclined 45° with the horizontal, as shown in Fig. 3-18(a). When the masses are released, what will be the tension T in the cord? Assume the coefficient of friction between the plane and the 14-kg mass is 0.25 and between the plane and the 7-kg mass is 0.375.

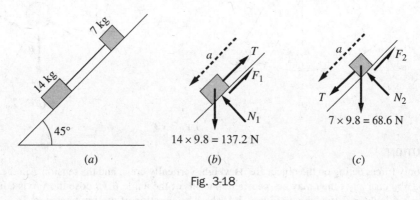

Fig. 3-18

SOLUTION

Free-body diagrams of the two masses are shown in Fig. 3-18(b) and (c). The equations of motion for the 14-kg mass are as follows, where the summations are parallel and perpendicular to the plane:

$$\sum F_{\parallel} = 137.2 \sin 45° - F_1 - T = 14a \tag{1}$$

$$\sum F_{\perp} = N_1 - 137.2 \cos 45° = 0 \tag{2}$$

The equations of motion for the 7-kg mass are as follows:

$$\sum F_{\parallel} = T + 68.6 \sin 45° - F_2 - 7a \tag{3}$$

$$\sum F_{\perp} = N_2 - 68.6 \cos 45° = 0 \tag{4}$$

Assume that both masses are moving. Of course, if the friction is great enough, the masses may not move. From Eq. (2),

$$N_1 = 137.2 \times 0.707 \qquad \text{then} \qquad F_1 = 0.25N_1 = 24.3 \text{ N}$$

From Eq. (4),

$$N_2 = 68.6 \times 0.707 \qquad \text{then} \qquad F_2 = 0.375N_2 = 18.2 \text{ N}$$

Substituting these values into Eqs. (1) and (3), the following equations result:

$$137.2 + 0.707 - 24.3 - T = 14a \tag{5}$$

$$T + 68.6 \times 0.707 - 18.2 = 7a \tag{6}$$

Multiply Eq. (6) by 2 and subtract from Eq. (5) to obtain $T = 4.0$ N.

3.17. An object of weight W is suspended on a cord of length ℓ as shown in Fig. 3-19(a). Determine the period and frequency of this simple pendulum, assuming a small angular displacement.

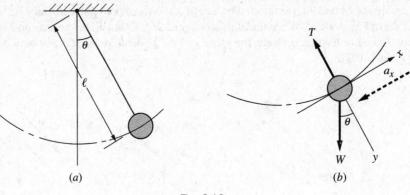

(a)   (b)

Fig. 3-19

SOLUTION

The only forces acting on the object are its weight vertically down and the tension T in the cord. The position of the object at any time t may be specified in terms of the angle θ. Choose the x axis tangent to the path of the object in the position shown in Fig. 3-19(b). The equation of motion becomes

$$\sum F_x = -W \sin \theta = \frac{W}{g} a_x$$

Thus, the acceleration is zero when θ is zero, i.e., at the lowest position of the object.

The above differential equation may be solved by noting that a_x is tangent to the path and may therefore be written

$$a_x = \ell \alpha$$

where α = angular acceleration of the cord and object. Then,

$$-W \sin \theta = \frac{W}{g} \ell \alpha = \frac{W}{g} \ell \frac{d^2 \theta}{dt^2} \qquad \text{or} \qquad -\frac{g}{\ell} \sin \theta = \frac{d^2 \theta}{dt^2}$$

The solution of this differential equation is simplified by using a series expansion for $\sin \theta$:

$$\sin \theta = \theta - \frac{\theta^3}{3!} + \frac{\theta^5}{5!} - \frac{\theta^7}{7!} + \cdots$$

For small angular displacements, $\sin \theta$ is approximately equal to θ, expressed in radians. Thus, for small displacements, the equation of motion becomes

$$-\frac{g}{\ell} \theta = \frac{d^2 \theta}{dt^2} \qquad \text{or} \qquad \frac{d^2 \theta}{dt^2} + \frac{g}{\ell} \theta = 0$$

The solution of this second-order ordinary differential equation is in the form of sines and cosines:

$$\theta = A \sin \sqrt{\frac{g}{\ell}} t + B \cos \sqrt{\frac{g}{\ell}} t$$

The constants A and B can be evaluated in a given problem by using the boundary conditions.

The frequencies are

$$\omega = \sqrt{\frac{g}{\ell}} \text{ rad/s} \qquad f = \frac{1}{2\pi} \sqrt{\frac{g}{\ell}} \text{ Hz (cycles/s)}$$

Incidentally, a cycle is the motion of the particle from a starting point through all possible positions back to the same point. A cycle is complete when the object moves, let us say, from its top left position to its right position and back to its top left position.

The time to complete one cycle is called the *period T*. This is then the reciprocal of the frequency f:

$$T = \frac{1}{f(\text{cycles/s})} = \frac{1}{f} (\text{s/cycle})$$

3.18. Consider the motion of an object of mass m resting on a frictionless horizontal plane, as shown in Fig. 3-20(a). It is attached to a spring that has a constant of K (N/m). Displace the mass a distance x_0 from its equilibrium position (spring tension or compression is zero at equilibrium position) and then release it with zero velocity. Find expressions for the displacement and velocity of the object.

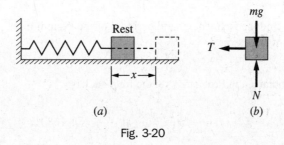

(a)   (b)

Fig. 3-20

SOLUTION

Figure 3-20(b) is a free-body diagram showing the object in a position a distance x from equilibrium. Acting on the object in the horizontal direction is the force T in the spring, which is stretched the distance x.

Within the elastic limit of the material, it is assumed that the tension in the spring is proportional to its change in length from the unstressed position. Then

$$T = Kx$$

where T = spring force
K = spring constant
x = change of length

A summation of forces horizontally yields

$$\sum F_x = -T = ma_x$$

Note that forces directed to the left, and distances to the left of the rest position, are assumed negative. In this case, the distance x is to the right; hence, a_x is written positive. Tension T is to the left, i.e., negative.

Substituting $T = Kx$ and $a_x = d^2x/dt^2$, the above equation becomes $-Kx = md^2x/dt^2$ (simple harmonic motion). This differential equation is similar to the one found in Problem 3.17. Its solution is in the form of sines and cosines:

$$x = A\sin\sqrt{\frac{K}{m}}t + B\cos\sqrt{\frac{K}{m}}t$$

The values of A and B can now be calculated. The value of x is x_0 when t is zero; hence,

$$x_0 = A\sin\sqrt{\frac{K}{m}}0 + B\cos\sqrt{\frac{K}{m}}0 = 0 + B \qquad \text{so} \qquad B = x_0$$

giving

$$x = A\sin\sqrt{\frac{K}{m}}t + x_0\cos\sqrt{\frac{K}{m}}t$$

To evaluate A, it is necessary to differentiate x with respect to time, since the other known condition is that the velocity v is zero when the time t is zero:

$$v = \frac{dx}{dt} = A\sqrt{\frac{K}{m}}\cos\sqrt{\frac{K}{m}}t - x_0\sqrt{\frac{K}{m}}\sin\sqrt{\frac{K}{m}}t$$

When $t = 0$, $v = 0$ and $\sin \sqrt{K/m}\, 0 = 0$. Then $0 = A\sqrt{K/m}\,\cos 0 - 0$, and $A = 0$. The displacement and velocity are, respectively,

$$x = x_0 \cos \sqrt{\frac{K}{m}}t \qquad v = -x_0 \sqrt{\frac{K}{m}} \sin \sqrt{\frac{K}{m}}t$$

Of course, if the initial velocity had some value other than zero, say v_0, then $A = v_0/\sqrt{K/m}$.

3.19. In Problem 3.18, find the frequency f and period T of the system if the mass m has a weight of 4 N and the spring constant is 2 N/m.

SOLUTION

Using the results obtained in Problem 3.18, the frequency is

$$f = \frac{1}{2\pi}\sqrt{\frac{K}{m}} = \frac{1}{2\pi}\sqrt{\frac{22}{4/9.8}} = 1.168 \text{ Hz}$$

$$T = \frac{1}{f} = \frac{1}{1.168 \text{ Hz}} = 0.856 \text{ seconds per cycle}$$

3.20. A particle of mass m rests on the top of a smooth sphere of radius r as shown in Fig. 3-21(a). Assuming that the particle starts to move from rest, at what point will it leave the sphere?

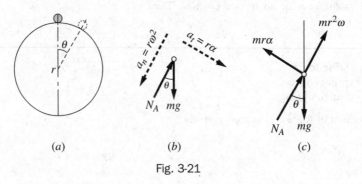

(a) $\qquad\qquad\qquad$ (b) $\qquad\qquad\qquad$ (c)

Fig. 3-21

SOLUTION

Let θ be the angular displacement at any time t during its travel. The free-body diagram indicates the only two forces acting on the particle, i.e., the plane reaction along the radius and the weight mg [see Fig. 3-21(b)].

   The equations of motion found by summing forces along the radius and the tangent are

$$\sum F_r = -N_A + mg \cos \theta = mr\omega^2 \tag{1}$$

$$\sum F_t = mg \sin \theta = mr\alpha \tag{2}$$

A third equation is necessary for a solution. This is the one expressing a relationship among θ, ω, and α; and for review it is derived here. Eliminating dt from $\omega = d\theta/dt$ and $\alpha = d\omega/dt$ yields

$$\alpha\, d\theta = \omega\, d\omega \tag{3}$$

From Eq. (2), $\alpha = (g/r) \sin \theta$. Substituting into (3), $(g/r) \sin \theta\, d\theta = \omega\, d\omega$. Then,

$$\int_0^\theta \frac{g}{r} \sin \theta\, d\theta = \int_0^\omega \omega\, d\omega \qquad \text{or} \qquad \frac{g}{r}(1 - \cos \theta) = \frac{\omega^2}{2} \tag{4}$$

At the position of departure from the sphere, the normal reaction N_A will be zero. Equation (1) now becomes

$$mg \cos \theta = mr\omega^2 \qquad \text{or} \qquad \frac{g}{r} \cos \theta = \omega^2 \tag{5}$$

Replacing ω^2 in Eq. (4) by its value in Eq. (5), we obtain

$$\frac{g}{r}(1 - \cos\theta) = \frac{g}{2r}\cos\theta \qquad \cos\theta = \frac{2}{3} \qquad \theta = 0.841 \text{ rad or } 48.2°$$

For an alternate solution, let Fig. 3-21(c) be the free-body diagram showing the two inertia forces $mr\alpha$ and $mr\omega^2$. The resulting equilibrium equations become

$$\sum F_v = N_A \cos\theta - mg + mr\omega^2 \cos\theta + mr\alpha\sin\theta = 0$$

$$\sum F_h = N_A \sin\theta + mr\omega^2 \sin\theta - mr\alpha\cos\theta = 0$$

When the particle leaves the sphere, $N_A = 0$. Eliminating ω^2 from the simultaneous equations gives $\alpha = (g/r)\sin\theta$. Eliminating α from the equations gives $\omega^2 = (g/r)\cos\theta$. Using Eq. (3) and substituting for α give Eq. (4). Substituting for ω^2 yields

$$\frac{g}{r}(1 - \cos\theta) = \frac{g}{2r}\cos\theta$$

This equation gives $\theta = 48.2°$.

3.21. A particle of mass m slides down the frictionless chute of Fig. 3-22 (a) and enters a "loop-the-loop" of diameter d. What should be the height h at the start in order that the particle just makes a complete circuit in the loop?

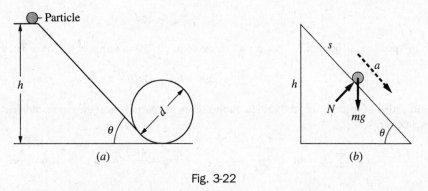

(a)   (b)

Fig. 3-22

SOLUTION

A free-body diagram of the particle at any time in its travel down is shown in Fig. 3-22(b). The equations of motion parallel and perpendicular to the plane are

$$\sum F_{\parallel} = mg\sin\theta = ma \tag{1}$$

$$\sum F_{\perp} = N - mg\cos\theta = 0 \tag{2}$$

From Eq. (1), the acceleration is $a = g\sin\theta$. This value of a is now substituted into the kinematics equation $a\,ds = v\,dv$, where s refers to displacement along the plane. Then $g\sin\theta\,ds = v\,dv$ and

$$\int_0^s g\sin\theta\,ds = \int_0^v v\,dv \qquad \text{or} \qquad g\sin\theta\,s = \frac{v^2}{2} \tag{3}$$

At the bottom of the plane the speed is found by substituting $h/\sin\theta$ for s in Eq. (3):

$$g\sin\theta\frac{h}{\sin\theta} = \frac{v^2}{2} \qquad \text{or} \qquad v^2 = 2gh \tag{4}$$

This indicates that the speed at the bottom of the smooth incline is independent of the slope of the incline and is the same as if the particle fell vertically downward.

Next draw a free-body diagram of the particle at the top of the loop (see Fig. 3-23). The forces acting are the gravitational force mg and the force N_A of the loop along the vertical radius. Since the minimum speed is the one with which we are concerned, the value of N_A in this case is zero. Expressed somewhat differently, as the speed v increases, the normal acceleration a_n increases. To provide this increasing value of a_n, the reaction N_A must increase also, because $\sum F_n = ma_n$.

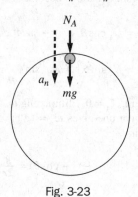

Fig. 3-23

In determining the speed at the top of the loop, use the fact just determined that the motion on a smooth path (such as the plane or the side of the loop) is equivalent to vertical motion under the acceleration of gravity only. Hence, the particle loses speed in moving up the side of the loop in an amount equal to the loss in straight-line vertical motion:

$$v_{top}^2 = v_{bot}^2 - 2gd$$

But the speed at the bottom v_{bot} has been expressed in Eq. (4) in terms of height h. Then

$$v_{top}^2 = 2gh - 2gd = 2g(h - d)$$

A summation of forces along the vertical radius when the particle is at the top results in the following equation, since $N_A = 0$:

$$\sum F_n = mg = \frac{mv_{top}^2}{d/2} \qquad \text{or} \qquad mg = \frac{4mg}{d}(h - d) \qquad \therefore h = \frac{5}{4}d$$

This means that the particle must leave at this height in order that a minimum speed be obtained. If $h < 5d/4$, the particle will not follow the circular path at the top, but will "jump across."

3.22. A flexible chain of length ℓ rests on a frictionless table with length c overhanging the rounded edge, as shown in Fig. 3-24(a). The system, originally at rest, is released. The chain weighs w N/m. Assume that the chain maintains contact with the table surfaces. Find the length of the overhang as a function of time.

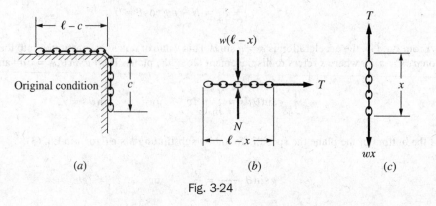

(a)   (b)   (c)

Fig. 3-24

SOLUTION

The free-body diagrams of the two pieces of the chain are drawn with the same tension T shown acting on each piece. See Fig. 3-24(b) and (c). The length of overhang is x. The equations of motion are

$$T = \frac{w(\ell - x)}{g}a = \frac{w(\ell - x)}{g}\frac{d^2x}{dt^2} \tag{1}$$

$$wx - T = \frac{wx}{g}a = \frac{wx}{g}\frac{d^2x}{dt^2} \tag{2}$$

Add (1) and (2) to obtain

$$wx = \frac{w\ell}{g}\frac{d^2x}{dt^2} \quad \text{or} \quad \frac{d^2x}{dt^2} = \frac{g}{\ell}x$$

The solution of this second-order differential equation is

$$x = Ae^{\sqrt{g/\ell}\,t} + Be^{-\sqrt{g/\ell}\,t}$$

To evaluate the constants A and B, note the initial conditions, namely, $x = c$ and $v = 0$. Substituting $x = c$ when $t = 0$, we get $c = A + B$. The velocity is found by differentiating x with respect to t:

$$v = \frac{dx}{dt} = A\sqrt{\frac{g}{\ell}}e^{\sqrt{g/\ell}\,t} - B\sqrt{\frac{g}{\ell}}e^{-\sqrt{g/\ell}\,t}$$

Substituting the condition $v = 0$ when $t = 0$, we get $A - B = 0$. Solving simultaneously

$$A - B = 0 \quad \text{and} \quad c = A + B \quad \therefore A = B = \frac{1}{2}c$$

The solution of the problem is

$$x = \frac{1}{2}ce^{\sqrt{g/\ell}\,t} + \frac{1}{2}ce^{-\sqrt{g/\ell}\,t}$$

The exponential functions could be evaluated for any given time t to determine the length x of the overhang.

3.23. An object of weight W falls from rest in a fluid with the resistance R proportional to the velocity. This is approximately true for slowly moving objects. Derive an expression for the vertical distance from a fixed plane as a function of time.

SOLUTION

The free-body diagram of the object shows the weight W acting down and the retarding force R acting up (see Fig. 3-25). Assuming that x, $\dot{x}$, and $\ddot{x}$ are positive in the downward direction, the differential equation of the motion is

$$W - k\dot{x} = \frac{W}{g}\ddot{x} = \frac{W}{g}\frac{d\dot{x}}{dt}$$

Rewrite this equation and then integrate:

$$\frac{d\dot{x}}{W/k - \dot{x}} = \frac{kg}{W}dt \quad \text{and} \quad -\ln\left(\frac{W}{k} - \dot{x}\right) = \frac{kg}{W}t + C_1$$

Assuming $\dot{x} = 0$ when $t = 0$, $-\ln(W/k) = C_1$ and

$$-\ln\left(\frac{W}{k} - \dot{x}\right) = \frac{kg}{W}t - \ln\left(\frac{W}{k}\right) \quad \text{or} \quad \frac{W/k - \dot{x}}{W/k} = e^{-(kg/W)t}$$

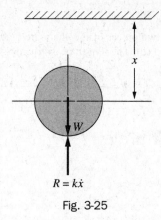

Fig. 3-25

from which

$$\dot{x} = \frac{W}{k}(1 - e^{-(kg/W)t})$$ (1)

Since the limiting velocity $\dot{x}_{max}$ occurs as $t \rightarrow \infty$,

$$\dot{x}_{max} = \frac{W}{k}$$ (2)

To determine the distance x as a function of time, write (1) as

$$dx = \frac{W}{k}(1 - e^{-(kg/W)t})\,dt$$

Integrating,

$$x = \frac{W}{k}t - \frac{W}{k}(-W/kg)\,e^{-(kg/W)t} + C_2$$

Assuming $x = 0$ when $t = 0$, $0 = 0 + W^2/k^2g + C_2$. Then,

$$x = \frac{W}{k}t + \frac{W^2}{k^2g}e^{-(kg/W)t} - \frac{W^2}{k^2g}$$

Using $W/k = \dot{x}_{max}$, this expression is regrouped as

$$x = \frac{W}{k}t - \frac{\dot{x}_{max}^2}{g}(1 - e^{-(kg/W)t})$$ (3)

3.24. An object of mass m falls from rest in a fluid with the resistance R proportional to the square of the speed. This is approximately true for high-speed objects. Derive an expression for the velocity as a function of time or position.

SOLUTION

The free-body diagram of the object (Fig. 3-26) shows the gravitational force mg acting down and the retarding force R acting up. Assuming that x, $\dot{x}$, and $\ddot{x}$ are positive in the downward direction, the differential equation of motion is

$$mg - k\dot{x}^2 = m\ddot{x} = m\frac{d\dot{x}}{dx}\frac{dx}{dt} = m\frac{d\dot{x}}{dx}\dot{x}$$

Rewrite this equation and integrate:

$$\frac{\dot{x}\,d\dot{x}}{mg/k - \dot{x}^2} = \frac{k}{m}dx \qquad \text{and} \qquad -\frac{1}{2}\ln\left(\frac{mg}{k} - \dot{x}^2\right) = \frac{k}{m}x + C_1$$

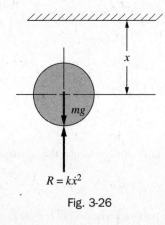

Fig. 3-26

Assuming $\dot{x} = 0$ when $x = 0$, $-\frac{1}{2}\ln(mg/k) = C_1$ and

$$-\frac{1}{2}\ln\left(\frac{mg}{k} - \dot{x}^2\right) = \frac{k}{m}x - \frac{1}{2}\ln\left(\frac{mg}{k}\right)$$

Then

$$-\frac{2k}{m}x = \ln\left(1 - \frac{k\dot{x}^2}{mg}\right) \quad \text{or} \quad \frac{mg}{k} - \dot{x}^2 = \frac{mg}{k}e^{-(2k/m)x} \quad \text{or} \quad 1 - \frac{k}{mg}\dot{x}^2 = e^{-(2k/m)x}$$

or

$$v^2 = \frac{mg}{k}(1 - e^{-(2k/m)x}) \tag{1}$$

The limiting velocity occurs as $x \to \infty$; thus,

$$v_{\max} = \sqrt{\frac{mg}{k}} \tag{2}$$

3.25. A projectile of mass m is given an initial velocity v_0 at an angle θ with the horizontal. Determine the range, the maximum height, and the time of flight, assuming that the projectile hits on the same horizontal plane from which it is fired. Neglect air resistance in this solution.

SOLUTION

The horizontal line is chosen as the x axis. The range is r. The projectile is shown in Fig. 3-27 at some point (x, y) along the path. The only force acting is its weight mg.

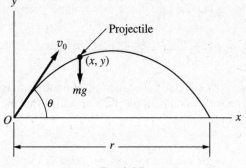

Fig. 3-27

The equations of motion in the x and y directions are

$$\sum F_x = 0 = m\ddot{x} \quad \text{and} \quad \sum F_y = -mg = m\ddot{y}$$

from which $\ddot{x} = 0$ and $\ddot{y} = -g$. Integrating $\dot{x} = C_1$ and $\dot{y} = -gt + C_2$. But $\dot{x}$ has constant value $v_0 \cos \theta$; and at $t = 0$, $\dot{y} = v_0 \sin \theta$. Hence,

$$\dot{x} = v_0 \cos \theta \qquad \text{and} \qquad \dot{y} = -gt + v_0 \sin \theta$$

Another integration yields

$$x = v_0 t \cos \theta + C_3 \qquad \text{and} \qquad y = v_0 t \sin \theta - \frac{1}{2} g t^2 + C_4$$

Since $x = 0$ and $y = 0$ when $t = 0$, $C_3 = C_4 = 0$, and the equations of motion are

$$x = v_0 t \cos \theta \qquad \text{and} \qquad y = v_0 t \sin \theta - \frac{1}{2} g t^2$$

To obtain the equation of the path in cartesian coordinates, eliminate t from the equations. This is most easily done by solving the x equation for t and substituting its value into the y equation. Substituting $t = x/(v_0 \cos \theta)$ into the y equation yields

$$y = v_0 \sin \theta \frac{x}{v_0 \cos \theta} - \frac{1}{2} g \frac{x^2}{v_0^2 \cos^2 \theta} = x \tan \theta - \frac{g}{2 v_0^2 \cos^2 \theta} x^2$$

which is the equation of a parabola.

To determine the time of flight, equate y to zero. The height is zero at the beginning of the motion and also when the projectile hits the ground.

$$y = 0 = v_0 t \sin \theta - \frac{1}{2} g t^2 = t \left(v_0 \sin \theta - \frac{1}{2} g t \right)$$

The values of t that satisfy this equation are $t = 0$ (beginning of flight) and

$$t = \frac{2 v_0 \sin \theta}{g} \qquad \text{(time of flight)}$$

obtained from $v_0 \sin \theta = \frac{1}{2} g t$.

The range may be calculated by substituting this value of t into the x equation of motion:

$$x = r = v_0 \cos \theta \left(\frac{2 v_0 \sin \theta}{g} \right) = \frac{v_0^2}{g} \sin 2\theta \qquad \text{(range)}$$

Note that the maximum range for a given initial velocity v_0 is achieved when $\sin 2\theta = 1$, or when $2\theta = 90°$. Thus θ should be 45° for maximum range for a given v_0.

The maximum height can be determined either by assuming that it occurs at one-half the time of flight (assuming on air resistance) or by solving for the time t when the y component of velocity is equated to zero. Assume that the time is one-half the time of flight. Substituting this value into the y equation results in

$$h = v_0 \sin \theta \left(\frac{v_0 \sin \theta}{g} \right) - \frac{1}{2} g \left(\frac{v_0 \sin \theta}{g} \right)^2 = \frac{v_0^2 \sin^2 \theta}{2g} \qquad \text{(maximum height)}$$

3.26. A particle of mass m moves under the attraction of another mass M [see Fig. 3-28(a)], which we assume to be at rest. Study the motion of the mass m, noting that the distance between the two particles is not necessarily constant.

SOLUTION

The particle m has radius vector $\mathbf{r}$ with respect to mass M. The unit vectors $\mathbf{e}_r$ and $\mathbf{e}_\phi$ for the polar coordinate system are shown. A free-body diagram of mass m contains only the attractive force $\mathbf{F}$, which is in the negative $\mathbf{e}_r$ direction [see Fig. 3-28(b)]. According to *Newton's law of gravitation*, this central force is

$$\mathbf{F} = -G \frac{Mm}{r^2} \mathbf{e}_r \qquad (1)$$

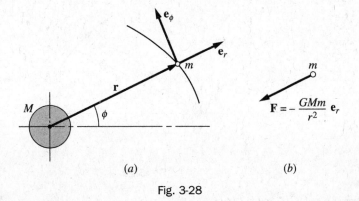

Fig. 3-28

where G is the universal gravitational constant:

$$G = 6.67 \times 10^{-11} \, \text{m}^3/(\text{kg} \cdot \text{s}^2)$$

In polar coordinates, the acceleration $\mathbf{a}$ is [see Eq. (34) in Chap. 2]

$$\mathbf{a} = (\ddot{r} - r\dot{\phi}^2)\mathbf{e}_r + (r\ddot{\phi} + 2\dot{r}\dot{\phi})\mathbf{e}_\phi \qquad (2)$$

Substituting from Eqs. (1) and (2) into

$$\mathbf{F} = m\mathbf{a} \qquad (3)$$

gives

$$-\frac{GMm}{r^2}\mathbf{e}_r = m(\ddot{r} - r\dot{\phi}^2)\mathbf{e}_r + m(r\ddot{\phi} + 2\dot{r}\dot{\phi})\mathbf{e}_\phi \qquad (4)$$

Equate coefficients of $\mathbf{e}_\phi$ and $\mathbf{e}_r$ to zero:

$$r\ddot{\phi} + 2\dot{r}\dot{\phi} = 0 \qquad (5)$$

$$\frac{-GM}{r^2} = \ddot{r} - r\dot{\phi}^2 \qquad (6)$$

Equation (5) is identical with

$$\frac{1}{r}\frac{d}{dt}(r^2\dot{\phi}) = 0 \qquad (7)$$

Integrating Eq. (7),

$$r^2\dot{\phi} = C \qquad (8)$$

Note that in Fig. 3-29 the radius vector $\mathbf{r}$ sweeps out an area dA as it rotates through the angle $d\phi$. The differential area dA is approximately equal to $\frac{1}{2}r(r\,d\phi) = \frac{1}{2}r^2 d\phi$. Dividing by dt,

$$\frac{dA}{dt} = \frac{1}{2}r^2\frac{d\phi}{dt} = \frac{1}{2}r^2\dot{\phi} = \frac{1}{2}C$$

Thus, the radius vector $\mathbf{r}$ sweeps out equal areas in equal times.

Next solve Eq. (6) to determine the path of mass m. For convenience, change variables and let $r = 1/u$; then $dr/du = -1/u^2$ and

$$\dot{r} = \frac{dr}{du}\frac{du}{d\phi}\frac{d\phi}{dt} = \frac{-1}{u^2}\frac{du}{d\phi}\dot{\phi}$$

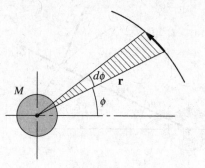

Fig. 3-29

But from Eq. (8), $\dot{\phi} = C/r^2 = Cu^2$; then

$$\dot{r} = -\frac{1}{u^2}\frac{du}{d\phi}Cu^2 = -C\frac{du}{d\phi}$$

and

$$\ddot{r} = \frac{d\dot{r}}{dt} = \frac{d\dot{r}}{d\phi}\frac{d\phi}{dt} = \left(-C\frac{d^2u}{d\phi^2}\right)Cu^2 = -C^2u^2\frac{d^2u}{d\phi^2}$$

Substituting into Eq. (6), we obtain

$$-GMu^2 = -C^2u^2\frac{d^2u}{d\phi^2} - C^2u^3$$

or

$$\frac{d^2u}{d\phi^2} + u = \frac{GM}{C^2} \tag{9}$$

As an intelligent guess, the solution of Eq. (9) might be of the form

$$u = A\cos\phi + B \tag{10}$$

This form must be substituted into Eq. (9) to determine the necessary conditions to make it a solution; thus,

$$\frac{du}{d\phi} = -A\sin\phi \qquad \frac{d^2u}{d\phi^2} = -A\cos\phi$$

and

$$-A\cos\phi + A\cos\phi + B = \frac{GM}{C^2}$$

Hence, $B = GM/C^2$, and the solution of Eq. (9) is

$$u = A\cos\phi + \frac{GM}{C^2} \qquad \text{where } u = \frac{1}{r} \tag{11}$$

Equation (11) represents a conic section, which from analytic geometry is the path of a point that moves so that the ratio of its distance from a fixed point M (a *focus*) to its perpendicular distance from a fixed line (the *directrix*) is a constant e (the *eccentricity*). Figure 3-30 illustrates the distances involved in the definition of e:

$$e = \frac{r}{d - r\cos\phi} \tag{12}$$

Equation (12) can be solved for $1/r$, yielding

$$\frac{1}{r} = \frac{1}{d}\cos\phi + \frac{1}{ed} \tag{13}$$

The following table lists the types of conics (curves) with their e values.

e	0	< 1	1	> 1
curve	circle	ellipse	parabola	hyperbola

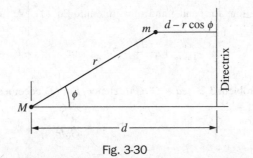

Fig. 3-30

Comparison of Eqs. (11) and (13) indicates that

$$\frac{1}{ed} = \frac{GM}{C^2} \qquad \text{or} \qquad ed = \frac{C^2}{GM} \tag{14}$$

The equation of motion is

$$u = \frac{1}{r} = \frac{1}{d}\cos\phi + \frac{GM}{C^2} \tag{15}$$

3.27. In the preceding problem the path of a planet m around the sun M is elliptical (eccentricity $e < 1$). The sun is at one of the foci of the ellipse. Show that the period T (time for a complete revolution of the planet around the sun) is given by

$$T = \frac{2\pi a^{3/2}}{\sqrt{GM}}$$

where a = semimajor axis of ellipse
   G = gravitational constant
   M = mass of sun

SOLUTION

The path is an ellipse with semimajor axis a and semiminor axis b. Figure 3-31 shows that

$$f + g = 2a \tag{1}$$

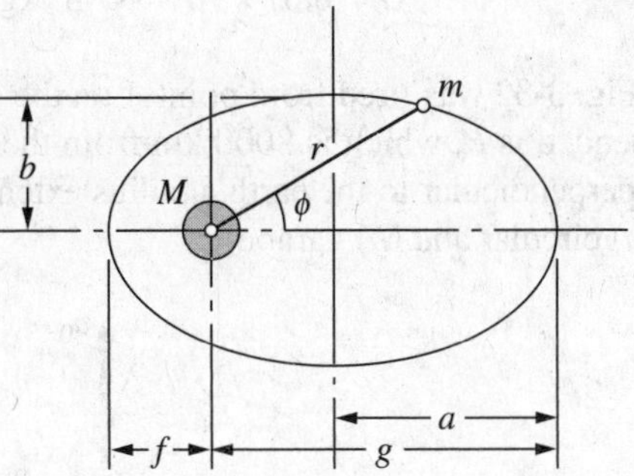

Fig. 3-31

Also, from Eq. (13) of Problem 3.26,

$$\frac{1}{r} = \frac{1}{d}\cos\phi + \frac{1}{ed} \tag{2}$$

When $\phi = 0°$, $r = g$; and when $\phi = 180°$, $r = f$. Thus, Eq. (2) becomes

$$\frac{1}{g} = \frac{1}{d}\cos 0° + \frac{1}{ed} = \frac{1+e}{ed} \qquad \text{and} \qquad \frac{1}{f} = \frac{1}{d}\cos 180° + \frac{1}{ed} = \frac{1-e}{ed}$$

Solve these two equations for f and g and substitute into Eq. (1). We find

$$\frac{ed}{1+e} + \frac{ed}{1-e} = 2a \qquad \text{or} \qquad 1 - e^2 = \frac{ed}{a} \tag{3}$$

But by Eq. (14) of Problem 3.26, $ed = C^2/GM$. Hence, Eq. (3) becomes

$$1 - e^2 = \frac{C^2}{GMa}$$

The area A of the ellipse is $A = \pi ab = \pi a^2\sqrt{1-e^2}$. Thus,

$$A = \pi a^2 \frac{C}{\sqrt{GMa}} = \frac{2\pi a^{3/2}}{\sqrt{GM}} \frac{C}{2} \tag{4}$$

Equation (4) gives the area swept out in one complete revolution, or in time T. This area is also the product of the constant rate dA/dt and time T: $A = (dA/dt)T$. Finally, substituting $dA/dt = C/2$ as determined in Problem 3.26, we have

$$A = \frac{C}{2}T = \frac{2\pi a^{3/2}}{\sqrt{GM}} \frac{C}{2} \qquad \text{or} \qquad T = \frac{2\pi a^{3/2}}{\sqrt{GM}}$$

3.28. Convert the gravitational constant $G = 6.67 \times 10^{-8}$ cm^3/g $\cdot$ s^2 into acceptable SI units.

SOLUTION

The units cm^3 (except length, area, and volume) and g are not acceptable in SI. So

$$G = 6.67 \times 10^{-8} \frac{\text{cm}^3}{\text{g} \cdot \text{s}^2} = 6.67 \times 10^{-8} \frac{(10^{-2}\text{m/cm})^3}{10^{-3}\text{kg/g}} = 6.67 \times 10^{-11} \text{m}^3/\text{kg} \cdot \text{s}^2$$

Since 1 N $= 1$ kg$\cdot$m/s^2, the units can be written in the preferred form

$$\frac{\text{m}^3}{\text{kg} \cdot \text{s}^2} = \frac{\text{kg} \cdot \text{m}^3}{\text{kg}^2 \cdot \text{s}^2} = \frac{(\text{N} \cdot \text{s}^2/\text{m})\text{m}^3}{\text{kg}^2 \cdot \text{s}^2} = \frac{\text{N} \cdot \text{m}^2}{\text{kg}^2}$$

Thus, $$G = 6.67 \times 10^{-11}\, \text{N} \cdot \text{m}^2/\text{kg}^2$$

3.29. The satellite shown in Fig. 3-32 was fired from point A on the surface of the earth. The burnout point at which all fuel is expended is P, which is 8000 km from the earth's center C. Assume the velocity at P is v_0 and that it is perpendicular to the earth's radius extended. Determine the value of v_0 so that the satellite's orbit is (*a*) circular and (*b*) parabolic.

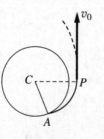

Fig. 3-32

SOLUTION

(*a*) From Eq. (8) in Problem 3.26,

$$C = r^2\dot{\phi} = r(r\dot{\phi}) = rv_0 = (8 \times 10^6)v_0 \tag{1}$$

The solution of the differential Eq. (9) in Problem 3.26 is given in Eq. (11) as

$$u = \frac{1}{r} = A\cos\phi + \frac{GM}{C^2} \tag{2}$$

The solution of Eq. (13) in the same problem is given as

$$\frac{1}{r} = \frac{1}{d}\cos\phi + \frac{1}{ed} \tag{3}$$

Hence, we see

$$A = \frac{1}{d} \qquad \text{and} \qquad \frac{1}{ed} = \frac{GM}{C^2} \tag{4}$$

which combine to give

$$e = \frac{AC^2}{GM} \tag{5}$$

where M, the mass of the earth, is 5.97×10^{24} kg.
   Substituting into Eq. (2) above, we find

$$\frac{1}{8 \times 10^6} = A\cos 0° + \frac{(6.67 \times 10^{-11})(5.97 \times 10^{24})}{(8 \times 10^6 v_0)^2} \tag{6}$$

or

$$A = \frac{1}{8 \times 10^6}\left(1 - \frac{4.98 \times 10^7}{v_0^2}\right) \tag{7}$$

Substitute $e = 0$ (for circular orbit) and the value of A from Eq. (7) into Eq. (5) to find

$$0 = \frac{1}{8 \times 10^6}\left(1 - \frac{4.98 \times 10^7}{v_0^2}\right)\frac{C^2}{GM} \tag{8}$$

Equation (8) can only be satisfied if

$$1 - \frac{4.98 \times 10^7}{v_0^2} = 0$$

Thus $v_0 = 7060$ m/s   or   25 400 km/h

(*b*) If the path is to be parabolic, then $e = 1$. Substituting $e = 1$ into Eq. (5), we get

$$1 = \frac{1}{8 \times 10^6}\left(1 - \frac{4.98 \times 10^7}{v_0^2}\right)\frac{C^2}{GM}$$

$$= \frac{1}{8 \times 10^6}\left(1 - \frac{4.98 \times 10^7}{v_0^2}\right)\frac{(8 \times 10^6)^2 v_0^2}{(6.67 \times 10^{-11})(5.97 \times 10^{24})}$$

This simplifies to

$$v_0 = 9980 \text{ m/s} \qquad \text{or} \qquad 35\,900 \text{ km/h}$$

Note that a v_0 greater than this last value will yield an eccentricity greater than 1, which means that the satellite will depart on a hyperbolic path and never return to earth.

3.30. A satellite with all fuel expended is at a point 800 km from the earth's surface and is traveling with a velocity of 11 000 m/s in a direction perpendicular to an earth radius extended. Determine the eccentricity of the orbit. Assume the earth has a radius of 6340 km.

SOLUTION

Equation (12) of Problem 3.26 can be written as $ed - er \cos \phi = r$. From this, we can write

$$r = \frac{ed}{1 + e \cos \phi} = \frac{C^2/GM}{1 + e \cos \phi}$$

Our initial conditions are (a) $r = 6340 + 800 = 7140$ km and (b) $\phi = 0$. Also, $C = r^2 \dot{\phi} = r(v_0) = 7140 \times 1000 v_0$, $G = 6.67 \times 10^{-11}$, and $M = 5.97 \times 10^{24}$. Hence, with $v_0 = 11\,000$ m/s, we can write Eq. (1) as

$$7.14 \times 10^6 = \frac{(7.14 \times 10^6 \times 11\,000)^2/(6.67 \times 10^{-11})(5.97 \times 10^{24})}{1 + e}$$

or $\qquad 1 + e = \dfrac{4440 \times 5280 \times (36\,000)^2}{(3.43 \times 10^{-8})(4.09 \times 10^{23})} = \dfrac{(11\,000)^2 \times 7.14 \times 10^6}{(6.67 \times 10^{-11})(5.97 \times 10^{24})}$

From this, the eccentricity of the orbit is $e = 2.17$, and the path is hyperbolic.

3.31. Given that the period T for the passage of the earth around the sun is 1 yr and that the earth at its perihelion is 147×10^6 km from the sun and at its aphelion is 152×10^6 km from the sun, determine the mass M of the sun. Ignore the effects of the other planets.

SOLUTION

Figure 3-33 shows the earth in both its near and far positions (perihelion and aphelion, respectively). The semimajor axis a of the elliptical orbit is obtained by adding the two distances and dividing by 2; this gives $a = 149.5 \times 10^6$ km. Using the formula derived in Problem 3-27,

$$M = \frac{4\pi^2 a^3}{GT^2} = \frac{4\pi^2(149.5 \times 10^9)^3}{6.67 \times 10^{-11}(365 \times 24 \times 3600)^2} = 2 \times 10^{30} \text{ kg}$$

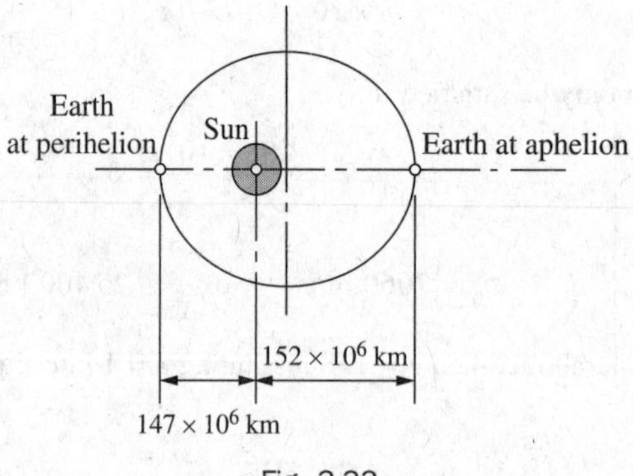

Fig. 3-33

3.32. An automobile weighing 1800 kg is accelerated at the rate of 1.2 m/s^2 along a horizontal roadway. What constant force F (parallel to the ground) is required to produce this acceleration?

Ans.   2160 N

3.33. A body is projected up a 25° plane with an initial velocity of 15 m/s. If the coefficient of friction between the body and the plane is 0.25, determine how far the body will move up the plane and the time required to reach the highest point.

   Ans. 17.7 m, 2.36 s

3.34. A meteorite weighing 450 kg is found buried 20 m in the earth. Assuming a striking velocity of 300 m/s, what was the average retarding force of the earth on the meteorite?

   Ans. $F = 1.01 \times 10^6$ N

3.35. Two particles of the same mass are released from rest on a 25° incline when they are 10 m apart. The coefficient of friction between the upper particle and the plane is 0.15, and between the lower one and the plane it is 0.25. Find the time required for the upper one to overtake the lower one.

   Ans. $t = 4.74$ s

3.36. Refer to Fig. 3-34. An automobile weighing 1270 kg and traveling at 50 km/h hits a depression in the road which has a radius of curvature of 15 m. What is the total force to which the springs are subjected?

   Ans. $N = 28.7$ kN

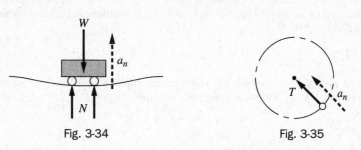

Fig. 3-34   Fig. 3-35

3.37. A small bob of 0.025-kg mass is whirled in an assumed horizontal circular path on a string 400 mm long as shown in Fig. 3-35. What is the tension in the string when the constant angular speed is 30 rad/s?

   Ans. $T = 9.0$ N

3.38. A 400-N weight has a velocity of 10 m/s horizontally on a smooth surface. Determine the value of the horizontal force that will bring the weight to rest in 4 s.

   Ans. 102 N

3.39. The ball of a conical pendulum weighing 50 N hangs at the end of a 3-m string and describes a circular path in a horizontal plane. If the weight is swung so that the string makes an angle of 30° with the vertical, what is the linear speed of the ball?

   Ans. 2.91 m/s

3.40. Two masses of 40 and 35 kg are attached by a cord that passes over a frictionless pulley. If the masses start from rest, find the distance covered by either mass in 6 s.

   Ans. 11.8 m

3.41. A 40-kg mass is dragged along the surface of a table by means of a cord (the cord pulls horizontally on the mass) which passes over a frictionless pulley at the edge of the table and is attached to a 12-kg mass. If the coefficient of friction between the 40-kg mass and the table is 0.15, determine the acceleration of the system and the tension in the cord.

   Ans. 1.13 m/s^2, 104 N

3.42. A block *A* has a mass of 8 kg and is at rest on a frictionless horizontal surface. A 4-kg mass *B* is attached to a rope as shown in Fig. 3-36. Determine the acceleration of mass *B* and the tension in the cord. The pulley is frictionless.

Ans. $a = 3.27 \text{ m/s}^2$ down, $T = 26.2 \text{ N}$

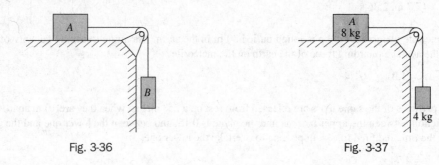

Fig. 3-36   Fig. 3-37

3.43. Two blocks are connected as shown in Fig. 3-37. The coefficient of friction between block *A* and the plane is 0.30. The pulley is frictionless. Determine the tension in the rope.

Ans. $T = 34 \text{ N}$

3.44. A block *B* rests on a block *A*, which is being pulled along a smooth horizontal surface by a horizontal force *P*. If the coefficient of friction between the two blocks is μ, determine the maximum acceleration before slipping occurs between *A* and *B*.

Ans. $a = \mu g$

3.45. A 10-kg box is dropped onto the body of a truck moving 50 km/h horizontally. If the coefficient of friction is 0.5, calculate how far the truck will move before the box stops slipping.

Ans. $s = 39.4 \text{ m}$

3.46. The 40- and 30-kg masses in Fig. 3-38 are attached to cords passing over pulleys. Neglecting the weight of the pulleys and cords and assuming no friction, compute the tensions in the cords. Note that the length of the cord passing over the pulleys is constant.

$$x_2 + (x_2 - c) + x_1 + 2 \text{ half-circumferences} = \text{constant}$$

Ans. $T_1 = 220 \text{ N}, T_2 = 440 \text{ N}$

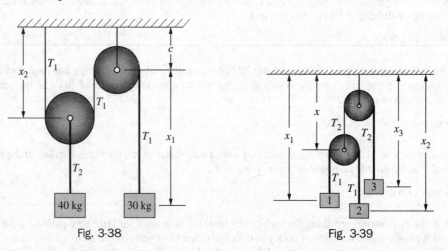

Fig. 3-38   Fig. 3-39

3.47. In the system of pulleys and weights shown in Fig. 3-39, let x_1, x_2, x_3 be the positions of the 1-, 2-, and 3-kg masses, respectively, after the system is released. Neglect the masses of the pulleys and the cords, and assume no friction. Determine the tensions T_1 and T_2.

Ans. $T_1 = 13.8 \text{ N}, T_2 = 27.6 \text{ N}$

3.48. Refer to Fig. 3-40. The masses A and B are 15 and 55 kg, respectively. Assume that the coefficient of friction between A and the plane is 0.25 and that between B and the plane is 0.10. What is the force between the two as they slide down the incline? In the free-body diagrams, P is the unknown force between A and B.

　　Ans.　$P = 13.3$ N

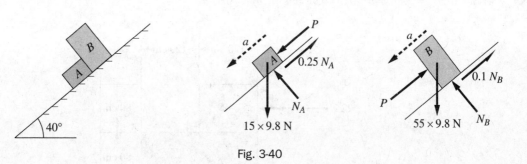

Fig. 3-40

3.49. Two solid blocks resting on a frictionless plane are connected by a string, as shown in Fig. 3-41. Determine the maximum force P that can be applied to the 8-kg block if the maximum strength of the string is 10 N. Consider the blocks as particles.

　　Ans.　$P = 47$ N

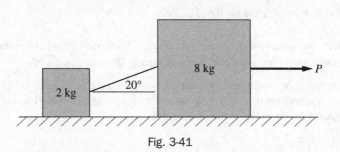

Fig. 3-41

3.50. Complete Problem 3.49 using D'Alembert's method.

　　Ans.　$P = 47$ N

3.51. The two blocks A and B in Fig. 3-42 are connected by a weightless pulley and cord. The coefficient of friction between the blocks and the turntable is 0.35. The weights of blocks A and B are 50 and 100 N, respectively. If the turntable rotates about a vertical axis with a constant angular speed, determine the speed at which the blocks begin to slide. Also find the tension in the string.

　　Ans.　$\omega = 2.27$ rad/s, $T = 49$ N

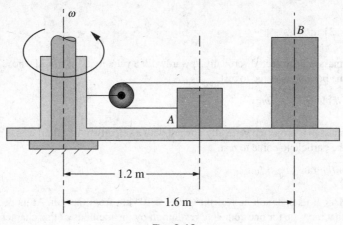

Fig. 3-42

3.52. While the two blocks shown in Fig. 3-43 are at rest, a 22-N force is applied to the top block. The coefficient of friction between the two blocks is 0.20, and the floor is frictionless. (*a*) Determine the acceleration of each block. (*b*) Determine the time that elapses before the right edge of the top block lines up with the right edge of the bottom block.

Ans.   (*a*) $a_A = 2.83$ m/s^2, $a_B = 0.98$ m/s^2 (*b*) $t = 0.92$ s

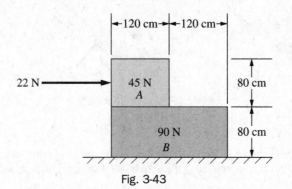

Fig. 3-43

3.53. In Problem 3.52, find the acceleration if the 22-N force is replaced by a 6.6-N force.

Ans.   Both blocks have acceleration 0.479 m/s

3.54. In Problem 3.18, the mass m is displaced a distance x_0 from equilibrium and released with initial velocity v_0. Show that the equation of motion is $x = v_0\sqrt{m/K}\,\sin\sqrt{m/K}\,t + x_0\cos\sqrt{m/K}\,t$.

3.55. At what height above the earth's surface must a 1.8-m pendulum be in order to have a period of 2.8 s? Assume $g = 9.8$ m/s^2 at the earth's surface and the gravitational force varies inversely as the square of the distance from the earth's center. Take the radius of the earth as 6450 km.

Ans.   257 km

3.56. In Problem 3.21, assume the mass is 14 kg and the diameter of the loop is 12 m. Determine the vertical reaction of the track on the mass when the mass is at the top of the loop for (*a*) $h = 15$ m and (*b*) $h = 18$ m.

Ans.   (*a*) $N = 0$, (*b*) $N_T = 137$ N

3.57. A chain 10 m long is stretched with one-half its length on a frictionless horizontal table and the other half hanging freely. If the chain starts from rest, find the time for the chain to leave the table. See Fig. 3-22.

Ans. $t = 1.33$ s

3.58. A body of mass 1.5 kg falls in a fluid where the resistance is $R = k\dot{x}$ and $k = 0.7$ N · s/m. What is the terminal velocity?

Ans. $\dot{x}_{max} = 21$ m/s

3.59. A particle of mass m is projected vertically upward with a velocity v_0 in a fluid whose resistance is kv. Determine the time for the particle to come to rest.

Ans. $t = (m/k)\ln(1 + kv_0/mg)$

3.60. A particle of mass m is projected vertically upward with a velocity v_0 in a fluid whose resistance is kv^2. Determine the time for the particle to come to rest.

Ans. $t = \sqrt{m/kg}\,\tan^{-1} v_0\sqrt{k/mg}$

3.61. The planet Mars at its aphelion in its orbit is 249×10^6 km from the sun. At its perihelion it is 207×10^6 km away. Show that the time for one complete revolution by the methods of this chapter is close to the actual value of 687 days.

3.62. The Soviet satellite Sputnik I had a mass of about 83 kg and orbited on a path around the earth that varied between 220 and 900 km above the surface of the earth. Using the radius of the earth as 6340 km, show that the time for a revolution is about 1.5 h.

3.63. The spherical plastic balloon satellite Echo I was orbited on a path that varied from 1521 to 1689 km from the earth's surface. Show that the time for a revolution was initially about 2 h.

3.64. A satellite with all fuel expended is at a point 483 km from the earth's surface and is traveling with a velocity of 9150 m/s in a direction perpendicular to an extended radius of the earth. Determine the type of path which the satellite will follow.

    Ans. $e = 0.434$; path is elliptical

3.65. At what point on a journey from the earth to the moon will the attractive forces of the two masses on the spaceship be equal? Take the mass of the moon as 0.012 times that of the mass of the earth, and the distance from the earth to the moon as 385 000 km.

    Ans. $d = 348\,000$ km from the earth's center

3.66. Given the period of the earth around the sun as 365 days and the perihelion and aphelion as 147×10^6 km and 152×10^6 km, respectively, determine the eccentricity of the earth's orbit.

    Ans. $e = 0.017$

3.67. A weather satellite is to be placed in a circular orbit around the earth at an altitude of 483 km. Its initial velocity in orbit is parallel to the earth's surface. What should this initial velocity be?

    Ans. $v_0 = 27\,400$ km/h

3.68. Referring to the satellite in Problem 3.64, find the maximum altitude and the velocity at that altitude.

    Ans. $h = 10\,900$ km, $v = 13\,000$ km/h

3.69. Communication satellites complete a circular orbit around the earth in one day. They are seen to be stationary relative to the earth. Determine the necessary altitude and velocity for a communication satellite.

    Ans. $h = 35\,900$ km, $v = 11\,100$ km/h

CHAPTER 4

Kinematics of a Rigid Body in Plane Motion

4.1 Plane Motion of a Rigid Body

Plane motion of a rigid body takes place if every point in the body remains at a constant distance from a fixed plane. In Fig. 4-1, it is assumed that the *XY* plane is the fixed reference plane. The lamina* shown is representative of all laminae that compose the rigid body. The *Z* distance to any point in the lamina remains constant as the lamina moves.

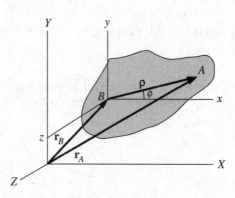

Fig. 4-1 A lamina contained in a rigid body.

It is customary to select an arbitrary point *B* in the body as the origin of the *xyz* reference frame. The position vector $\mathbf{r}_A$ of any point *A* in the lamina may now be written in terms of the position vector $\mathbf{r}_B$ of *B* and the vector **BA**, which is labeled $\boldsymbol{\rho}$. Thus,

$$\mathbf{r}_A = \mathbf{r}_B + \boldsymbol{\rho} \tag{1}$$

If, as a rigid body implies, *AB* is constant, the time rate of change of $\boldsymbol{\rho}$ is that given in Eq. (32) in Chapter 2. Thus, we can write

$$\mathbf{v}_A = \dot{\mathbf{r}}_A = \dot{\mathbf{r}}_B + \dot{\boldsymbol{\rho}} = \dot{\mathbf{r}}_B + \rho\omega\mathbf{e}_\phi \tag{2}$$

---

*The lamina is a thin slice of a rigid body that has motion in the *XY* plane only.

where $\dot{\mathbf{r}}_B$ = linear velocity of pole B relative to fixed axes X, Y, and Z
$\qquad \omega$ = magnitude of angular velocity of $\boldsymbol{\rho}$ about any line parallel to Z axis
$\qquad \mathbf{e}_\phi$ = unit vector perpendicular to $\boldsymbol{\rho}$ and in direction of increasing ϕ (as indicated by right-hand rule, this is counterclockwise about z axis)

The acceleration $\mathbf{a}_A$ may next be found by applying Eq. (36) in Chapter 2:

$$\mathbf{a}_A = \dot{\mathbf{v}}_A = \ddot{\mathbf{r}}_A = \ddot{\mathbf{r}}_B - \rho\omega^2\mathbf{e}_r + \rho\alpha\mathbf{e}_\phi \qquad (3)$$

where $\ddot{\mathbf{r}}_B$ = linear acceleration of pole B relative to fixed axes X, Y, and Z
$\qquad \mathbf{e}_r$ = unit vector along $\boldsymbol{\rho}$ directed from B toward A
$\qquad \mathbf{e}_\phi$ = unit vector as specified in Eq. (2)
$\qquad \alpha$ = magnitude of angular acceleration of $\boldsymbol{\rho}$ about any line parallel to Z axis

An alternative method of writing Eqs. (2) and (3) may be developed as follows:

$$\boldsymbol{\omega} = \dot{\phi}\mathbf{k} = \omega\mathbf{k} \qquad \text{and} \qquad \boldsymbol{\alpha} = \ddot{\phi}\mathbf{k} = \dot{\omega}\mathbf{k} = \alpha\mathbf{k} \qquad (4)$$

where ω and α will be positive if they are in a counterclockwise direction about the z axis (the right-hand rule).

Then, in Eq. (2) the term $\rho\omega\mathbf{e}_\phi$ may be replaced by the cross product $\boldsymbol{\omega} \times \boldsymbol{\rho}$, which is identical with it (see Problem 4.1). Likewise, in Eq. (3) the component of the acceleration $\rho\alpha\mathbf{e}_\phi$ is equivalent to $\boldsymbol{\alpha} \times \boldsymbol{\rho}$. The component $-\rho\omega^2\mathbf{e}_r$ is identical with $\boldsymbol{\omega} \times (\boldsymbol{\omega} \times \boldsymbol{\rho})$. These substitutions give

$$\mathbf{v}_A = \mathbf{v}_B + \boldsymbol{\omega} \times \boldsymbol{\rho} \qquad (5)$$

$$\mathbf{a}_A = \mathbf{a}_B + \boldsymbol{\omega} \times (\boldsymbol{\omega} \times \boldsymbol{\rho}) + \boldsymbol{\alpha} \times \boldsymbol{\rho} \qquad (6)$$

Equations (5) and (6) may also be written

$$\mathbf{v}_A = \mathbf{v}_B + \mathbf{v}_{A/B} \qquad (7)$$

$$\mathbf{a}_A = \mathbf{a}_B + \mathbf{a}_{A/B} \qquad (8)$$

where $\mathbf{v}_{A/B}$ and $\mathbf{a}_{A/B}$ are the *relative velocity* and *relative acceleration* that A possesses as it rotates around B, which is moving with respect to the X, Y, and Z reference frame.

4.2 Translation

Translation is motion in which the line $\boldsymbol{\rho}$ from B to A does not rotate. Thus, as the lamina moves, every straight line in the lamina is always parallel to its original direction.

4.3 Rotation

Rotation is motion in which the base point B is fixed. Extending this to the rigid body, there is a line through B parallel to the Z axis that is fixed in the X, Y, and Z coordinate system. To describe this motion, the velocity $\mathbf{v}_B$ and acceleration $\mathbf{a}_B$ in Eqs. (2) and (3) or in Eqs. (5) and (6) are equated to zero.

4.4 Instantaneous Axis of Rotation

The instantaneous *axis of rotation* is that line in a body in plane motion that is the locus of points of zero velocity. It is perpendicular to the plane of motion (parallel to the Z axis in our system). All other points in the rigid body rotate about that line at that instant. It is important to realize that the position of this line of zero velocity in general changes continuously. To locate the instant center I for a lamina that has an angular velocity $\boldsymbol{\omega}$, write the velocity expressions for any two points A and C in the body in terms of I as the base point (see Fig. 4-2). Thus,

$$\mathbf{v}_A = \mathbf{v}_I + \boldsymbol{\omega} \times \boldsymbol{\rho}_A$$

$$\mathbf{v}_C = \mathbf{v}_I + \boldsymbol{\omega} \times \boldsymbol{\rho}_C \tag{9}$$

But $\mathbf{v}_I$ is zero because I is the instant center. Hence, $\mathbf{v}_A = \boldsymbol{\omega} \times \boldsymbol{\rho}_A$ and $\mathbf{v}_C = \boldsymbol{\omega} \times \boldsymbol{\rho}_C$. These equations mean $\boldsymbol{\rho}_A$ is perpendicular to $\mathbf{v}_A$ (I is on $\boldsymbol{\rho}_A$) and $\boldsymbol{\rho}_C$ is perpendicular to $\mathbf{v}_C$ (I is on $\boldsymbol{\rho}_C$). Therefore the instant center I is the intersection of the perpendiculars to $\mathbf{v}_A$ and $\mathbf{v}_C$.

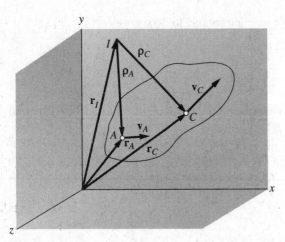

Fig. 4-2   The instant center I of a lamina of a rigid body.

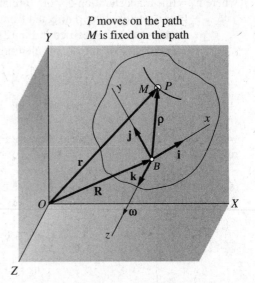

Fig. 4-3   A rotating lamina of a rigid body.

4.5   The Coriolis' Acceleration

In Sec. 4.1, the velocity and acceleration of a point in the rigid body were presented when the point was not moving relative to the xyz frame. Figure 4-3 shows the lamina, in plane motion, where pole B is the origin of a rotating reference frame. Point P is moving along the path in the plane of the lamina. The acceleration of P is given by

$$\mathbf{a}_P = \mathbf{a}_{P/\text{path}} + \mathbf{a}_M + 2\boldsymbol{\omega} \times \mathbf{v}_{P/\text{path}} \tag{10}$$

where $\mathbf{a}_{P/\text{path}}$ = acceleration of P relative to path considered as fixed—use components tangent and normal to path

$\mathbf{a}_M$ = acceleration of the point M which is coincident with P

$\mathbf{v}_{P/\text{path}}$ = velocity of P relative to point M (this velocity can only be tangent to path along which the particle is moving in the body)

$\boldsymbol{\omega}$ = angular velocity of body (or lamina)

$2\boldsymbol{\omega} \times \mathbf{v}_{P/\text{path}}$ = the Coriolis' acceleration, the direction of which is obtained using the right-hand rule

It can be seen in Eq. (10) that referring the acceleration to rotating axes gives rise to a third term in the acceleration equation if $\mathbf{v}_{P/\text{path}}$ is nonzero. The velocity equation is unchanged. This third term is called the *Coriolis acceleration*. It arises from the fact that $\boldsymbol{\rho}$, measured in the rotating reference frame, is defined with respect to $\mathbf{ijk}$. The unit vectors $\mathbf{i}, \mathbf{j}, \mathbf{k}$ change directions and hence have derivatives.

The proof of Coriolis' law is as follows. In the lamina shown parallel to the fixed XY plane in Fig. 4-3, let

$\mathbf{R}$ = position vector of base point B

$\mathbf{r}$ = position vector of point P, which is moving along a path on lamina

$\boldsymbol{\rho}$ = radius vector of P with respect to base point B

$\boldsymbol{\omega}$ = angular velocity of lamina about Z axis

Let **i** and **j** be unit vectors along the x and y axes, which are fixed in the lamina and hence rotate with angular velocity $\omega\mathbf{k}$ (note ω is the same about either z axis of Z axis). Since **i** and **j** rotate, there is a time change in these unit vectors. In Sec. 2.3, it was shown that the time rate of change of a unit vector is a vector that is at right angles to the unit vector and has a magnitude equal to the angular speed ω. Thus,

$$\dot{\mathbf{i}} = \omega\mathbf{j} = \boldsymbol{\omega} \times \mathbf{i} \qquad \text{and} \qquad \dot{\mathbf{j}} = -\omega\mathbf{i} = \boldsymbol{\omega} \times \mathbf{j} \tag{11}$$

Now write

$$\mathbf{r} = \mathbf{R} + \boldsymbol{\rho} \tag{12}$$

Next take the time derivative of Eq. (12), obtaining

$$\dot{\mathbf{r}} = \dot{\mathbf{R}} + \dot{\boldsymbol{\rho}} \tag{13}$$

But in terms of the moving axes, P is located by

$$\boldsymbol{\rho} = x\mathbf{i} + y\mathbf{j} \tag{14}$$

and

$$\dot{\boldsymbol{\rho}} = \dot{x}\mathbf{i} + \dot{y}\mathbf{j} + x\dot{\mathbf{i}} + y\dot{\mathbf{j}} = \mathbf{v}_{P/\text{path}} + \boldsymbol{\omega} \times \boldsymbol{\rho} \tag{15}$$

where $\dot{x}\mathbf{i} + \dot{y}\mathbf{j} = \mathbf{v}_{P/\text{path}}$ and $x\dot{\mathbf{i}} + y\dot{\mathbf{j}} = x(\boldsymbol{\omega} \times \mathbf{i}) + y(\boldsymbol{\omega} \times \mathbf{j}) = \boldsymbol{\omega} \times (x\mathbf{i} + y\mathbf{j}) = \boldsymbol{\omega} \times \boldsymbol{\rho}$.

Thus, we can rewrite Eq. (13) as

$$\dot{\mathbf{r}} = \dot{\mathbf{R}} + \boldsymbol{\omega} + \boldsymbol{\rho} + \mathbf{v}_{P/\text{path}} \tag{16}$$

Note that the first two terms on the right-hand side of Eq. (16) give the absolute velocity of the point M that is fixed on the path but coincides with P at the instant [refer to Eq. (5)]. Equation (15) may be expressed as

$$\mathbf{v}_P = \mathbf{v}_M + \mathbf{v}_{P/\text{path}} \tag{17}$$

To derive the expression for the acceleration of P, take the time derivative of Eq. (13) but express $\dot{\boldsymbol{\rho}}$ as $\dot{x}\mathbf{i} + \dot{y}\mathbf{j} + \boldsymbol{\omega} \times \boldsymbol{\rho}$:

$$\frac{d\dot{\mathbf{r}}}{dt} = \frac{d\dot{\mathbf{R}}}{dt} + \frac{d}{dt}(\dot{x}\mathbf{i} + \dot{y}\mathbf{j} + \boldsymbol{\omega} \times \boldsymbol{\rho}) \tag{18}$$

Thus,

$$\ddot{\mathbf{r}} = \ddot{\mathbf{R}} + (\ddot{x}\mathbf{i} + \ddot{y}\mathbf{j}) + (\dot{x}\dot{\mathbf{i}} + \dot{y}\dot{\mathbf{j}}) + \dot{\boldsymbol{\omega}} \times \boldsymbol{\rho} + \boldsymbol{\omega} \times \dot{\boldsymbol{\rho}}$$

$$= \ddot{\mathbf{R}} + \mathbf{a}_{P/\text{path}} + \boldsymbol{\omega} \times (\dot{x}\mathbf{i} + \dot{y}\mathbf{j}) + \boldsymbol{\alpha} \times \boldsymbol{\rho} + \boldsymbol{\omega} \times (\dot{x}\mathbf{i} + \dot{y}\mathbf{j} + \boldsymbol{\omega} \times \boldsymbol{\rho})$$

$$\mathbf{a}_P = \ddot{\mathbf{R}} + \mathbf{a}_{P/\text{path}} + \boldsymbol{\omega} \times \mathbf{v}_{P/\text{path}} + \boldsymbol{\alpha} \times \boldsymbol{\rho} + \boldsymbol{\omega} \times \mathbf{v}_{P/\text{path}} + \boldsymbol{\omega} \times (\boldsymbol{\omega} \times \boldsymbol{\rho})$$

$$= \ddot{\mathbf{R}} + \boldsymbol{\alpha} \times \boldsymbol{\rho} + \boldsymbol{\omega} \times (\boldsymbol{\omega} \times \boldsymbol{\rho}) + \mathbf{a}_{P/\text{path}} + 2\boldsymbol{\omega} \times \mathbf{v}_{P/\text{path}} \tag{19}$$

The first three terms on the right-hand side of Eq. (19) give the absolute acceleration of the point M that is fixed on the path and that coincides with P at the instant [refer to Eq. (6)]. Equation (19) can be written as

$$\ddot{\mathbf{r}} = \mathbf{a}_M + \mathbf{a}_{P/\text{path}} + 2\boldsymbol{\omega} \times \mathbf{v}_{P/\text{path}} \tag{20}$$

which is equivalent to Eq. (10).

SOLVED PROBLEMS

As in previous chapters, vectors in the diagrams are identified only by their magnitudes when the directions are evident by inspection.

4.1. Determine the linear velocity **v** of any point P in the rigid body rotating with angular velocity $\boldsymbol{\omega}$ about an axis, as shown in Fig. 4-4.

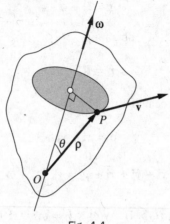

Fig. 4-4

SOLUTION

Select *any* reference point O on the axis of rotation. The radius vector $\boldsymbol{\rho}$ for point P relative to O is shown. The velocity vector **v** for point P is tangent to a circle that is in the plane perpendicular to the axis of rotation. The magnitude of the vector **v** is the product of the radius of the circle and the angular speed ω. It is apparent from the figure that the radius of the circle is $\rho \sin \theta$. Hence, the magnitude of **v** is $\rho \omega \sin \theta$.

By the definition of the cross product, $\boldsymbol{\omega} \times \boldsymbol{\rho}$ has the same magnitude $\rho \omega \sin \theta$ and is perpendicular to the plane containing $\boldsymbol{\rho}$ and $\boldsymbol{\omega}$. Hence, we conclude

$$\mathbf{v} = \boldsymbol{\omega} \times \boldsymbol{\rho}$$

In the plane motion of the lamina shown at the beginning of this chapter, $\boldsymbol{\omega}$ is perpendicular to the plane of the lamina and $\boldsymbol{\rho}$ is in the plane of the lamina; hence, $\mathbf{v} = \boldsymbol{\omega} \times \boldsymbol{\rho}$ is in the plane of the lamina and is perpendicular to $\boldsymbol{\rho}$.

4.2. A flywheel 500 mm in diameter is brought uniformly from rest up to a speed of 300 rpm in 20 s. Find the velocity and acceleration of a point on the rim 2 s after starting from rest.

SOLUTION

This problem illustrates the application of the equations of rotation.

First determine the magnitude of the angular acceleration α:

$$\alpha = \frac{\omega - \omega_0}{t} = \frac{2\pi(300/60) - 0}{20} = 1.57 \text{ rad/s}^2$$

Next determine the angular speed of the wheel 2 s after starting:

$$\omega_1 = \omega_0 + \alpha t = 0 + (1.57 \text{ rad/s}^2)(2 \text{ s}) = 3.14 \text{ rad/s}$$

The speed of a point on the rim is

$$v = r\omega = (0.25 \text{ m})(3.14 \text{ rad/s}) = 0.785 \text{ m/s}$$

The magnitude of the normal acceleration of a point on the rim is

$$a_n = r\omega^2 = (0.25 \text{ m})(3.14 \text{ rad/s})^2 = 2.46 \text{ m/s}^2$$

The magnitude of the tangential acceleration of a point on the rim is

$$a_t = r\alpha = (0.25 \text{ m})(1.57 \text{ rad/s}^2) = 0.39 \text{ m/s}^2$$

The magnitude of the total acceleration of a point on the rim is

$$a = \sqrt{a_n^2 + a_t^2} = \sqrt{(2.46)^2 + (0.39)^2} = 2.49 \text{ m/s}^2$$

The angle between the total acceleration vector and the radius to the point is

$$\theta = \cos^{-1}\frac{a_n}{a} = \cos^{-1}\frac{2.46}{2.49} = 0.155 \text{ rad or } 8.9°$$

Figure 4-5 indicates the acceleration result.

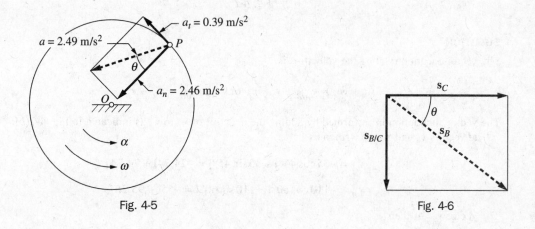

Fig. 4-5   Fig. 4-6

4.3. A ball rolls 2 m across a flat car in a direction perpendicular to the path of the car. In the same time interval during which the ball is rolling, the car moves at a constant speed on a horizontal straight track for a distance of 2.5 m. What is the absolute displacement of the ball?

SOLUTION

The vector equation for the absolute displacement $\mathbf{s}_B$ of the ball B in terms of the absolute displacement $\mathbf{s}_C$ of the car C is

$$\mathbf{s}_B = \mathbf{s}_{B/C} + \mathbf{s}_C$$

The displacement $\mathbf{s}_{B/C}$ of the ball relative to the car is 2 m at right angles to the track. The absolute displacement of the car is 2.5 m along the track. Figure 4-6 indicates these relations.

The absolute displacement $\mathbf{s}_B$ of the ball is the sum of the two given vectors. Its magnitude is

$$s_B = \sqrt{(s_{B/C})^2 + (s_C)^2} = \sqrt{(2)^2 + (2.5)^2} = 3.2 \text{ m}$$

The angle that the vector $\mathbf{s}_B$ makes with the track is

$$\theta = \tan^{-1}\left(\frac{s_{B/C}}{s_C}\right) = \tan^{-1}\left(\frac{2}{2.5}\right) = 0.675 \text{ rad or } 38.7°$$

4.4. Automobile A is traveling 35 km/h along a straight road headed northwest. Automobile B is traveling 110 km/h along a straight road headed 60° south of west [see Fig. 4-7(a)]. What is the relative velocity of A to B? Of B to A?

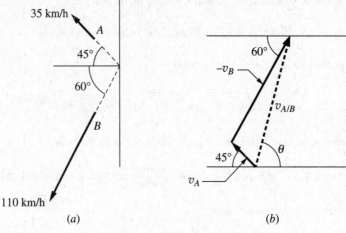

Fig. 4-7

SOLUTION

The vector equation relating the velocities is

$$\mathbf{v}_A = \mathbf{v}_{A/B} + \mathbf{v}_B \qquad \text{or} \qquad \mathbf{v}_{A/B} = \mathbf{v}_A - \mathbf{v}_B$$

The vector subtraction is performed by adding the negative of $\mathbf{v}_B$ to $\mathbf{v}_A$ as indicated in Fig. 4-7(b).
The vectors $\mathbf{v}_A$ and $\mathbf{v}_B$ are written as

$$\mathbf{v}_A = -35\cos 45°\mathbf{i} + 35\sin 45°\mathbf{j} = -24.74\mathbf{i} + 24.74\mathbf{j}$$

$$\mathbf{v}_B = -110\cos 60°\mathbf{i} - 110\sin 60°\mathbf{j} = -55\mathbf{i} - 95.26\mathbf{j}$$

The vector $\mathbf{v}_{A/B}$ is then

$$\mathbf{v}_{A/B} = \mathbf{v}_A - \mathbf{v}_B = (-24.74 + 55)\mathbf{i} + (24.74 + 95.26)\mathbf{j} = 30.3\mathbf{i} + 120\mathbf{j}$$

The vector $\mathbf{v}_{B/A}$ is equal in magnitude but opposite in direction as shown in Fig. 4-8 and as follows:

$$\mathbf{v}_{B/A} = \mathbf{v}_B - \mathbf{v}_A = (-55 + 24.74)\mathbf{i} + (-95.26 - 24.74)\mathbf{j} = -30.3\mathbf{i} - 120\mathbf{j}$$

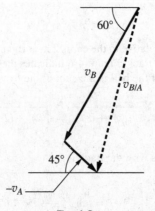

Fig. 4-8

4.5. An airplane pilot determines from a sectional map that the true course over the ground should be 295° (relative to north) to a destination 160 km away. There is a south wind of 32 km/h (blowing from the south). Airspeed (speed relative to the air which is moving north 32 km/h) is 192 km/h. What should be the pilot's true heading to accomplish the task and how long should it take, assuming no change in the wind?

SOLUTION

Figure 4-9 shows the true course (T.C.) of 295° measured clockwise from the true north line. The wind speed 32 km/h is drawn to the north from O, the point of departure. A sketch is perhaps helpful. Sketch a vector from the arrow end of the wind vector, with a radius 192 km/h long to point C. This intersects the true course line at 295° in point C.

It is seen from the velocity triangle that $\mathbf{OC} = \mathbf{OW} + \mathbf{WC}$. This is merely the statement that the velocity of the plane along the true course line (its magnitude is called *ground speed* or *absolute speed*) is equal to the plane velocity relative to the wind (its magnitude is called *airspeed*) plus the velocity of the wind relative to the ground.

From the law of cosines, the triangle provides

$$192^2 = 32^2 + C^2 - 2 \times 32C \cos 65° \qquad \text{or} \qquad C^2 - 27C - 35\,840 = 0$$

This provides $C = 203.3$ km/h. Then the angle θ can be found from

$$203.3^2 = 192^2 + 32^2 - 2 \times 192 \times 32 \cos\theta \qquad \therefore \theta = 106.3°$$

The true heading (T.H.) is then $106.3 + 180 = 286°$, the direction along which the fore and aft line of the plane should be headed.

The estimated time of flight is 160 km divided by 203.3 km/h, or 47 min.

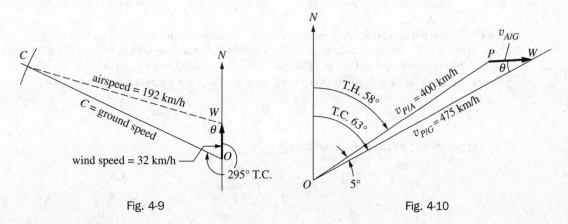

Fig. 4-9   Fig. 4-10

4.6. Correcting for magnetic deviation and variation, a pilot calculates the plane's true heading to be 58°. Airspeed is 400 km/h. Several checkpoints indicate that the plane is making good a true course of 63° at a ground speed of 475 km/h. Determine the wind direction and magnitude.

SOLUTION

Sketch the true heading line at an angle of 58° with the true north line, as shown in Fig. 4-10. Along it, sketch $\mathbf{OP}$ as 400 km/h. The true course line makes an angle of 63° with the true north line through O. Along it, $\mathbf{OW}$ is the ground speed 475 km/h. But $\mathbf{OW} = \mathbf{OP} + \mathbf{PW}$. This is interpreted as follows: the absolute velocity of the plane equals its velocity relative to the air plus the velocity of the air (wind vector) or $\mathbf{v}_{P/G} = \mathbf{v}_{P/A} + \mathbf{v}_{A/G}$. The law of cosines provides

$$v_{A/G}^2 = 475^2 + 400^2 - 2 \times 475 \times 400 \cos 5° \qquad \therefore v_{A/G} = 84.1 \text{ km/h}$$

The angle θ is found from

$$400^2 = 475^2 + 84.1^2 - 2 \times 475 \times 84.1 \cos\theta \qquad \therefore \theta = 24.5°$$

4.7. The rod of length ℓ moves so that the velocity of point A is of constant magnitude and directed to the left. Determine the angular velocity ω and angular acceleration α of the rod when it makes an angle θ with the vertical, as shown in Fig. 4-11.

SOLUTION

The velocity $\mathbf{v}_A$ of point A is expressed in terms of the relative velocity of A to B ($\mathbf{v}_{A/B}$) because this term will introduce the desired quantity ω:

$$\mathbf{v}_A = \mathbf{v}_{A/B} + \mathbf{v}_B$$

This vector equation is sketched in Fig. 4-12. The velocity v_A is horizontal to the left, v_B is vertical down, and $v_{A/B}$ is perpendicular to the rod with magnitude $\ell\omega$. From the triangle we see that

$$\ell\omega\cos\theta = v_A \qquad \therefore \omega = \frac{v_A}{\ell\cos\theta}$$

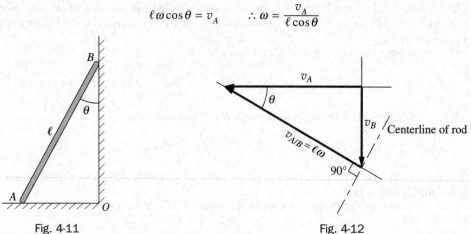

Fig. 4-11   Fig. 4-12

Since $\mathbf{v}_{A/B}$ is directed up to the left, A must turn clockwise about B; that is ω is clockwise.

Now determine the angular acceleration α. The magnitude of the tangential component of A relative to B is $\ell\alpha$ and the normal component is $\ell\omega^2$. The equation is $\mathbf{a}_A = (\mathbf{a}_{A/B})_t + (\mathbf{a}_{A/B})_n + \mathbf{a}_B$.

Sketch the acceleration polygon with a_B down ($a_A = 0$). From the triangle shown in Fig. 4-13,

$$\tan\theta = \frac{\ell\alpha}{\ell\omega^2} = \frac{\alpha}{\omega^2} \qquad \therefore \alpha = \omega^2 \tan\theta$$

Using ω from above, the angular acceleration is

$$\alpha = \frac{v_A^2 \tan\theta}{\ell^2 \cos^2\theta}$$

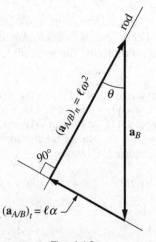

Fig. 4-13

4.8.   The ladder of length ℓ makes an angle θ with the vertical wall, as shown in Fig. 4-14(a). The foot of the ladder moves to the right with constant speed v_A. Determine $\dot\theta$ and $\ddot\theta$ in terms of v_A, ℓ, and θ. Use vector relationships.

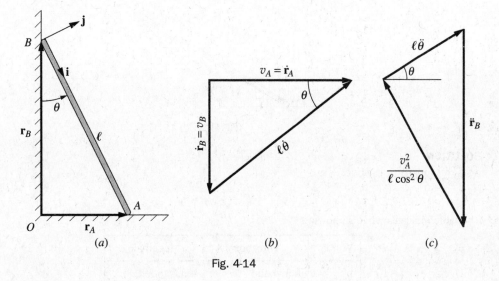

Fig. 4-14

SOLUTION

Let **i** and **j** be unit vectors along and perpendicular to the ladder; they move with the ladder. In vector notation, the three vectors in Fig. 4-14(a) are related by

$$\mathbf{r}_A = \mathbf{r}_B + \ell\mathbf{i}$$

Taking the time derivative, we have

$$\dot{\mathbf{r}}_A = \dot{\mathbf{r}}_B + \ell\dot{\mathbf{i}}$$

In the above equation, $\dot{\mathbf{r}}_A$ is $\mathbf{v}_A$, which is known completely. Also, $\dot{\mathbf{i}} = \theta\mathbf{j}$ [see Eq. (11)], and $\dot{\mathbf{r}}_B$ can only be vertical. Figure 4-14(b) shows this relationship. Thus,

$$\cos\theta = \frac{v_A}{\ell\theta} \qquad \text{or} \qquad \dot\theta = \frac{v_A}{\ell\cos\theta} \tag{1}$$

Since $\ell\dot\theta\mathbf{j}$ is in the positive **j** direction, $\dot\theta$ is positive (ladder is moving counterclockwise).
  The derivative of Eq. (1) with respect to time is

$$\ddot\theta = \frac{v_A\sin\theta}{\ell\cos^2\theta}\,\dot\theta = \frac{v_A^2\sin\theta}{\ell^2\cos^3\theta} = \frac{v_A^2\tan\theta}{\ell^2\cos^2\theta} \tag{2}$$

The angular acceleration $\ddot\theta$ could also be obtained by differentiating $\dot{\mathbf{r}}_A = \dot{\mathbf{r}}_B + \ell\dot\theta\mathbf{j}$, which yields

$$\ddot{\mathbf{r}}_A = \ddot{\mathbf{r}}_B + \ell\ddot\theta\mathbf{j} - \ell\dot\theta^2\mathbf{i} \tag{3}$$

In Eq. (3), $\ddot{\mathbf{r}}_A$ is zero because $\dot{\mathbf{r}}_A$ is constant, $\ddot{\mathbf{r}}_B$ can only be vertical, the **j** component (magnitude $\ell\ddot\theta$) is positive, and the **i** component [magnitude $\ell\dot\theta^2 = v_A^2/(\ell\cos^2\theta)$] is negative. Figure 4-14(c) shows these relations. Then

$$\tan\theta = \frac{\ell\ddot\theta}{v_A^2/(\ell\cos^2\theta)} \qquad \text{or} \qquad \ddot\theta = \frac{v_A^2\tan\theta}{\ell^2\cos^2\theta}$$

which is identical with Eq. (2). Since $\ell\ddot\theta\mathbf{j}$ is in the positive **j** direction, $\ddot\theta$ is positive (the ladder is accelerating counterclockwise).
  Note that $\dot\theta$ in Eq. (1) agrees (as it should) with the magnitude of $\boldsymbol\omega$ in Problem 4.7. Also, the $\ddot\theta$ in equation (2) equals the magnitude of $\boldsymbol\alpha$ in Problem 4.7.

4.9. A rod 2.5 m long slides down the plane shown in Fig. 4-15 with $\mathbf{v}_A = 4$ m/s to the left and $\mathbf{a}_A = 5$ m/s^2 to the right. Determine the angular velocity ω and the angular acceleration α of the rod when $\theta = 30°$.

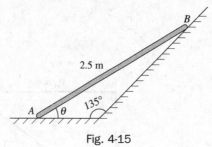

Fig. 4-15

SOLUTION

As in Problem 4.7,

$$\mathbf{v}_A = \mathbf{v}_{A/B} + \mathbf{v}_B$$

The table indicates what is known.

Vector	Direction	Magnitude
$\mathbf{v}_A$	horizontal	4 m/s to left
$\mathbf{v}_{A/B}$	$\perp$ rod	$\ell\omega$ (ω unknown)
$\mathbf{v}_B$	along the 45° line	unknown

Sketch the velocity polygon to fit the vector equation given above (see Fig. 4-16). An analysis of the angles shows that the three angles in the triangle are 45°, 60°, and 75°. The law of sines gives

$$\frac{4}{\sin 75°} = \frac{v_{A/B}}{\sin 45°} \qquad \therefore v_{A/B} = 2.93 \text{ m/s}$$

$$v_{A/B} = 2.5\omega \qquad \therefore \omega = 1.17 \text{ rad/s}$$

To determine α, use the vector equation $\mathbf{a}_A = (\mathbf{a}_{A/B})_n + (\mathbf{a}_{A/B})_t + \mathbf{a}_B$ with the following table.

Vector	Direction	Magnitude
$\mathbf{a}_A$	horizontal	5 m/s^2 to right
$(\mathbf{a}_{A/B})_n$	along the rod from A to B	$\ell\omega^2$ $2.5(1.17)^2 = 3.42$ m/s^2
$(\mathbf{a}_{A/B})_t$	$\perp$ rod	$\ell\alpha$ (α unknown)
$\mathbf{a}_B$	along the 45° line	unknown

Sketch the acceleration polygon to fit the above table. First draw $\mathbf{a}_A$ (see Fig. 4-17). Then through the tail of this vector draw $(\mathbf{a}_{A/B})_n$. Through the head of the vector $\mathbf{a}_A$ draw a line along a 45° line, and through the head of the vector $(\mathbf{a}_{A/B})_n$ draw a perpendicular to the rod in the 30° plane.

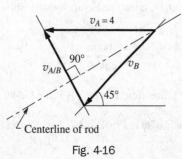

Fig. 4-16

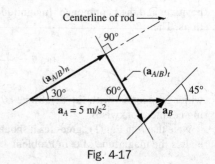

Fig. 4-17

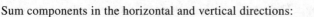

Sum components in the horizontal and vertical directions:

$$3.42 \cos 30° + 2.5\alpha \cos 60° + a_B \cos 45° = 5$$
$$3.42 \sin 30° - 2.5\alpha \sin 60° + a_B \sin 45° = 0 \qquad \therefore \alpha = 1.38 \text{ rad/s}^2$$

4.10. In the slider crank mechanism shown in Fig. 4-18, the crank is rotating at a constant speed of 480 rpm. The connecting rod is 600 mm long, and the crank is 100 mm long. For an angle of 30°, determine the absolute velocity of the crosshead *P*.

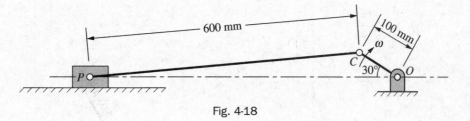

Fig. 4-18

SOLUTION

The angular velocity of the crankpin *C* is 120 rpm, so $\omega = 2\pi(480/60) = 16\pi$ rad/s. The linear speed of the crankpin *C* is therefore,

$$v = r\omega = (0.1)(16\pi) = 5.03 \text{ m/s}$$

Figure 4-19 indicates this velocity in a direction perpendicular to the crank.

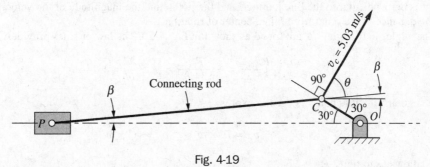

Fig. 4-19

The component of this velocity along the connecting rod *PC* will now be determined. Angle θ must be found first. The figure indicates that $\theta = 90° - 30° - \beta$. But β may be found by using the sine law in the triangle *PCO*:

$$\frac{100}{\sin \beta} = \frac{600}{\sin 30°}$$

from which $\beta = 4.8°$ and $\theta = 60° - 4.8° = 55.2°$. Hence, the component of the velocity of *C* along the connecting rod is 5.03 cos 55.2° = 2.87 m/s.

But all points on the connecting rod must have the same velocity along the rod, hence point *P*, which is a point on the rod, has a velocity component of 2.87 m/s along the rod. However, its velocity is along the line of travel of the crosshead. Then

$$v_p = \frac{2.87}{\cos \beta} = \frac{2.87}{0.9965} = 2.88 \text{ m/s}$$

4.11. In Problem 4.10, determine the velocity of the crosshead by use of instant centers.

SOLUTION

The instant center of the connecting rod relative to the frame is the point about which all points in the rod appear to rotate at that instant. Point *C* is a point on the crank and on the rod. Its absolute velocity (relative to the frame) is the same whether it is a point on the crank or the rod. However, as a point on the crank, its velocity is perpendicular to the crank. Therefore, since its velocity is the same when it is considered as a point on the rod, the instant center for the rod is somewhere along the crank extended. (The velocity of a point in rotation is perpendicular to the radius drawn to it from the center of rotation.)

Similarly, point P is a point on the crosshead and on the connecting rod. As a point on the crosshead, its velocity is horizontal. Hence, as a point on the rod, its velocity is the same (i.e., horizontal). The instant (instantaneous) center for the rod is therefore on a perpendicular to the line of travel of the crosshead, i.e., on a vertical line through P.

The instant center I is at the intersection of the vertical line through P and the crank extended, as shown in Fig. 4-20.

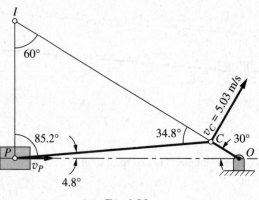

Fig. 4-20

Since I is the center of rotation for all points on the rod, it follows that the linear velocity of a particular point is perpendicular to the line joining I and the point, and the magnitude of the velocity is proportional to the distance of the point from I, i.e., center of rotation.

The angles in triangle PIC are found as shown in Fig. 4-20. The law of sines provides

$$\frac{IC}{\sin 85.2°} = \frac{IP}{\sin 34.8°} \qquad \therefore \frac{IC}{IP} = 1.746$$

The angular velocity about I is used to express

$$v_P = IP \times \omega \qquad v_C = IC \times \omega$$

This allows us to find v_P:

$$\frac{v_P}{v_C} = \frac{IP}{IC} = 1.746^{-1} \qquad \therefore v_P = \frac{5.03}{1.746} = 2.88 \text{ m/s}$$

4.12. Determine the angular velocity of the connecting rod, referring to Problem 4.10. Sketch the velocity polygon (triangle) represented by

$$\mathbf{v}_P = \mathbf{v}_{P/C} + \mathbf{v}_C$$

SOLUTION

Referring to Fig. 4-21, the law of cosines gives

$$v_{P/C}^2 = 5.03^2 + 2.88^2 - 2(5.03)(2.88)\cos 60° \qquad \therefore v_{P/C} = 4.37 \text{ m/s}$$

Then the angular velocity of the rod is found by dividing $v_{P/C}$ by the length of the rod.

$$\omega_{PC} = \frac{v_{P/C}}{\ell} = \frac{4.37}{0.600} = 7.28 \text{ rad/s}$$

ω_{PC} is counterclockwise.

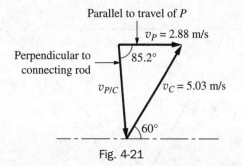

Fig. 4-21

4.13. Referring to Problem 4.10, find the acceleration of the crosshead in the slider crank mechanism. Use $\omega_{PC} = 7.28$ rad/s from Problem 4.12.

SOLUTION

The acceleration of P, which is horizontal, is determined by the following vector equation:

$$\mathbf{a}_P = \mathbf{a}_{P/C} + \mathbf{a}_C$$

The acceleration $\mathbf{a}_{P/C}$ of point P relative to point C is one due to rotation. It is good to write it in terms of its tangential and normal components, which are, respectively, perpendicular and parallel to the connecting rod. The equation now becomes

$$\mathbf{a}_P = (\mathbf{a}_{P/C})_n + (\mathbf{a}_{P/C})_t + \mathbf{a}_C$$

Sketch the acceleration polygon, as shown in Fig. 4-22 starting with $\mathbf{a}_C$. Since the angular velocity of the crank is constant, the acceleration of point C consists of only the normal component directed toward the center O. Its magnitude is $r\omega^2$:

$$(a_C)_n = r\omega^2 = 0.1 \times (16\pi)^2 = 253 \text{ m/s}^2$$

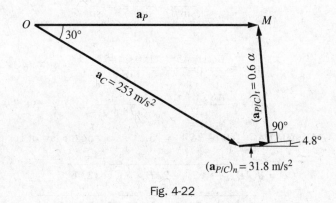

Fig. 4-22

The normal component $(\mathbf{a}_{P/C})_n$ is directed toward C with magnitude

$$(a_{P/C})_n = r\omega_{PC}^2 = 0.6 \times 7.28^2 = 31.8 \text{ m/s}^2$$

The tangential component $(\mathbf{a}_{P/C})_t$ is perpendicular to $\overline{PC}$ and meets the acceleration $\mathbf{a}_P$ on the horizontal. The magnitude of $(\mathbf{a}_{P/C})_t$ is $0.6\alpha_{PC}$. So only two unknowns exist: α and a_P. They are found from the two component equations as follows:

$$\sum a_x: 253\cos 30° + 31.8\cos 4.8° - 0.6\alpha\sin 4.8° = a_P$$

$$\sum a_y: 253\sin 30° - 31.8\sin 4.8° - 0.6\alpha\cos 4.8° = 0$$

$$\alpha = 207 \text{ rad/s}^2 \quad \text{and} \quad a_P = 240 \text{ m/s}^2$$

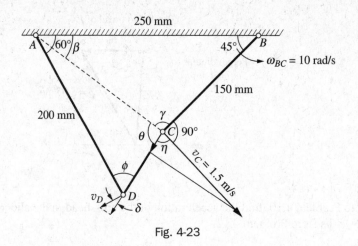

Fig. 4-23

4.14 If the angular velocity of BC is as shown in the quadric crank mechanism in Fig. 4-23, determine the angular velocity of AD and the velocity of point D for the phase indicated.

SOLUTION

The velocity of point C as a point on BC is perpendicular to BC, and its magnitude is found to be

$$v_C = BC \times \omega_{BC} = (0.15)(10) = 1.5 \text{ m/s}$$

To determine the velocity of D by resolution of velocities, first analyze Fig. 4-23 to find angles. By the law of cosines,

$$AC = \sqrt{(AB)^2 + (BC)^2 - 2AB \times BC \cos 45°} = \sqrt{(250)^2 + (150)^2 - 2 \times 250 \times 150 \times \cos 45°} = 178.8 \text{ mm}$$

By the law of sines,

$$\frac{BC}{\sin \beta} = \frac{AC}{\sin 45°} = \frac{AB}{\sin \gamma}$$

Hence,

$$\sin \beta = \frac{BC \sin 45°}{AC} = \frac{150 \sin 45°}{178.8} \qquad \therefore \beta = 36.4°$$

$$\sin \gamma = \frac{AB \sin 45°}{AC} = \frac{250 \sin 45°}{178.8} \qquad \therefore \gamma = 98.7$$

In triangle ADC, angle $DAC = 60° - 36.4° = 23.6°$. Applying the law of cosines in triangle ADC,

$$CD = \sqrt{(AD)^2 + (AC)^2 - 2AD \times AC \cos 23.6°}$$

$$= \sqrt{(200)^2 + (178.8)^2 - 2(200)(178.8) \cos 23.6°} = 80.2 \text{ mm}$$

By the law of sines,

$$\frac{CD}{\sin 23.6°} = \frac{AD}{\sin \theta} = \frac{AC}{\sin \phi} \qquad \text{or} \qquad \frac{80.2}{\sin 23.6°} = \frac{200}{\sin \theta} = \frac{178.8}{\sin \phi}$$

Solving, $\theta = 86.7°$ and $\phi = 63.2°$.

It is now evident that the angle η between CD and the velocity $\mathbf{v}_c$ is

$$\eta = 360° - (90° + \theta + \gamma) = 360° - (90° + 86.7° + 98.7°) = 84.6°$$

The component of this velocity along bar CD is $1.5 \cos 84.6° = 0.141$ m/s. Note that this component is directed from C toward D. This is also the component of the velocity of D along CD.

The angle δ between velocity vector of D and bar CD is $\delta = 180° - (90° + 63.2°) = 26.8°$. The magnitude of the velocity of D is

$$v_D = \frac{0.141}{\cos 26.8°} = 0.158 \text{ m/s}$$

Note that this velocity is directed such that the arm AD turns clockwise whereas the arm BC turns counterclockwise. The angular speed of AD is

$$\omega_{AD} = \frac{v_D}{AD} = \frac{(0.158)}{0.2} = 0.79 \text{ rad/s clockwise}$$

4.15. Solve Problem 4.14 by use of instant centers.

SOLUTION

Refer to Fig. 4-24. Points C and D are points on the arm CD. The instant center of CD relative to the frame is the intersection of the lines drawn perpendicular to the absolute velocities of C and D. The latter velocities, however, are perpendicular to BC and AD. Hence, the instant center I is at the intersection of BC and AD. The velocity of C is shown perpendicular to BC and of magnitude 1.5 m/s.

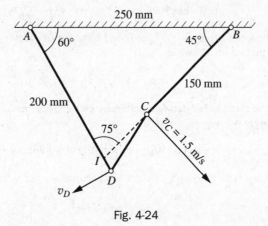

Fig. 4-24

Now, let's find the distances IC and ID from the law of sines on triangle ABI, we find

$$\frac{BI}{\sin 60°} = \frac{250}{\sin 75°} \qquad \therefore BI = 224.1 \text{ mm} \qquad \text{and} \qquad IC = 224.1 - 150 = 74.1 \text{ mm}$$

$$\frac{AI}{\sin 45°} = \frac{250}{\sin 75°} \qquad \therefore AI = 183.0 \text{ mm} \qquad \text{and} \qquad ID = 200 - 183 = 17 \text{ mm}$$

The angular velocity of CD is then $\omega_{CD} = v_C/IC = 1.5/0.741 = 20.2 \text{ rad/s}$. The velocity of D is then

$$v_D = \omega_{CD} \times ID = 20.2 \times 0.017 = 0.34 \text{ m/s}$$

Note: This is slightly different from the value found in Problem 4.14 due to round-off error.

4.16. Figure 4-25 shows a quadric crank mechanism with various lengths given (or calculated). If the crank AB is rotating 3 rad/s counterclockwise, determine the velocities of points B and C and the angular velocities of BC and DC.

SOLUTION

Since point B is on the rotating crank AB, its velocity is

$$\mathbf{v}_B = \boldsymbol{\omega}_{AB} \times \boldsymbol{\rho}_{AB} \qquad \text{or} \qquad 3\mathbf{k} \times (0.3 \times 0.5\mathbf{i} + 0.3 \times 0.866\mathbf{j}) = 0.45\mathbf{j} - 0.779\mathbf{i} \text{ m/s}$$

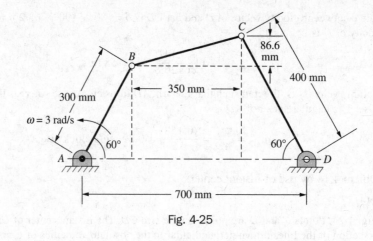

Fig. 4-25

Note that the magnitude of $\mathbf{v}_B$ can be found directly from $v_B = r\omega = 0.3(3) = 0.9$ m/s. Since this vector is perpendicular to the crank, it is directed to the left and up. It can be written

$$0.9(-\cos 30°\mathbf{i} + \sin 30°\mathbf{j}) = 0.45\mathbf{j} - 0.779\mathbf{i}\ \text{m/s}$$

This is, of course, the same expression as that found by using the vector cross product.

To determine the motion of BC, we use

$$\mathbf{v}_C = \mathbf{v}_B + \mathbf{v}_{C/B} = \mathbf{v}_B + \boldsymbol{\omega}_{BC} \times \boldsymbol{\rho}_{BC} \tag{1}$$

To use Eq. (1), assume that DC is rotating counterclockwise. This means that point C moves to the left and down. Since a 30° angle is involved, we write

$$\mathbf{v}_C = -v_C \cos 30°\mathbf{i} - v_C \sin 30°\mathbf{j} = -0.866v_C\mathbf{i} - 0.5v_C\mathbf{j}$$

Next, assume that BC is rotating counterclockwise; hence, we can write

$$\boldsymbol{\omega}_{BC} = \omega_{BC}\mathbf{k}$$

Also note that

$$\boldsymbol{\rho}_{BC} = 0.35\mathbf{i} + 0.0866\mathbf{j}$$

Make these substitutions in Eq. (1) to get

$$-0.866v_C\mathbf{i} - 0.5v_C\mathbf{j} = -0.779\mathbf{i} + 0.45\mathbf{j} + \omega_{BC}\mathbf{k} \times (0.35\mathbf{i} + 0.0866\mathbf{j})$$

or
$$-0.866v_C\mathbf{i} - 0.5v_C\mathbf{j} = -0.779\mathbf{i} + 0.45\mathbf{j} + 0.35\omega_{BC}\mathbf{j} - 0.0866\omega_{BC}\mathbf{i} \tag{2}$$

Equating the **i** terms and then the **j** terms will yield

$$-0.866v_C = -0.779 - 0.0866\omega_{BC} \tag{3}$$

$$-0.5v_C = 0.45 + 0.35\omega_{BC} \tag{4}$$

Multiply Eq. (3) by 0.35/86.6 and add to Eq. (4). This yields $v_C = 0.675$ m/s, and ω_{BC} is then found to be -2.25 rad/s. This means that BC is rotating clockwise instead of counterclockwise as assumed at the beginning.

To find ω_{DC}, we note that

$$\mathbf{v}_C = \boldsymbol{\omega}_{DC} \times \boldsymbol{\rho}_{DC} = \omega_{DC}\mathbf{k} \times (-0.2\mathbf{i} + 0.3464\mathbf{j}) \quad\quad \text{or} \quad\quad \mathbf{v}_C = -0.675 \times 0.866\mathbf{i} - 0.675 \times 0.5\mathbf{j}$$

$$= -0.2\omega_{DC}\mathbf{j} - 0.3464\omega_{DC}\mathbf{i} \tag{5}$$

Equate either the **i** terms or the **j** terms to find $\omega_{DC} = 1.69$ rad/s (counterclockwise as assumed).

Note: The following is a different method that can be used to solve this problem. It involves use of the law of sines to solve the relative velocity part of Eq. (1) or

$$\mathbf{v}_C = \mathbf{v}_{C/B} + \mathbf{v}_B$$

We know that $\mathbf{v}_B$ has a magnitude of 0.9 m/s directed to the left and up at an angle of 30° with the horizontal. We also know that $\mathbf{v}_C$ has a magnitude of 0.4 m/s, with ω_{DC} unknown in this second solution of the problem. Thus, $\mathbf{v}_C$ is along a line making an angle of 30° with the horizontal, although its sense is unknown. Also known is the fact that $\mathbf{v}_{C/B}$ is perpendicular to *BC*, with a magnitude equal to the product of ω_{BC} and the length *BC*. Length *BC* is equal to $\sqrt{(350)^2 + (86.6)^2} = 361$ mm. The velocity of *C* to *B* (since it is perpendicular to *BC*) makes an angle with respect to the vertical given by $\theta = \tan^{-1}(86.6/350) = 13.9°$.

Figure 4-26(*a*) shows $\mathbf{v}_B$, which is completely known, drawn from an arbitrary point *O* to point *R*, with a length of 0.9 m/s and at an angle of 30°. From *R* draw a line, making an angle of 13.9° with the vertical. Then draw a line from *O*, making an angle of 30° with the horizontal. These lines meet at point *S*. The figure is redrawn in Fig. 4-26(*b*) to make the law of sines easier to apply. From Fig. 4-26(*b*), we can see that

$$\frac{v_{C/B}}{\sin 60°} = \frac{v_C}{\sin 46.1°} = \frac{0.9}{\sin 73.9°}$$

This yields $v_C = 0.675$ m/s, and leads to $\omega_{DC} = 1.69$ rad/s clockwise. Also, $v_{C/B} = 0.812$ m/s and leads to $\omega_{BC} = 2.25$ rad/s clockwise.

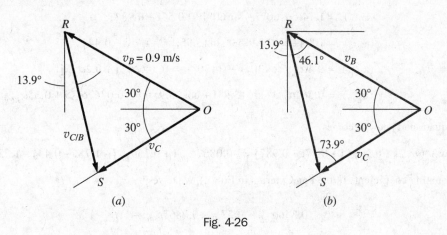

(*a*)   (*b*)

Fig. 4-26

4.17. In Problem 4.16, determine the linear accelerations of points *B* and *C* and the angular accelerations of *BC* and *CD*.

SOLUTION

The mechanism is shown in Fig. 4-27, with components of accelerations some of which are known completely and some of which are known only in direction. The acceleration of *C* as a point on *BC* is expressed as

$$\mathbf{a}_C = (\mathbf{a}_{C/B})_t + (\mathbf{a}_{C/B})_n + \mathbf{a}_B$$

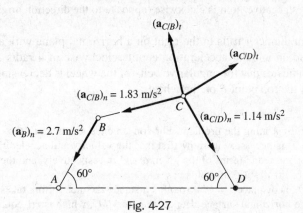

Fig. 4-27

Note that $\mathbf{a}_B$ has only a normal component. If there were an α_{AB}, there would be a tangential component to add to the equation.

Of course, C is a point on DC, and its acceleration is then written $\mathbf{a}_C = (\mathbf{a}_{C/D})_t + (\mathbf{a}_{C/D})_n$. Hence, we write

$$(\mathbf{a}_{C/D})_t + (\mathbf{a}_{C/D})_n = (\mathbf{a}_{C/B})_t + (\mathbf{a}_{C/B})_n + \mathbf{a}_B \tag{1}$$

Three of the terms in Eq. (1) are known completely. They are the normal components of the accelerations, because all angular velocities were found in Problem 4.16.

To determine the normal acceleration of B, use Eq. (6) or $\boldsymbol{\omega}_{AB} \times \boldsymbol{\omega}_{AB} \times \boldsymbol{\rho}_{AB}$ or use $AB \times (\omega_{AB})^2$. B has an acceleration of 2.7 m/s² directed from B toward A as shown.

Similarly, $(\mathbf{a}_{C/B})_n$ is directed from C toward B and has a magnitude $BC \times (\omega_{BC})^2 = 0.361(2.25)^2 = 1.83$ m/s². Also, $(\mathbf{a}_{C/D})_n$ is directed from C toward D and has a magnitude $DC \times (\omega_{DC})^2 = 0.4(1.69)^2 = 1.14$ m/s². The $(\mathbf{a}_{C/B})_t$ is perpendicular to BC and is assumed to act to the left and up. A positive value would mean that α_{BC} is counterclockwise. The $(\mathbf{a}_{C/D})_t$ is perpendicular to DC and is assumed to act to the right and up. A positive value would mean α_{DC} is clockwise.

The five acceleration vectors in Eq. (1) are

$$\mathbf{a}_B = 2.7(-\cos 60° \mathbf{i} - \sin 60° \mathbf{j}) = -1.35\mathbf{i} - 2.34\mathbf{j}$$

$$(\mathbf{a}_{C/D})_n = 1.14(\cos 60° \mathbf{i} - \sin 60° \mathbf{j}) = 0.57\mathbf{i} - 0.987\mathbf{j}$$

$$(\mathbf{a}_{C/B})_n = 1.83(-\cos 13.9° \mathbf{i} - \sin 13.9° \mathbf{j}) = -1.78\mathbf{i} - 0.44\mathbf{j}$$

$$(\mathbf{a}_{C/D})_t = 0.4\alpha_{DC}(\cos 30° \mathbf{i} + \sin 30° \mathbf{j}) = 0.346\alpha_{DC}\mathbf{i} + 0.2\alpha_{DC}\mathbf{j}$$

$$(\mathbf{a}_{C/B})_t = 0.361\alpha_{BC}(-\sin 13.9° \mathbf{i} + \cos 13.9° \mathbf{j}) = -0.0867\alpha_{BC}\mathbf{i} + 0.35\alpha_{BC}\mathbf{j}$$

Equation (1) now becomes

$$0.346\alpha_{DC}\mathbf{i} + 0.2\alpha_{DC}\mathbf{j} + 0.57\mathbf{i} - 0.987\mathbf{j} = -0.0867\alpha_{BC}\mathbf{i} + 0.35\alpha_{BC}\mathbf{j} - 1.78\mathbf{i} - 0.44\mathbf{j} - 1.35\mathbf{i} - 2.34\mathbf{j} \tag{2}$$

Equating coefficients of the $\mathbf{i}$ and $\mathbf{j}$ terms in Eq. (2), we have

$$0.346\alpha_{DC} + 0.57 = -0.0867\alpha_{BC} - 1.78 - 1.35 \tag{3}$$

$$0.2\alpha_{DC} - 0.987 = 0.35\alpha_{BC} - 0.44 - 2.34 \tag{4}$$

These simplify to

$$0.346\alpha_{DC} + 0.0867\alpha_{BC} = -3.7 \qquad \therefore \alpha_{DC} = -10.5 \text{ rad/s}^2 \tag{3'}$$

$$0.2\alpha_{DC} - 0.35\alpha_{BC} = -1.79 \qquad \alpha_{BC} = -0.87 \text{ rad/s}^2 \tag{4'}$$

Since the sign of α_{DC} is negative, it is counterclockwise (opposite to the direction originally assumed). Also, the sign of α_{BC} is negative, so it is clockwise (opposite to the direction originally assumed).

4.18. A wheel 3 m in diameter rolls to the right on a horizontal plane with an angular velocity of 8 rad/s (clockwise) and an angular acceleration counterclockwise of 4 rad/s², as shown in Fig. 4-28. The latter merely indicates that the angular velocity of the wheel is decreasing. Determine the velocity and acceleration of the top point B on the wheel.

SOLUTION

Draw Fig. 4-29 illustrating the problem. For convenience, the center O is chosen as the base point in this relative motion. It is necessary to show first that the velocity and acceleration of the center O of a rolling wheel may be expressed in terms of the given ω and α, respectively, and the distance from O to the surface on which the wheel rolls (1.5 m in this case).

In Fig. 4-29, O is displaced to O' through a distance s. Since rolling takes place, MM'' on the wheel contacts MM' on the horizontal surface. Hence, the arc MM'', which is $r\theta$, equals MM', which is equal to OO'

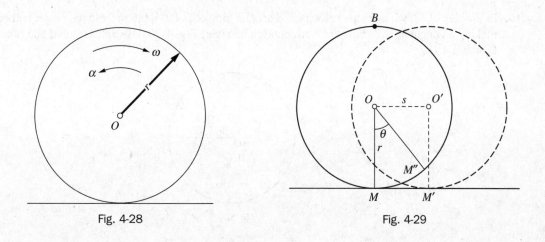

Fig. 4-28   Fig. 4-29

or s. Therefore, $s = r\theta$, where s is the magnitude of the linear displacement of center O, r is the radius of the wheel, and θ is the magnitude of the angular displacement of the wheel. Differentiation yields

$$v_O = \frac{ds}{dt} = r\frac{d\theta}{dt} = r\omega \qquad \text{and} \qquad a_O = \frac{d^2 s}{dt^2} = r\frac{d\omega}{dt} = r\alpha$$

These facts may now be applied to the present problem:

$$v_O = (1.5)(8) = 12 \text{ m/s}$$

$$a_O = (1.5)(4) = 6 \text{ m/s}^2$$

where $\mathbf{v}_O$ is directed to the right and $\mathbf{a}_O$ is directed to the left.

The velocity vector equation that will be used is

$$\mathbf{v}_B = \mathbf{v}_{B/O} + \mathbf{v}_O$$

The velocity of B relative to O is perpendicular to the radius OB and to the right (since OB is moving clockwise). Its magnitude is

$$v_{B/O} = OB \times \omega = (1.5)(8) = 12 \text{ m/s}$$

The velocity $\mathbf{v}_B$ is therefore made up of two components ($\mathbf{v}_{B/O}$ and $\mathbf{v}_O$), each horizontal to the right and each 12 m/s. Hence,

$$v_B = 24 \text{ m/s} \qquad \text{(to the right)}$$

To determine the absolute acceleration $\mathbf{a}_B$, apply the vector equation

$$\mathbf{a}_B = (\mathbf{a}_{B/O})_t + (\mathbf{a}_{B/O})_n + \mathbf{a}_O$$

The relative acceleration $(\mathbf{a}_{B/O})_t$ is directed horizontally to the left (since the angular acceleration of OB is counterclockwise). Its magnitude is equal to OB times the magnitude of the angular acceleration α, or

$$(a_{B/O})_t = (1.5)(4) = 6 \text{ m/s}^2 \qquad \text{(directed left)}$$

The component $(\mathbf{a}_{B/O})_n$ is directed toward O at the instant considered and is equal in magnitude to OB times the square of the angular speed ω:

$$(a_{B/O})_n = (1.5)(8)^2 = 96 \text{ m/s}^2 \qquad \text{(directed down)}$$

To find the acceleration of B, note that the horizontal component is $6 + 6 = 12$ m/s^2 to the left. The vertical component is 96 m/s^2 down. Therefore,

$$a_B = \sqrt{(a_h)^2 + (a_v)^2} = \sqrt{(12)^2 + (96)^2} = 96.7 \text{ m/s}^2$$

and $\tan \phi = 12/96$ or $\phi = 0.124$ rad or $7.12°$, where ϕ is measured relative to the vertical.

4.19. In Problem 4.18, what are the velocity and acceleration of point A, 0.6 m from the center and on a line making an angle of 30° above the horizontal radius? See Fig. 4-30. Use values stated and obtained in Problem 4.18.

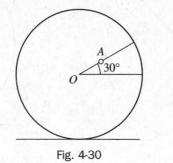

Fig. 4-30

SOLUTION

The usual vector equations apply.

$$\mathbf{v}_A = \mathbf{v}_{A/O} + \mathbf{v}_O \tag{1}$$

$$\mathbf{a}_A = (\mathbf{a}_{A/O})_t + (\mathbf{a}_{A/O})_n + \mathbf{a}_O \tag{2}$$

The velocity equation (1) is represented by Fig. 4-31. The magnitude of the horizontal component of $\mathbf{v}_A$ is

$$(v_A) = 4.8 \sin 30° + 12 = 14.4 \text{ m/s}$$

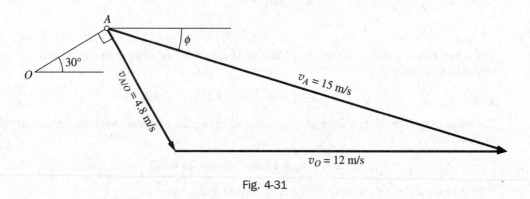

Fig. 4-31

The magnitude of the vertical component of $\mathbf{v}_A$ is

$$(v_A) = -4.8 \cos 30° = -4.16 \text{ m/s}$$

Hence,

$$v_A = \sqrt{(14.4)^2 + (4.16)^2} = 15 \text{ m/s} \qquad \text{and} \qquad \phi = \tan^{-1}\left(\frac{4.16}{14.4}\right) = 0.281 \text{ rad or } 16.1°$$

Vector	Direction	Magnitude
$\mathbf{a}_A$	?	?
$(\mathbf{a}_{A/O})_t$	$\perp OA$	$OA \times \alpha = 0.6 \text{ m} \times 4 \text{ rad/s}^2 = 2.4 \text{ m/s}^2$
$(\mathbf{a}_{A/O})_n$	along OA	$OA \times \omega^2 = 0.6 \text{ m} \times (8 \text{ rad/s})^2 = 38.4 \text{ m/s}^2$
$\mathbf{a}_O$	horizontal	6 m/s^2 to the left

Figure 4-32 represents Eq. (2). To obtain the magnitude of $\mathbf{a}_A$, note that the magnitudes of its components horizontally and vertically are

$$(a_A)_h = -6 - 0.6 \times 8^2 \cos 30° - 0.6 \times 4 \sin 30° = -40.5 \text{ m/s}^2$$

$$(a_A)_v = -0.6 \times 8^2 \sin 30° + 0.6 \times 4 \cos 30° = -17.1 \text{ m/s}^2$$

Hence,

$$a_A = \sqrt{(-40.5)^2 + (-17.1)^2} = 44.0 \text{ m/s}^2 \qquad \text{with } \theta = \tan^{-1}\left(\frac{-17.1}{-40.5}\right) = 0.4 \text{ rad or } 22.9°$$

Note that ϕ is below the negative x axis.

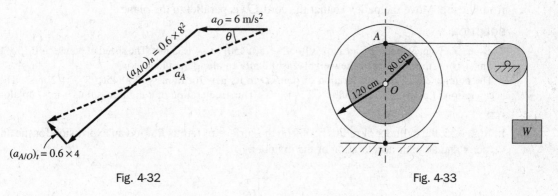

Fig. 4-32   Fig. 4-33

4.20. A cylinder and axle roll under the influence of a weight W, as shown in Fig. 4-33. What is the displacement s_O of the center of the cylinder when the weight is displaced 4 m down? The pulley is assumed to have frictionless bearings.

SOLUTION

Point I is the instant center between the cylinder and the surface on which it is rolling. Hence, the magnitude of the absolute displacement of A, which is equal to that of the displacement of W, may be expressed as 5θ, where the radius $AI = 2$ m and θ is the angle or rotation. Since $2.0\theta = 4$, $\theta = 2$ rad and

$$s_O = 1.2\theta = 2.4 \text{ m} \qquad \text{(directed to right)}$$

4.21. Figure 4-34 shows a weight A on a frictionless plane. The weight is attached to a cord that is wrapped around the small step of pulley B, which turns on frictionless bearings. Another cord, which is wrapped around the large step of pulley B, is in turn wrapped around the smaller of two cylinders C, which are integrally connected. The center of cylinder C is moving with a speed of 2 m/s and an acceleration of 3 m/s^2, both down the plane. Determine the velocity and acceleration of weight A. Assume that the cords are parallel to the planes as shown.

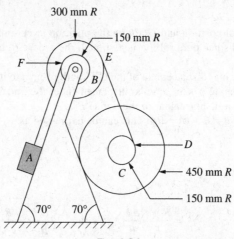

Fig. 4-34

SOLUTION

Point E on cord DE has the same speed as point D, which is on the cylinder. As a point on the cylinder, D has a speed of $[(450 + 150)/450] \times 2 = 2.67$ m/s. The mating point on the pulley B has the same speed as E (2.67 m/s). Then the speed of point F is $(150/300) \times 2.67 = 1.33$ m/s. This, then, is the speed of weight A: 1.33 m/s, up along the plane.

The component of the acceleration of point D that is parallel to the plane is $(600/450) \times 3 = 4$ m/s^2. This, then, is also the magnitude of the tangential component of the acceleration of point E. The tangential component of the acceleration of point F is then $(150/300) \times 4 = 2$ m/s^2. Thus, the acceleration of weight A is 2 m/s^2 up along the plane.

4.22. Solve Problem 4.21 if the cord ED is unwrapping from the bottom of the smaller cylinder instead of from the top. Move the pulley so that the cord ED is parallel to the plane.

SOLUTION

The speed of point D in its new location is $[(450 - 150)/450] \times 2 = 1.33$ m/s. The speed of point F is $0.15 \times 1.33 = 0.2$ m/s. Thus, the velocity of the weight A is 0.2 m/s up along the plane.

The component of the acceleration of point D in its new location is $(300/450) \times 3 = 2$ m/s^2. Then the acceleration of point F is $(150/300) \times 2 = 1$ m/s^2. The acceleration of weight A is thus 1 m/s^2 up along the plane.

4.23. In Fig. 4-35, the cylinder of radius r rolls on the surface of radius R. Find an expression for the angular velocity and angular acceleration of the cylinder.

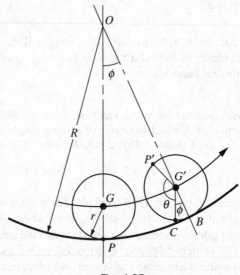

Fig. 4-35

SOLUTION

Let OGP be the original position and $OG'B$ be the position after some time has elapsed. Point P has then moved to point P', and since pure rolling is assumed, the arc BCP' of the cylinder must equal the arc PB on the surface.

The angle θ is the angular displacement of the $G'P'$ with respect to its original position GP (or $G'C$, which is parallel to GP). The angle ϕ is the angular displacement of the line OGP in the same time interval. Hence, $\theta + \phi$ is the total angular displacement of the line $G'P'$.

Arc BCP' = arc PB, or $r(\theta + \phi) = R\phi$. This can be expressed as

$$\theta = \frac{R - r}{r} \phi$$

Hence,

$$\frac{d\theta}{dt} = \frac{R - r}{r} \frac{d\phi}{dt} \quad \text{and} \quad \frac{d^2\theta}{dt^2} = \frac{R - r}{r} \frac{d^2\phi}{dt^2}$$

If the velocity of point G is v_G and the magnitude of its acceleration tangent to its circular path (of radius $R - r$) is a_G, the angular velocity $d\phi/dt$ and angular acceleration $d^2\phi/dt^2$ of G are found as follows (note again that G moves on a circular path of radius $R - r$):

$$\frac{d\phi}{dt} = \frac{v_G}{R - r} \qquad \text{and} \qquad \frac{d^2\phi}{dt^2} = \frac{(a_G)_t}{R - r}$$

The angular velocity $d\theta/dt$ and angular acceleration $d^2\theta/dt^2$ of any point on the cylinder relative to its center are then

$$\frac{d\theta}{dt} = \frac{R - r}{r}\frac{d\phi}{dt} = \frac{v_G}{r} \qquad \text{and} \qquad \frac{d^2\theta}{dt^2} = \frac{(a_G)_t}{r}$$

The velocity and the acceleration of any point on the cylinder may now be found by first referring the motion of the point to the center and then adding the motion of the center. For example, the velocity $\mathbf{v}_B$ of the contact point B equals the sum of the velocity of B relative to the center (magnitude is $r\,d\theta/dt$) and the velocity of the center $\mathbf{v}_G$. If the wheel is rolling up the plane, $d\theta/dt$ is clockwise, and $r\,d\theta/dt$ is thus tangent to the cylinder and directed down to the left. Its magnitude is $r\,d\theta/dt$ or $r(v_G/r) = v_G$. Add this to the velocity of G, which is of the same magnitude, parallel to it but directed up to the right. The absolute velocity of B is found to be zero; i.e., B is the instant center.

4.24. In an epicyclic gear train, the arm is moving 6 rad/s clockwise and has an acceleration of 10 rad/s^2 counterclockwise. Determine the velocity and acceleration of point B in the position shown in Fig. 4-36.

SOLUTION

In the previous problem, it was shown that the magnitude of the angular velocity ω and angular acceleration α of the small wheel about its own center are given by

$$\omega = \frac{d\theta}{dt} = \frac{R - r}{r}\frac{d\phi}{dt} \qquad \text{and} \qquad \frac{d^2\theta}{dt^2} = \frac{R - r}{r}\frac{d^2\phi}{dt^2}$$

where ϕ is the angular change of the arm OG. Hence, $\omega = [(75 - 25)/25] \times 6 = 12$ rad/s (counterclockwise) and $\alpha = [(75 - 25)/25] \times 10 = 20$ rad/s^2 (clockwise). By inspection, clockwise motion of the arm means counterclockwise motion of the small gear.

The velocity $\mathbf{v}_B$ of point B is the vector sum of its relative velocity to G and the velocity of G: $\mathbf{v}_B = \mathbf{v}_{B/G} + \mathbf{v}_G$. The magnitudes of the vectors $\mathbf{v}_G$ and $\mathbf{v}_{B/G}$ are found to be

$$v_G = (R - r)\omega = (0.075 - 0.025) \times 6 = 0.3 \text{ m/s}$$
$$v_{B/G} = r\omega = 0.025 \times 12 = 0.3 \text{ m/s}$$

The vector sum is shown in Fig. 4-37. The magnitude of the velocity at B is:

$$v_B = \sqrt{0.3^2 + 0.3^2} = 0.424 \text{ m/s}.$$

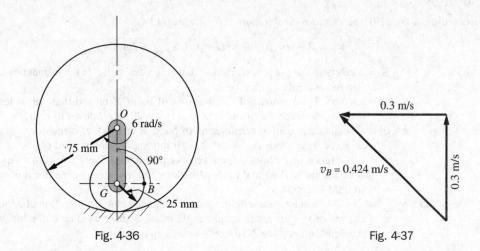

Fig. 4-36   Fig. 4-37

The magnitudes of the various vectors are:

$$(a_{B/G})_t = r\alpha = 0.025 \times 20 = 0.5 \text{ m/s}^2 \qquad \text{(down)}$$

$$(a_{B/G})_n = r\omega^2 = 0.025 \times 12^2 = 3.6 \text{ m/s}^2 \qquad \text{(left)}$$

$$(a_G)_t = (R - r)10 = 0.05 \times 10 = 0.5 \text{ m/s}^2 \qquad \text{(right)}$$

$$(a_G)_n = (R - r)6^2 = 0.05 \times 36 = 1.8 \text{ m/s}^2 \qquad \text{(up)}$$

The acceleration $\mathbf{a}_B$ of point B is the vector sum of its relative acceleration of G (two components: normal and tangent) and the acceleration of G (two components also):

$$\mathbf{a}_B = (\mathbf{a}_{B/G})_t + (\mathbf{a}_{B/G})_n + (\mathbf{a}_G)_t + (\mathbf{a}_G)_n$$

A vector diagram need not be drawn. The vertical summation is a net amount of $1.8 - 0.5 = 1.3 \text{ m/s}^2$ up. The horizontal summation is a net amount of 3.1 m/s^2 to the left. The resultant absolute acceleration of B is thus to the left and up of magnitude $a_B = \sqrt{1.3^2 + 3.1^2} = 3.36 \text{ m/s}^2$.

Note: The absolute velocity $\mathbf{v}_B$ may be determined also by using the instant center, the common point of tangency of the two circles (see preceding problem). Although the absolute velocity of the instant center is zero, its absolute acceleration is not.

4.25. In Fig. 4-38, the washer is sliding outward on the rod with a velocity of 4 m/s when its distance from point O is 2 m. Its velocity along the rod is increasing at the rate of 3 m/s^2. The angular velocity of the rod is 5 rad/s counterclockwise, and its angular acceleration is 10 rad/s^2 clockwise. Determine the absolute acceleration of point P on the washer.

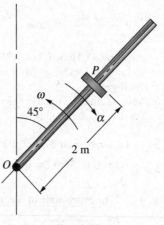

Fig. 4-38

SOLUTION

According to Eq. (19), the absolute acceleration of P is expressed as

$$\mathbf{a}_P = \ddot{\mathbf{R}} + \boldsymbol{\alpha} \times \boldsymbol{\rho} + \boldsymbol{\omega} \times (\boldsymbol{\omega} \times \boldsymbol{\rho}) + \mathbf{a}_{P/\text{rod}} + 2\boldsymbol{\omega} \times \mathbf{v}_{P/\text{rod}}$$

where

$\ddot{\mathbf{R}}$ = acceleration of point M on the rod that coincides with P at the instant involved; $\ddot{\mathbf{R}} = 0$ (the rod is rigid)

$\boldsymbol{\alpha} \times \boldsymbol{\rho}$ = tangential component of acceleration of point M on rod that coincides with P at instant involved; magnitude $r\alpha = 2(10) = 20 \text{ m/s}^2$, down to right

$\boldsymbol{\omega} \times (\boldsymbol{\omega} \times \boldsymbol{\rho})$ = normal component of acceleration of point M on rod that coincides with P at instant involved; magnitude $r\omega^2 = 2(5^2) = 50 \text{ m/s}^2$, along rod toward O

$(\mathbf{a}_{P/\text{rod}})_t$ = acceleration of P along its path relative to rod, i.e., 3 m/s^2 outward along rod

$(\mathbf{a}_{P/\text{rod}})_n$ = acceleration of P normal to its path along rod, i.e., zero here since it is moving on a straight-line path

$2\boldsymbol{\omega} \times \mathbf{v}_{P/\text{rod}}$ = Coriolis' acceleration with magnitude indicated and in a direction obtained by rotating vector $\mathbf{v}_{P/\text{rod}}$ through a right angle in same sense as $\boldsymbol{\omega}$ (counterclockwise in this problem); magnitude $2(5)(4) = 40 \text{ m/s}^2$, up to left

The vector diagram indicates each of these accelerations and their vector sum $\mathbf{a}_P$ (see Fig. 4-39). Add components in the x and y directions. Thus, $a_P = 51$ m/s^2, with $\theta = 22°$.

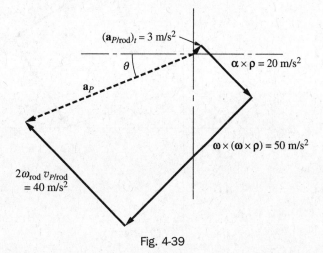

Fig. 4-39

4.26. One vane of an impeller wheel has its center of curvature C located as shown in Fig. 4-40 at a given instant. A particle P, 200 mm from the center O, has a velocity 0.25 m/s and an acceleration 0.5 m/s^2 directed outward and tangent to the vane. Determine the acceleration of P if the angular velocity of the wheel is 2 rad/s counterclockwise and the angular acceleration is 3 rad/s^2 clockwise.

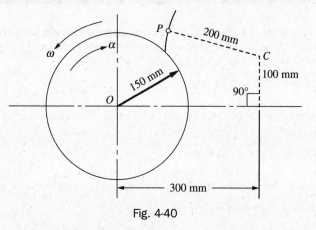

Fig. 4-40

SOLUTION

According to Eq. (19), the absolute acceleration of P is

$$\mathbf{a}_P = {}_0\ddot{\mathbf{R}} + \boldsymbol{\alpha} \times \boldsymbol{\rho} + \boldsymbol{\omega} \times (\boldsymbol{\omega} \times \boldsymbol{\rho}) + \mathbf{a}_{P/\text{vane}} + 2\boldsymbol{\omega} \times \mathbf{v}_{P/\text{vane}}$$

where

$(\mathbf{a}_{P/\text{vane}})_t = 20$ m/s^2 outward and $\perp$ to PC

$(\mathbf{a}_{P/\text{vane}})_n = (v_{P/\text{vane}})^2/PC = (10)^2/8 = 12.5$ m/s^2 directed from P to C

$\boldsymbol{\alpha} \times \boldsymbol{\rho} =$ tangential component of acceleration of point M on the vane that coincides with point P at the instant; magnitude $= \rho \times \alpha = 0.2 \times 3 = 0.6$ m/s^2, directed down to right and $\perp$ to $\rho (\rho = OP)$

$\boldsymbol{\omega} \times (\boldsymbol{\omega} \times \boldsymbol{\rho}) =$ normal component of acceleration of point M on vane that coincides with P at the instant; magnitude $= \rho \times \omega^2 = 0.2(2)^2 = 0.8$ m/s^2, directed from P to O

$(\mathbf{v}_{P/\text{vane}}) = 0.25$ m/s, outward $\perp$ to PC

$(\boldsymbol{\omega}_{\text{vane}}) = 2$ rad/s, counterclockwise

$2\boldsymbol{\omega}_{\text{vane}} \times \mathbf{v}_{P/\text{vane}} = 2(2)(0.25) = 1.0$ m/s^2 directed from C toward P; direction is obtained by rotating $\mathbf{v}_{P/\text{vane}}$, which is $\perp$ to CP and outward, through 90° in sense of $\boldsymbol{\omega}_{\text{vane}}$, i.e., counterclockwise in plane of paper

The vector sum of these components yields $a_P = 0.525$ m/s^2 and $\theta_x = 215°$. See Fig. 4-41.

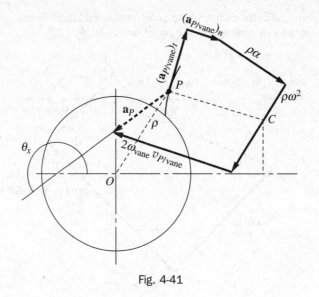

Fig. 4-41

SUPPLEMENTARY PROBLEMS

4.27. A rigid body is rotating 12 rad/s about an axis through the origin and has direction cosines 0.421, 0.365, and 0.831 with respect to the x, y, and z axes, respectively. What is the velocity of a point in the body defined by the position vector (with respect to the origin) $\mathbf{r} = -2\mathbf{i} + 3\mathbf{j} - 4\mathbf{k}$ m?

Ans. $\mathbf{v} = -47.4\mathbf{i} + 0.26\mathbf{j} + 23.9\mathbf{k}$ m/s

4.28. A rigid body is rotating 200 rpm about the line $\mathbf{i} - 3\mathbf{j} + 4\mathbf{k}$. The origin is on the line. What is the velocity of the point $P(3, 3, -1)$ m?

Ans. $\mathbf{v} = -37.0\mathbf{i} + 53.6\mathbf{j} + 49.3\mathbf{k}$ m/s

4.29. Determine the angular velocities, in rad/s, of the second and minute hands of an old-fashioned watch.

Ans. 0.105 rad/s; 0.0018 rad/s

4.30. A rigid body is rotating at a rate of 60 rpm about a line from the origin to the point $(3, 0, 5)$, where the coordinates are in meters. Determine the velocity of the point $(1, -2, 2)$ in the body.

Ans. $10.8\mathbf{i} - 1.08\mathbf{j} - 6.47\mathbf{k}$ m/s

4.31. Refer to Fig. 4-42. The equal bars AB and CD are free to rotate about pins in the frame. The bar BD is equal in length to the distance AC. The constant angular velocity of AB is 10 rpm counterclockwise. Determine the motion of BD.

Ans. All points on BD have $v = 0.314$ m/s to the right; $a = 0.329$ m/s^2 up

4.32. At a given time, a shaft is rotating at 50 rpm about a fixed axis; 20 s later, it is rotating at 1050 rpm. What is the average angular acceleration α in rad/s^2?

Ans. $\alpha = 5.23$ rad/s^2

4.33. A flywheel with diameter 500 mm starts from rest with constant angular acceleration 2 rad/s^2. Determine the tangential and normal components of the acceleration of a point on the rim 3 s after the motion began.

Ans. $a_t = 0.5$ m/s^2, $a_n = 9$ m/s^2

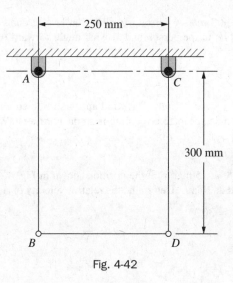

Fig. 4-42

4.34. A particle is at rest on a phonograph turntable at a radius r from the center. Assuming that the turntable starts from rest and accelerates with a uniform angular acceleration α_0, determine the magnitudes of the tangential and normal components of the acceleration of the particle at time t.

Ans. $a_n = \alpha_0^2 / r t^2, a_t = \alpha_0 r$

4.35. A rotor decreases uniformly from 1800 rpm to rest in 320 s. Determine the angular deceleration and the number of radians before coming to rest.

Ans. $\alpha = -0.589$ rad/s^2, $\theta = 30\ 100$ rad

4.36. A bar pivoted at one end moving 5 rad/s clockwise is subjected to a constant angular deceleration. After a certain time interval the bar has an angular displacement of 8 rad counterclockwise and has moved through a total angle of 20.5 rad. What is the angular velocity at the end of the time interval?

Ans. $\omega = 7.58$ rad/s

4.37. The drum shown in Fig. 4-43 is used to hoist the weight W a distance 2 m. The drum accelerates uniformly from rest to 15 rpm in 1.5 s and then moves at a constant speed of 15 rpm. What is the total elapsed time?

Ans. $t = 7.12$ s

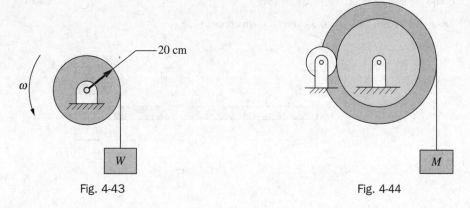

Fig. 4-43   Fig. 4-44

4.38. Refer to Fig. 4-44. The hoisting mechanism consists of a drum 1200 mm in diameter around which is wrapped the cable. Integral with the drum is a gear with pitch diameter 900 mm. This gear is driven by a pinion with a 300-mm pitch diameter. The acceleration of the mass M is 6 m/s^2 up. What is the angular acceleration of the pinion?

Ans. $\alpha = 30$ rad/s^2 clockwise

4.39. Rain is falling vertically at 2 m/s. A person is walking on level ground with a speed of 1.6 m/s. What is the velocity of the rain relative to the person and at what angle forward of the vertical should an umbrella be held?

Ans. 2.56 m/s, 38.7°

4.40. A person is walking due east at 3 km/h. The wind appears to be coming from the north. When the pace is dropped to 1 km/h, the wind appears to be coming from the northwest. What is the speed of the wind?

Ans. $v = 3.61$ km/h

4.41. Mass A moves to the right at 15 m/s from the position shown in Fig. 4-45. Mass B starts at the same instant, moving vertically upward at 20 m/s. Determine the relative velocity of A to B $\frac{1}{2}$ s after motion begins.

Ans. $v_{A/B} = 25$ m/s, $\theta_x = 307°$

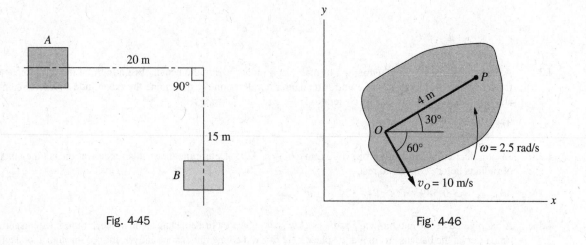

Fig. 4-45   Fig. 4-46

4.42. Refer to Fig. 4-46. Points O and P are on a thin lamina that has motion in the xy plane. Point O is known to have the velocity shown. Determine the velocity of point P.

Ans. $v_P = 0$

4.43. The iron bar in Fig. 4-47 is attracted by two large but unequal magnets in such a way that end A has a vertical acceleration of 4 m/s² and end B has a vertical acceleration of 6 m/s². Determine the acceleration of the center C and the angular acceleration of the bar.

Ans. $a_C = 5$ m/s² up, $\alpha = 0.333$ rad/s² counterclockwise

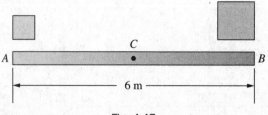

Fig. 4-47

4.44. A ladder of length ℓ leans against a vertical wall. The bottom moves away from the wall along a horizontal floor with a constant velocity v_0. Determine the velocity and acceleration of the top of the ladder. (*Hint:* $x = v_0 t$.)

Ans. $\dot{y} = \dfrac{-v_0 t}{\sqrt{\ell^2 - v_0^2 t^2}}$, $\ddot{y} = \dfrac{-v_0^2 \ell^2}{(\ell^2 - v_0^2 t^2)^{3/2}}$

4.45. A ladder of length ℓ makes an angle θ with the horizontal floor and leans against a vertical wall. Show that the center of the ladder moves on a circular path of radius $\frac{1}{2}\ell$.

4.46. The bar AB in Fig. 4-48 slides so that its bottom point A has a velocity of 400 mm/s to the left along the horizontal plane. What is the angular velocity of the bar in the position shown? Use the instantaneous center method.

Ans. $\omega = 0.293$ rad/s clockwise

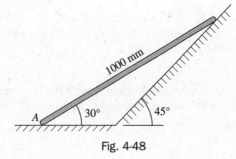

Fig. 4-48

4.47. The top of a ladder 3-m long slides down a smooth wall while its bottom slides along a smooth plane perpendicular to the wall. Show that the velocity of its midpoint is directed along the ladder when the ladder makes an angle of 45° with the horizontal plane.

4.48. In Fig. 4-49, the velocity of point A is 5 m/s to the right, and its acceleration is 8 m/s² to the right. What are the velocity and acceleration of point B?

Ans. $v_B = 8.66$ m/s down, $a_B = 33.9$ m/s² down

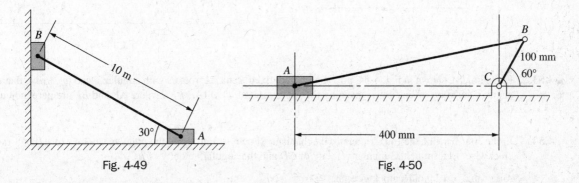

Fig. 4-49   Fig. 4-50

4.49. The crank CB of the slider crank mechanism is rotating at a constant 30 rpm clockwise. Determine the velocity of the crosshead A in the position shown in Fig. 4-50. Use two methods of solution.

Ans. $v_A = 0.242$ m/s to the right

4.50. In the slider crank mechanism shown in Fig. 4-51, the crank is turning clockwise at 120 rpm. What is the velocity of the crosshead when the crank is in the 60° phase? Use two methods of solution.

Ans. 3.09 m/s toward the right

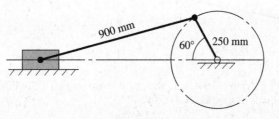

Fig. 4-51

4.51. The linear velocity of the crosshead (slider) in Fig. 4-52 is 2.4 m/s to the left along the horizontal plane. Using the instant center method, determine the angular velocity of the crank *AB* in the phase shown.

Ans. $\omega = 2.44$ rad/s counterclockwise

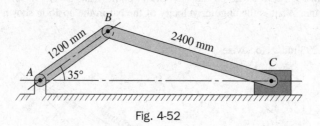

Fig. 4-52

4.52. A bar slides on a vertical post and is attached to a block *A* that is moving to the right with a constant velocity v_C. See Fig. 4-53. Determine the angular velocity $\dot\theta$ of the bar.

Ans. $\dot\theta = -(v_C/a)\sin^2\theta$

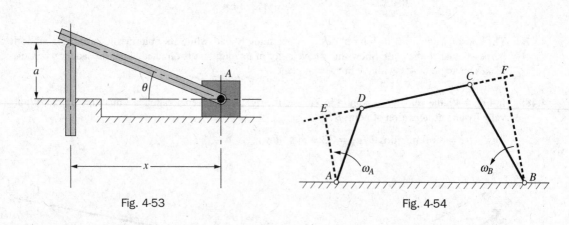

Fig. 4-53   Fig. 4-54

4.53. In the linkage shown in Fig. 4-54, pins *A* and *B* are fixed. Link *AD* rotates with angular speed ω_A. Prove that the angular speed ω_B of link *BC* is given by the expression $\omega_B = \omega_A(AE/BF)$, where *AE* and *BF* are perpendicular to *DC*.

4.54. In the four-bar linkage (quadric crank mechanism) shown in Fig. 4-55, the angular velocity of *AB* is 8 rad/s clockwise. Determine the angular velocity of *CD* and the angular velocity of *BC*.

Ans. $\omega_{CD} = 12.0$ rad/s clockwise, $\omega_{BC} = 0$

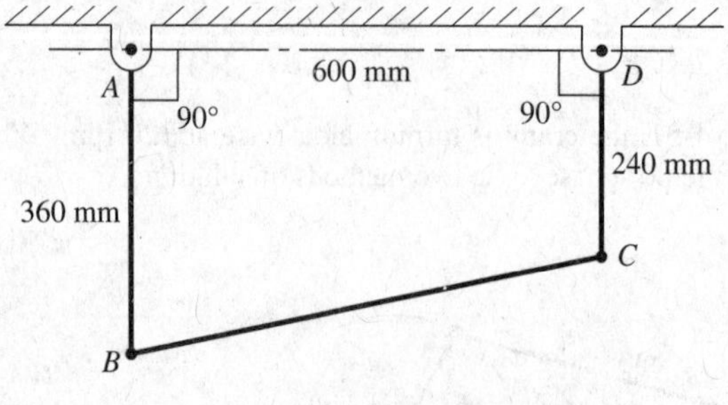

Fig. 4-55 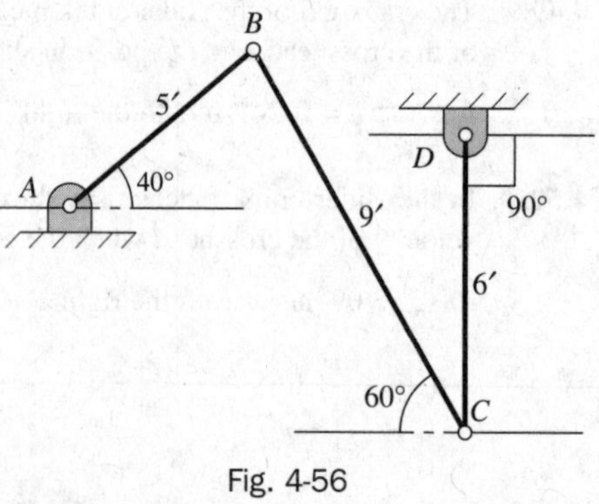   Fig. 4-56

4.55. Crank *AB* is moving 2 rpm clockwise in the phase shown in Fig. 4-56. What is the angular velocity of the arm *CD*? Use two methods of solution.

Ans. $\omega_{CD} = 3.28$ rpm counterclockwise

4.56. In the linkage shown in Fig. 4-57, bar *AB* is constrained to move horizontally and bar *CD* rotates about point *D*. If bar *AB* has a velocity of 0.6 m/s to the left and an acceleration of 1 m/s^2 to the right, what are the angular velocity and acceleration of *CD*?

Ans. $\omega = 5.3$ rad/s clockwise, $\alpha = 72$ rad/s^2 clockwise

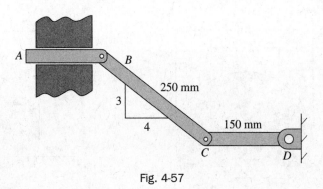

Fig. 4-57

4.57. A 3-m-diameter wheel rolls without slipping. The velocity and acceleration of the center of the wheel are 8 m/s and 5 m/s^2 to the right. What is the acceleration of the top point?

Ans. $a = 43.8$ m/s^2, $\theta_x = 283°$

4.58. A wheel 300 mm in diameter rolls to the right without slipping on a horizontal plane. Its angular speed is 30 rpm. What is the velocity of (*a*) the top point on the wheel and (*b*) the point on the front end of the horizontal diameter?

Ans.   (*a*) $v = 0.942$ m/s, $\theta_x = 0°$; (*b*) $v = 0.666$ m/s, $\theta_x = 315°$

4.59. The block in Fig. 4-58 moves on two 10-cm-diameter rollers as shown. If the block has a velocity of 0.9 m/s and an acceleration of 0.6 m/s^2, both to the right, determine the velocity and acceleration of the center of one of the rollers.

Ans. $v = 0.45$ m/s to the right, $a = 0.3$ m/s^2 to the right

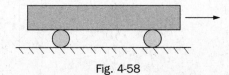

Fig. 4-58

4.60. The disk in Fig. 4-59 rolls without slipping on the horizontal plane. Its center has an acceleration of 1.2 m/s^2 directed horizontally to the left. At the instant that its center has a velocity of 0.9 m/s to the right, determine the acceleration of point *P*.

Ans. $a = 2.2$ m/s^2, $\theta_x = 183°$

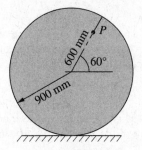

Fig. 4-59

4.61. A composite wheel rolls without slipping with angular velocity 30 rpm clockwise (see Fig. 4-60). Determine the velocities of points A and B. Use two methods of solution.

Ans. $v_A = 2.22$ m/s, $\theta_x = -45°$; $v_B = 4.71$ m/s to the right

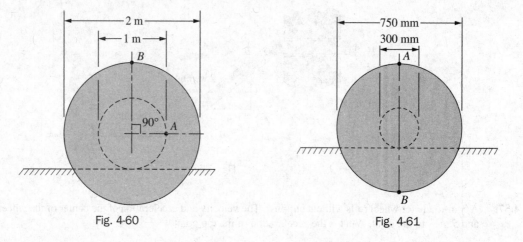

Fig. 4-60   Fig. 4-61

4.62. The composite wheel shown in Fig. 4-61 rolls without slipping on the horizontal plane. If the angular velocity of the wheel is 20 rad/s clockwise, determine the velocities of the top and bottom points of the wheel.

Ans. $v_A = 10.5$ m/s to the right, $v_B = 4.5$ m/s to the left

4.63. The wheel in Fig. 4-62 moves such that its center has a velocity of 4 m/s horizontally to the right. The angular velocity of the wheel is 4 rad/s clockwise. Determine the velocity of the points P and Q.

Ans. $v_P = 5.32$ m/s, $\theta_x = 25.0°$; $v_Q = 0$

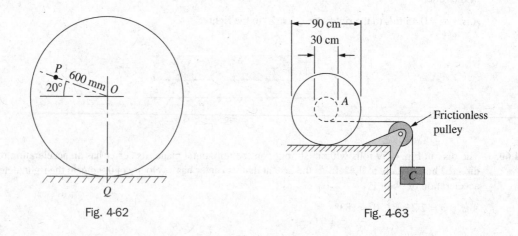

Fig. 4-62   Fig. 4-63

4.64. The composite wheel A in Fig. 4-63 rolls without slipping on the horizontal plane. The cord wrapped around the axle is attached to the weight C as shown. The velocity of C changes uniformly from 0.6 m/s downward to 1.8 m/s downward in 2 s. Determine the angular displacement of A during the interval.

Ans. $\theta = 8$ rad clockwise

4.65. The cylinder C shown in Fig. 4-64 is 500 mm in diameter and rolls without slipping on the horizontal plane. The pulley B is frictionless. If the displacement of A is 100 mm down, what is the angular displacement of C?

Ans. $\theta = 0.4$ rad clockwise

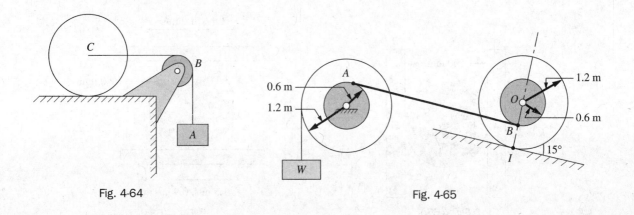

Fig. 4-64   Fig. 4-65

4.66. In the preceding problem, the velocity and acceleration of A are 100 mm/s and 50 mm/s^2, respectively. What are the angular velocity and angular acceleration of the cylinder C?

Ans. $\omega = 0.4$ rad/s clockwise, $\alpha = 0.2$ rad/s^2 clockwise

4.67. Weight W in Fig. 4-65 is suspended from the pulley, which turns in frictionless bearings. As the pulley turns, the cord AB from the axle of the cylinder is wrapped upon the pulley. The weight descends from rest with a constant acceleration of 4.8 m/s^2. Determine the displacement, velocity, and acceleration of the center O of the cylinder after 3 s. Cord AB is parallel to the inclined plane.

Ans. $s_0 = 21.6$ m, $v_0 = 14.4$ m/s, $a_0 = 4.8$ m/s^2

4.68. Solve Problem 4.67 if the cord AB is unwrapping from the top of the axle instead of the bottom. Move the pulley so that AB is still parallel to the plane.

Ans. $s_0 = 3.6$ m, $v_0 = 4.8$ m/s, $a_0 = 1.6$ m/s^2

4.69. In Fig. 4-66, valve A is actuated by the eccentric rotating 30 rpm counterclockwise. Express the velocity and acceleration of the valve in terms of the angle ϕ.

Ans. $\dot{x} = 0.20 \sin \phi$ m/s to the left, $\ddot{x} = 0.61 \cos \phi$ m/s^2 to the left

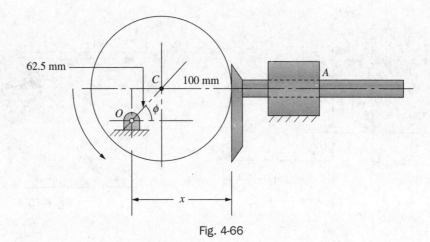

Fig. 4-66

4.70. Refer to Fig. 4-67. The gear rack A is stationary whereas gear rack B has a velocity of 600 mm/s down and an acceleration of 450 mm/s^2 down. Determine the velocity of the gear center O, the acceleration of the gear center, and the acceleration of the contact point C on the gear.

Ans. $v_0 = 0.3$ m/s down, $a_0 = 0.225$ m/s^2 down, $a_C = 1.0$ m/s^2 at $\theta_x = 207°$

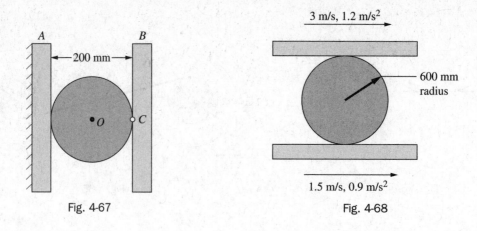

Fig. 4-67   Fig. 4-68

4.71. In Fig. 4-68, the upper plate is moving to the right with a speed of 3 m/s and a linear acceleration of 1.2 m/s^2. The lower plate is also moving to the right, with a speed of 1.5 m/s and a linear acceleration of 0.9 m/s^2. What are the angular velocity and acceleration of the 1.2-m-diameter disk. There is no slip between the disk and the plates.

Ans. $\omega = \frac{5}{4}$ rad/s clockwise, $\alpha = \frac{1}{4}$ rad/s^2 clockwise

4.72. Solve Problem 4.71 if the lower plate moves to the left with a speed of 1.5 m/s and an acceleration of 0.9 m/s^2.

Ans. $\omega = \frac{15}{4}$ rad/s clockwise, $\alpha = \frac{7}{4}$ rad/s^2 clockwise

4.73. In Fig. 4-69, the disk rolls without slipping on the horizontal plane with an angular velocity of 10 rpm clockwise and an angular acceleration of 6 rad/s^2 counterclockwise. The bar *AB* is attached as shown. The line *OA* is horizontal. Point *B* moves along the horizontal plane shown. Determine the velocity and acceleration of point *B* for the position shown.

Ans. $v_B = 1.1$ m/s to the right, $a_B = 7.47$ m/s^2 to the left

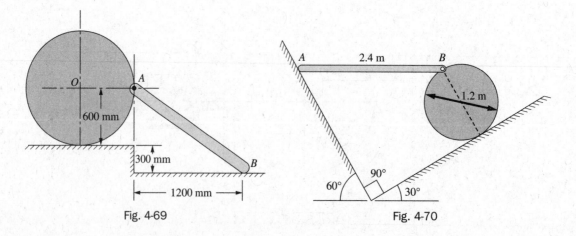

Fig. 4-69   Fig. 4-70

4.74. The 2.4 m bar *AB* in Fig. 4-70 is connected by a frictionless pin at *B* to the 1.2-m-diameter cylinder, which is rolling with constant center velocity 3.6 m/s down the plane inclined 30° with the horizontal. The end *A* slides on the frictionless plane, which makes an angle of 60° with the horizontal. For the position shown with the bar horizontal, determine the angular velocity and acceleration of the bar.

Ans. $\omega = 6.0$ rad/s clockwise, $\alpha = 62.4$ rad/s^2 counterclockwise

4.75. The crank *OB* in Fig. 4-71 is rotating with constant angular velocity 6 rad/s clockwise. Disk *C* rolls without slipping inside the fixed circle. Determine (*a*) the angular velocity of disk *C* and (*b*) the acceleration of point *P* in the position shown, where crank *OB* is horizontal and *P* is the top point on disk *C*.

Ans. (*a*) ω_C = 18 rad/s counterclockwise, (*b*) a_P = 102 m/s^2 at θ_x = 252°

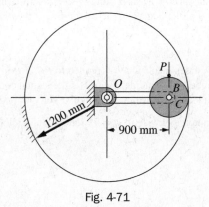

Fig. 4-71

4.76. A bead moves along a rod such that $r = 0.5t^2$. At the same time, the rod moves such that $\theta = t^2 + t$. Determine the radial and transverse components of the velocity and acceleration of the bead at time *t* = 2 s. Assume that *r* is in meters.

Ans. v_r = 2 m/s, v_θ = 10 m/s, a_r = − 49 m/s^2, a_θ = 24 m/s^2

4.77. A rod is rotating in a horizontal plane at a constant rate of 2 rad/s clockwise about a vertical axis through one end. A washer is sliding along the rod with a constant speed of 1.2 m/s. Determine the radial and transverse components of the acceleration at the instant when the washer is 600 mm from the vertical axis.

Ans. a_r = −2.4 m/s^2, a_θ = 4.8 m/s^2

4.78. In Fig. 4-72, the ball *P* moves with a constant velocity 2 m/s down along the smooth slot cut as shown in the disk rotating with angular velocity 3 rad/s clockwise and angular acceleration 8 rad/s^2 counterclockwise. Determine the acceleration of *P* in the position shown.

Ans. a = 14.9 m/s^2, with θ_x = 171°

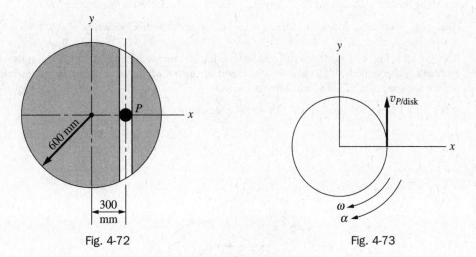

Fig. 4-72   Fig. 4-73

4.79. A particle moves with speed 600 mm/s around the circumference of a 1200-mm-diameter disk that is rotating in the opposite direction with angular velocity of 2 rad/s (clockwise). See Fig. 4-73. The disk has angular acceleration 4 rad/s^2 clockwise. Determine the acceleration of the particle in the position shown.

Ans. a = 2.47 m/s^2 with θ_x = 256°

4.80. A bug moves with constant speed v along the circumference of a disk of radius r. The disk rotates in the opposite direction with constant angular velocity ω rad/s. What is the absolute acceleration of the bug?

Ans. $(v - r\omega)^2/r$

4.81. A bug moves with constant speed v along a radius r of a disk rotating with constant angular velocity ω. What is the absolute acceleration of the bug as it reaches the rim of the disk?

Ans. $a = \omega\sqrt{r^2\omega^2 + 4v^2}$

4.82. In Fig. 4-74, member AD slides inside a collar pinned to PC at P, which is at a fixed distance from C. If M is the fixed point on the member AD coincident with P at the instant considered, then the velocity and acceleration of P relative to M are horizontal. Determine the angular velocity and acceleration of member CP.

Ans. $\omega_{CP} = 10$ rad/s counterclockwise, $\alpha_{CP} = 75$ rad/s^2 counterclockwise

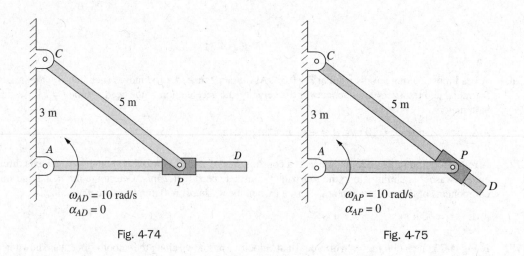

Fig. 4-74   Fig. 4-75

4.83. In Fig. 4-75, member CD slides inside a collar pinned to AP at P, which is a fixed distance from A. In this problem, M is the point in CD that is coincident with P; thus, the velocity and acceleration of P relative to M are along the line CD. Determine the angular velocity and acceleration of member CD.

Ans. $\omega_{CD} = 6.4$ rad/s counterclockwise, $\alpha_{CD} = 13.4$ rad/s^2 counterclockwise

4.84. The rod AB in Fig. 4-76 is pinned at B and rests on the 200-mm-radius wheel. The wheel rolls without slipping with an angular velocity of 12 rad/s clockwise and an angular acceleration of 3 rad/s^2 counterclockwise. What are the angular velocity and acceleration of rod AB?

Ans. $\omega = 1.6$ rad/s clockwise, $\alpha = 19.7$ rad/s^2 counterclockwise

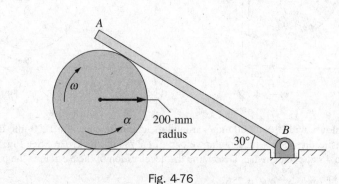

Fig. 4-76

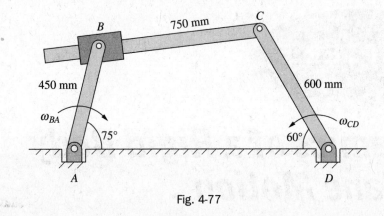

Fig. 4-77

4.85. Two rods *AB* and *CD* in Fig. 4-77 rotate about *A* and *D*, respectively. The 750-mm rod is free to slide in a collar at *B* and is pinned at *C*. If the angular speed of *AB* is constant at 10 rad/s clockwise and the angular speed of *CD* is constant at 8 rad/s counterclockwise, what are the angular speed and acceleration of the 750-mm rod?

Ans. $\omega = 0.353$ rad/s clockwise, $\alpha = 0.78$ rad/s^2 counterclockwise

CHAPTER 5

Dynamics of a Rigid Body in Plane Motion

5.1 Vector Equations of Plane Motion

In Section 4.1, plane motion of a rigid body was defined as that motion in which every point in the body remains at a constant distance from a fixed plane. In the case of kinetics of a rigid body, there is a further condition, namely, the body shall have a plane of symmetry. This is more restrictive than necessary, but it does simplify the moment equation. The vector equations of plane motion can be written

$$\sum \mathbf{F} = m\bar{\mathbf{a}} \qquad \text{(forces)} \tag{1}$$

$$\sum \mathbf{M}_O = (I_O \alpha + m\bar{x}a_{Oy} - m\bar{y}a_{Ox})\mathbf{k} \qquad \text{(moments)} \tag{2}$$

where
$\sum \mathbf{F}$ = resultant of external forces acting on body
$\sum \mathbf{M}_O$ = resultant of external moments acting on body
m = mass of body
$\bar{\mathbf{a}}$ = acceleration of mass center of body
α = angular acceleration of body
I_O = moment of inertia of body relative to reference point O
$\bar{x}, \bar{y}$ = coordinates of mass center relative to reference point O
a_{Ox}, a_{Oy} = magnitude of components of acceleration of reference point O along x and y axes

A right-hand set of coordinate axes is assumed in the above vector equations. This means that if the x and y axes are chosen positive to the right and up, then counterclockwise rotation must be chosen positive to be consistent with the right-hand system. Problem 5.1 illustrates this point.

5.2 Scalar Equations of Plane Motion

The moment equation in the preceding section may be simplified by the proper choice of the reference point O. One such choice is to use the mass center as O. Then $\bar{x}$ and $\bar{y}$ are zero. With this choice, the scalar equations of plane motion are

$$\sum F_x = m\bar{a}_x \qquad \sum F_y = m\bar{a}_y \qquad \sum \bar{M} = \bar{I}\alpha \tag{3}$$

where $\sum F_x, \sum F_y$ = sums of magnitudes of components of external forces along x and y axes
m = mass of body
$\bar{a}_x, \bar{a}_y$ = components of linear acceleration of mass center in x and y directions, respectively

128

$\sum \bar{M}$ = sum of moments of external forces about mass center
$\bar{I}$ = moment of inertia of body about mass center
α = angular acceleration of body

Note that the moment equation can also be written as $\sum M_O = I_O \alpha$, provided that the point O is a point whose acceleration either is zero or is directed through the mass center of the body.

Note also that the component of a force acting in the same direction as that chosen for the mass center acceleration is assigned a positive sign (if opposite in sense, it should be assigned a negative sign). Similarly, moments of forces are considered positive if they have the same sense as that assigned to the angular acceleration α.

It will be seen later that translation and rotation are special cases of plane motion.

5.3   Summary of the Equations

A pictorial representation of the equations can be used to emphasize that plane motion is a combination of translation and rotation. The drawing in Fig. 5-1 shows an object with all external forces acting on it equated to the object with the effective forces $m\bar{a}$ and with the moment $\bar{I}\alpha$ of the effective forces. It is apparent that $\sum F_x = m\bar{a}_x$, $\sum F_y = m\bar{a}_y$, and $\sum \bar{M} = \bar{I}\alpha$. Keep in mind that moments are taken relative to the mass center G.

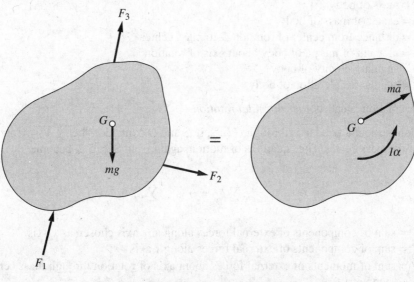

Fig. 5-1   An object being subjected to forces.

5.4   Translation of a Rigid Body

Translation of a rigid body is defined as the motion in which all particles of the body have the same acceleration. Then, the scalar equations of Section 5.2 become

$$\sum F_x = ma_x \qquad \sum F_y = ma_y \qquad \sum \bar{M} = 0 \tag{4}$$

where $\sum F_x$, $\sum F_y$ = sums of components of external forces in x and y directions
m = mass of body
a_x, a_y = components of acceleration of body in x and y directions
$\sum \bar{M}$ = sum of moments about mass center of body

5.5 Rotation of a Rigid Body

Rotation of a rigid body about a fixed axis is defined as the motion in which all particles along the fixed axis are at rest and all other particles of the body move on circular paths with centers along the axis of rotation.

(*a*) If a body has a plane of symmetry and rotates about a fixed axis perpendicular to this plane, then, from Sec. 5.2, the scalar equations of motion of the body under the action of an unbalanced force system are

$$\sum F_n = m\bar{r}\omega^2$$
$$\sum F_t = m\bar{r}\alpha \tag{5}$$
$$\sum M_O = I_O\alpha$$

where $\sum F_n$ = sum of components of all external forces (which are applied forces, gravitational force on body, and reaction **R** of the axis on the body) along n axis, which is line drawn between center of rotation O and mass center G; note that the positive sense is from G toward O because $\bar{a}_n = \bar{r}\omega^2$ has that sense

$\sum F_t$ = sum of components of external forces along t axis, which is perpendicular to n axis at O; note that the positive sense along this axis agrees with that of $\bar{a}_t = \bar{r}\alpha$

$\sum M_O$ = sum of moments of external forces about axis of rotation through O; note that positive sense agrees with assumed sense of angular acceleration α

m = mass of body

G = center of mass of body

$\bar{r}$ = distance from center of rotation O to mass center G

I_O = moment of inertia of body about axis of rotation

ω = angular velocity of body

α = angular acceleration of body

This type of rotation is called *noncentroidal rotation*.

(*b*) If the rotation is about a fixed axis through G (i.e., if G and O coincide), then $\sum \mathbf{F} = 0$ and moments are taken about the mass center. The equations of motion, again from Sec. 5.2, become

$$\sum F_x = 0 \qquad \sum F_y = 0 \qquad \sum \bar{M} = \bar{I}\alpha \tag{6}$$

where $\sum F_x$ = sum of components of external forces along any axis chosen as x axis

$\sum F_y$ = sum of components of external forces along y axis

$\sum M$ = sum of moments of external forces about axis of rotation through mass center G (axis of symmetry)

$\bar{I}$ = moment of inertia of body about axis of rotation through mass center G

α = angular acceleration of body

This type of rotation is called *centroidal rotation*.

5.6 Center of Percussion

The *center of percussion* is that point P on the n axis in Fig. 5-2 through which the resultant of the effective forces acts. It is at a distance q from the center of rotation O. The distance q is given by

$$q = \frac{k_O^2}{\bar{r}} \tag{7}$$

where k_O^2 = square of radius of gyration of body with respect to axis of rotation through O; note that $k_O^2 = I_O/m$, I_O being the mass moment of inertia of body about O and m its total mass

$\bar{r}$ = distance from center of rotation O to mass center G

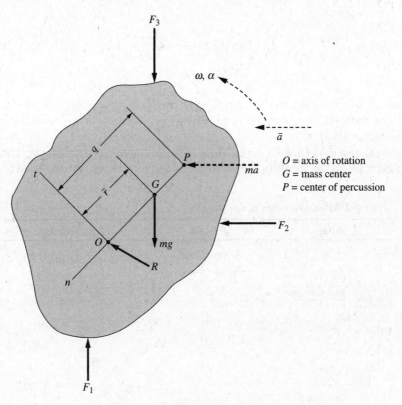

Fig. 5-2   The center of percussion *P*.

5.7   Inertia Force Method for Rigid Bodies

In Chapter 3, D'Alembert's principle was applied to *particles* in motion. Similarly, D'Alembert's principle can be applied to *rigid bodies* in motion. In the case of rigid body motion, not only must a force equal and opposite to $m\bar{a}$ be applied to the body at the center of mass, but also a couple equal and opposite to $\bar{I}\alpha$ must be applied to the free body. In Fig. 5-3, then, the reversed effective force and the reversed effective couple will balance out the external forces and couples.

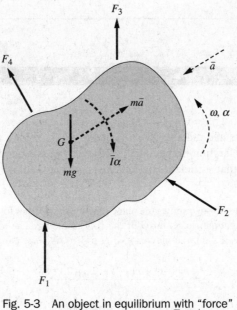

Fig. 5-3   An object in equilibrium with "force"
($-m\bar{a}$) and "moment" ($-\bar{I}\alpha$).

Hence,

$$\sum \mathbf{F} - m\bar{\mathbf{a}} = 0$$

$$\sum \mathbf{M} - I\alpha \mathbf{k} = 0 \tag{8}$$

The advantage of the inertia force method, based on D'Alembert's principle, is that it converts a dynamics problem into an equivalent problem in equilibrium. This allows moments to be conveniently taken about any axis and not only centroidal axes.

The mass moments of inertia I are displayed in Table 5-1 for three common shapes. These are the shapes most often used in problems. Moments of inertia for areas and masses are also included in Appendix B.

Table 5-1 Mass Moments of Inertia

Shape	Mass	Mass Moment
Rod	$m = \rho A L$	$I_x \cong 0,\ I_y = mL^2/3$
		$I_{y_C} = mL^2/12$
Cylinder	$m = \rho \pi R^2 h$	$I_y = mR^2/2$
		$I_{x_C} = m(3R^2 + h^2)/12$
		$I_x = m(3R^2 + 4h^2)/12$
Sphere	$m = 4\rho \pi R^3/3$	$I_C = 2mR^2/5$

SOLVED PROBLEMS

General Plane Motion

5.1. A ring of negligible mass and radius r has attached to it three small masses, as shown in Fig. 5-4(a). What is the angular acceleration immediately after the ring is placed on a horizontal plane? Assume no angular velocity and that sufficient friction exists so that there is no slipping.

SOLUTION

The free-body diagram in Fig. 5-4(b) shows the plane reactions (a normal force N and a frictional component F). A set of axes through the geometric center O of the ring is shown. The total gravitational force, which acts through the mass center G of the three masses in Fig. 5-4(a), is $4mg$. Note that the coordinates of G are

$$\bar{x} = \frac{mg(-r) + mg(0) + 2mg(0.866r)}{4mg} = 0.183r$$

$$\bar{y} = \frac{m(0) + mg(-r) + 2mg(0.5r)}{4mg} = 0$$

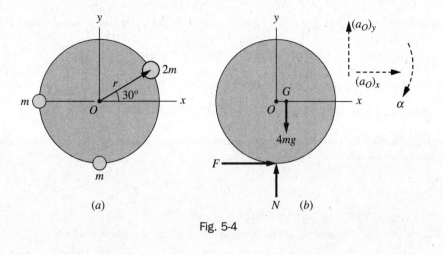

Fig. 5-4

No slipping is assumed, and thus the geometric center O has acceleration components

$$a_{Ox} = r\alpha \quad \text{and} \quad a_{Oy} = 0$$

To be consistent with the sense of a_O, shown acting to the right in Fig. 5-4(b), α must be shown acting clockwise. The moment equation is given in vector form in Sec. 5.1. In scalar form, this becomes

$$\sum M_O = I_O\alpha + m\bar{x}(a_O) - m\bar{y}(a_O)$$

This equation is based on a right-hand set of axes with α positive in the counterclockwise direction. In using the equation, we must substitute $-\alpha$ for α because we assumed its sense to be clockwise.

Using the free-body diagram, the equations of motion are

$$\sum F_x = m\bar{a}_{Ox} \quad \text{or} \quad F = 4mr\alpha$$

$$\sum F_y = m\bar{a}_{Oy} \quad \text{or} \quad N - 4mg = 0$$

The moment equation becomes

$$Fr - 4mg(0.183r) = 4mr^2(-\alpha) + 4m(0.183r)(0) - 4m(0)r\alpha$$

Using $F = 4mr\alpha$, the moment equation becomes

$$(4mr\alpha)r - 0.732mgr = -4mr^2\alpha$$

Thus $\alpha = (0.0915g/r)$ rad/s^2. If $g = 9.8$ m/s^2, then $\alpha = (0.897/r)$ rad/s^2.

5.2. A 300-kg wheel with a 750-mm diameter rolls without slipping down a plane inclined at an angle of 25° with the horizontal. Determine the friction force F and the acceleration of the mass center.

SOLUTION

Figure 5-5 shows the force system acting on the wheel. Considerable difficulty is usually experienced in indicating the direction of the friction force F. (Force F may have any value between $-\mu N$ and μN.) In this case, the friction must act up the plane; otherwise, the wheel would slip down the plane. Also, friction is the only force that has a moment about the mass center, and therefore it is the force causing the angular acceleration ($\sum \bar{M} = \bar{I}\alpha$).

Choose the x axis parallel to the plane, with the positive direction down. The y axis is positive up. The moment of inertia—from Table 5-1—is

$$\bar{I} = \frac{1}{2}mr^2 = \frac{1}{2}(300)(0.375)^2 = 21.1 \text{ kg·m}^2$$

The equations of motion are $\sum F_x = m\bar{a}_x$, $\sum F_y = m\bar{a}_y$, and $\sum \bar{M} = \bar{I}\alpha$. Substituting values, these become

$$300(9.8)\sin 25° - F = 300\bar{a}_x \tag{1}$$

$$N - 300(9.8)\cos 25° = 300\bar{a}_y = 0 \tag{2}$$

$$F(0.375) = 21.1\alpha \tag{3}$$

These three equations involve four unknown quantities; hence, another equation in the unknowns is needed. Since this is an example of rolling, we shall use

$$\bar{a}_x = r\alpha = 0.375\alpha$$

Substitute $\alpha = 2.67\bar{a}_x$ into Eq. (3) to obtain $F = 150.2$.
Substitute $F = 150.2\bar{a}_x$ into Eq. (1) to obtain $\bar{a}_x = 2.76 \text{ m/s}^2$ $\therefore F = 415 \text{ N}$

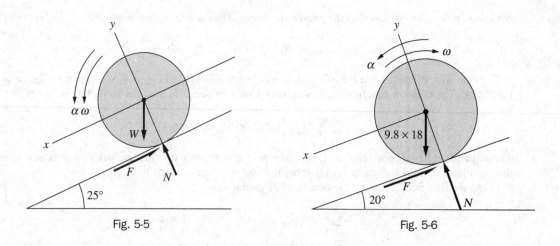

Fig. 5-5   Fig. 5-6

5.3. The center of a wheel having a mass of 18 kg and 600 mm in diameter is moving at a certain instant with a speed of 3 m/s up a plane inclined 20° with the horizontal (see Fig. 5-6). How long will it take to reach the highest point of its travel?

SOLUTION

The free-body diagram shows the friction F acting up the plane. Here, as in Problem 5.2, friction is the only force with a moment about the mass center. It therefore causes the angular acceleration, which must be counterclockwise. Note that the angular velocity ω is clockwise until the wheel stops at its highest point. Of course, on the way down the α and ω will be in the same direction—counterclockwise.

The equations of motion are $\sum F_x = m\bar{a}_x$, $\sum F_y = m\bar{a}_y$, and $\sum \bar{M} = \bar{I}\alpha$. Substitute values to obtain the following equations (assume the wheel is a cylinder, so that $\bar{I} = \frac{1}{2}mr^2$):

$$9.8 \times 18 \sin 20° - F = 18\bar{a}_x \tag{1}$$

$$N - 9.8 \times 18 \cos 20° = 18\bar{a}_y = 0 \tag{2}$$

$$F \times 0.3 = \frac{1}{2}18(0.3)^2\alpha \tag{3}$$

As in Problem 5.1, $\bar{a}_x = r\alpha = 0.3\alpha$.
From (3), $F = 9\bar{a}_x$. Substitute into Eq. (1) to obtain $\bar{a}_x = 2.23 \text{ m/s}^2$. The value of $\bar{a}_x$ is positive, i.e., down the plane.
To determine the time to come to rest, i.e., to reach its highest point after the initial speed of 3 m/s was observed, apply the kinematics scalar equation $v = v_0 + at$.

Keep in mind that the downward direction is positive. The final speed v is zero, the initial speed v_0 is up the plane and is therefore -3 m/s. The acceleration $\bar{a}_x$ is down the plane and is therefore 2.23 m/s^2. Then

$$v = v_0 + at \qquad 0 = -3 + 2.23t \qquad \therefore t = 1.35 \text{ s}$$

5.4. Analyze the motion of a homogeneous cylinder of radius R and mass m that is acted upon by a horizontal force P applied at various positions along a vertical axis, as shown in Fig. 5-7. Assume movement upon a horizontal plane.

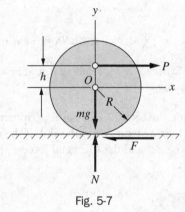

Fig. 5-7

SOLUTION

The free-body diagram shows the force P applied at a distance h above the center. Assume F acts to the left. The equations of motion are

$$\sum F_x = P - F = m\bar{a}_x \tag{1}$$

$$\sum F_y = N - mg = 0 \tag{2}$$

$$\sum \bar{M}_O = P \times h + F \times R = \frac{1}{2}mR^2\alpha \tag{3}$$

Note that P must be greater than F for motion to ensue to the right. Is it possible for motion to occur with friction F equal to zero?

Substitute $\bar{a}_x = R\alpha$ into Eq. (3) to obtain

$$P \times h + F \times R = \frac{1}{2}mR\bar{a}_x \tag{4}$$

Divide Eq. (4) by $\frac{1}{2}R$ to obtain

$$\frac{2Ph}{R} + 2F = m\bar{a}_x \tag{5}$$

Equating the left-hand sides of Eqs. (5) and (1),

$$\frac{2Ph}{R} + 2F = P - F \qquad \text{or} \qquad 3F = P\left(1 - \frac{2h}{R}\right) \tag{6}$$

It is evident that F will be zero if the term $1 - 2h/R$ is zero, i.e., if $h = \frac{1}{2}R$. Thus, if the force P is applied at a point one-half the radius above the center, the frictional force F is zero.

Also note that if $h = R$, the equation becomes $3F = P(1 - 2R/R) = -P$. The friction has now reversed itself and acts to the right. With $F = -\frac{1}{3}P$, Eq. (3) becomes, for the case where $h = R$,

$$PR - \frac{1}{3}PR = \frac{1}{2}mR^2\alpha \qquad \text{or} \qquad \frac{2}{3}P = \frac{1}{2}mR\alpha \tag{7}$$

This indicates that α is positive or the cylinder rolls to the right.

Next assume that h becomes smaller until finally P is applied through the mass center, where h is zero. Under these conditions, $3F = P[1 - (2 \times 0)/R] = P$.

Naturally, at any time in the previous discussion, if P becomes too large, F will tend to increase. As soon as the maximum possible value of the friction is exceeded, the cylinder will slip. A new assumption must then be made, namely, that the friction F is now equal to the product of the coefficient of friction and the normal force N. The equations of motion are then

$$\sum F_x = P - \mu N = m\bar{a}_x \tag{8}$$

$$\sum F_y = N - mg = 0 \tag{9}$$

$$\sum \bar{M}_O = Ph + \mu NR = \frac{1}{2}mR^2\alpha \tag{10}$$

These equations indicate the existence of both sliding (linear acceleration $\bar{a}_x$) and rolling (angular acceleration α). Refer to Problem 5.6 and 5.22.

5.5. A homogeneous sphere with a mass of 20 kg has a peripheral slot cut in it, as shown in Fig. 5-8. A force of 40 N is exerted on a string wrapped in the slot. If the sphere rolls without slipping, determine the acceleration of its mass center and the frictional force F. Neglect the effect of the slot.

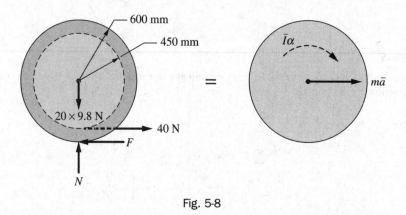

Fig. 5-8

SOLUTION

Assume rolling to the right. The angular acceleration will be clockwise, and the mass center acceleration will be to the right. Assuming that friction acts to the left, the equations of motion are

$$\sum F = m\bar{a} \qquad \text{or} \qquad 40 - F = 20\bar{a} \tag{1}$$

$$\sum M = I\alpha \qquad \text{or} \qquad F \times 0.6 - 40 \times 0.45 = \frac{2}{5}(20)(0.6)^2\alpha \tag{2}$$

Since $\bar{a} = 0.6\alpha$, the second equation may be rewritten

$$F \times 0.6 - 40 \times 0.45 = 4.8\bar{a}$$

Combine this with Eq. (1) to get

$$\bar{a} = 0.37 \text{ m/s}^2 \qquad \text{and} \qquad F = 7.14 \text{ N} \qquad \text{(to the left as assumed)}$$

5.6. The coefficient of friction between a plane and the homogeneous sphere, of mass 8 kg is 0.10, as shown in Fig. 5-9. Determine the angular acceleration of the sphere and the linear acceleration of its mass center.

SOLUTION

Draw a free-body diagram showing the frictional force F as unknown. At the beginning, it is not known whether the sphere will slip; therefore determine F to see whether it is greater than μN, that is, $0.10N$. If it is greater, this means that not enough friction is available and both rolling and sliding occur.

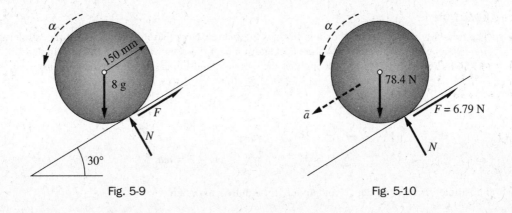

Fig. 5-9   Fig. 5-10

The equations of motion are (obtain $\bar{I}$ from Table 5-1)

$$\sum F = m\bar{a} \qquad \text{or} \qquad 8(9.8)\sin 30° - F = 8\bar{a} \qquad (1)$$

$$\sum \bar{M} = \bar{I}\alpha \qquad \text{or} \qquad F \times 0.15 = \frac{2}{5}(8)(0.15)^2\alpha \qquad (2)$$

Note that α is assumed counterclockwise; hence, $\bar{a}$ must be positive down the plane. Also, $\bar{a} = r\alpha = 0.15\alpha$. From (1), $39.2 - F = 8\bar{a}$; from (2), $F = 3.2\bar{a}$. Hence, $\bar{a} = 3.5$ m/s^2 and $F = 11.2$ N.

By inspection $N = 78.4 \cos 30° = 67.9$ N. Maximum friction available is $\mu N = 0.10(67.9) = 6.79$ N. But the F to prevent sliding (or cause rolling) is 11.2 N. This means that the problem must be reworked using the maximum friction available, namely 6.79 N instead of F. The relation $\bar{a} = r\alpha$ will no longer hold. Incidentally, we are also assuming that the coefficients of static and kinetic friction are equal.

Draw a new free-body diagram (see Fig. 5-10). The equations of motion are

$$78.4 \sin 30° - 6.79 = 8\bar{a} \qquad (3)$$

$$6.79 \times 0.15 = \frac{2}{5}(8)(0.15)^2\alpha \qquad (4)$$

Hence, $\bar{a} = 4.05$ m/s^2 and $\alpha = 14.1$ rad/s^2. The sphere will both roll and slip.

5.7. The cord passes over a massless and frictionless pulley, as shown in Fig. 5-11(a), carrying a mass m at one end and wrapped around a cylinder of mass $2m$ that rolls on a horizontal plane. What is the acceleration of the cylinder?

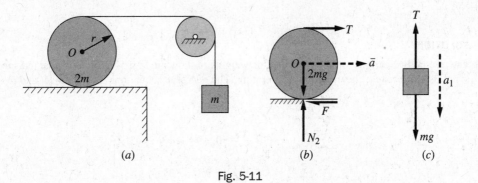

(a)   (b)   (c)

Fig. 5-11

SOLUTION

The free-body diagrams are shown in Fig. 5-1(b) and (c). Note that the magnitude of the acceleration a_1 of m does not equal the magnitude of the acceleration $\bar{a}$ of the center of gravity of the cylinder.

The following Eqs. (1) and (2) apply to the cylinder, and Eq. (3) applies to the mass m_1:

$$\sum F_h = m\bar{a} \qquad \text{or} \qquad T - F = 2m\bar{a} \tag{1}$$

$$\sum M_O = \bar{I}\alpha \qquad \text{or} \qquad (F + T)r = \frac{1}{2}(2m)r^2\alpha \tag{2}$$

$$\sum F_v = ma_1 \qquad \text{or} \qquad mg - T = ma_1 \tag{3}$$

Substitute $\bar{a}/r$ for α in Eq. (2) and divide through by r to obtain

$$F + T = m\bar{a} \tag{4}$$

Add (4) to (1) to obtain

$$T = \frac{3}{2}m\bar{a} \tag{5}$$

This may be solved simultaneously with (3) if the relation between $\bar{a}$ and a_1 is obtained. The horizontal component of the acceleration of the top point of the cylinder is equal to the sum of the acceleration $\bar{a}$ of the center and the product $r\alpha$. In this case, assuming pure rolling, $r\alpha$ is equal to $\bar{a}$ by kinematic considerations. Hence, the acceleration of the top point, which is the same as the acceleration a_1 of the mass m, is $\bar{a} + r\alpha = 2\bar{a}$.

From (5), $T = \frac{3}{2}m\left(\frac{1}{2}\alpha_1\right)$. Substituting into (3) and solving,

$$a_1 = \frac{4}{7}g$$

5.8. A wheel with a groove cut in it as shown in Fig. 5-12(a) is pulled up a rail inclined 30° with the horizontal by a rope passing over a pulley and supporting a 40-kg object. The wheel has a mass of 50 kg and has a moment of inertia $\bar{I}$ equal to 6 kg · m². How long will it take the mass center to attain a speed of 6 m/s starting from rest?

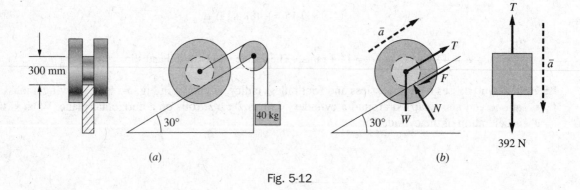

(a)   (b)

Fig. 5-12

SOLUTION

The free-body diagrams for the wheel and the weight are shown in Fig. 5-12(b). Note that tension T in the rope is not 392 N, because the weight is accelerating. The necessary equations for the object and the wheel are

$$\sum F_{object} = 392 - T = 40\bar{a} \tag{1}$$

$$\sum F_{wheel} = T - F - 50 \times 9.8 \sin 30° = 50\bar{a} \tag{2}$$

$$\sum \bar{M} = F \times 0.15 = 6\alpha \tag{3}$$

Since $\bar{a} = r\alpha = 0.15\alpha$, Eq. (3) may be written

$$F \times 0.15 = 6 \times 6.67\bar{a} = 40\bar{a} \quad \text{or} \quad F = 267\bar{a}$$

Putting F in (2) gives

$$T - 267\bar{a} - 245 = 50\bar{a} \quad \text{or} \quad T = 317\bar{a} + 245$$

Putting T in (1) gives

$$392 - 317\bar{a} - 245 = 40\bar{a} \quad \text{or} \quad \bar{a} = 0.21 \text{ m/s}^2$$

To find the time required to attain a speed of 6 m/s from rest, apply the kinematics equation:

$$v = v_0 + \bar{a}t \quad 6 = 0 + 0.21 \quad \therefore t = 28.6$$

5.9. Assuming the pulley in Fig. 5-13 to be massless and frictionless, determine the smallest coefficient of static friction that will cause the cylinder to roll. For the cylinder, $M = 70$ kg and $k_O = 400$ mm. Refer to Fig. 5-13(a).

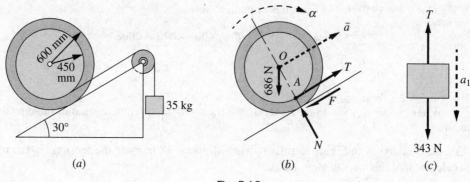

(a)   (b)   (c)

Fig. 5-13

SOLUTION

Draw free-body diagrams of the cylinder and weight as shown in Fig. 5-13(b) and (c). Assume that the cylinder is rolling up the plane. Since pure rolling is called for, the relation $\bar{a} = r\alpha$ holds. (Acceleration a_1 is the same as the component of the acceleration of point A parallel to the plane.) By kinematic considerations, $\mathbf{a}_A = \mathbf{a}_{A/O} + \bar{\mathbf{a}}$. Dealing only with components parallel to the plane, the component of $a_{A/O} = OA \times \alpha = 0.45\alpha$ down the plane since α is clockwise. The value of $\bar{a} = 0.6\alpha$ up the plane. Hence, $a_1 = 0.15\alpha$ up the plane.

The equations of motion may now be written. The moment of inertia for the cylinder is $\bar{I} = mk_O^2 = 70 \times 0.4^2 = 11.2$ kg·m^2:

$$\sum F = ma_1 \quad \text{or} \quad 343 - T = 35a_1 \tag{1}$$

$$\sum F = m\bar{a} \quad \text{or} \quad T - F - 686\sin 30° = 70\bar{a} \tag{2}$$

$$\sum \bar{M} = \bar{I}\alpha \quad \text{or} \quad F \times 0.6 - T \times 0.45 = 11.2\alpha \tag{3}$$

But $\bar{a} = 0.6\alpha$ and $a_1 = 0.15\alpha$, as has been shown. Eliminate F between (2) and (3) to obtain

$$0.25T - 686 \times 0.5 = 60.67\alpha$$

Solve this simultaneously with (1) to find $\alpha = -4.15$ rad/s^2 and $T = 365$ N.

Hence, $F = 196$ N. By inspection $N = 686\cos 30° = 594$ N. Thus, the required coefficient of friction $\mu = 196/594 = 0.33$.

Note that in this problem the cylinder will roll down the plane and the 35-kg mass will ascend since both α and a_1 are negative.

5.10. In Fig. 5-14, the homogeneous 200-kg sphere rolls without slipping on a horizontal surface. Determine the acceleration of the mass center and the friction needed.

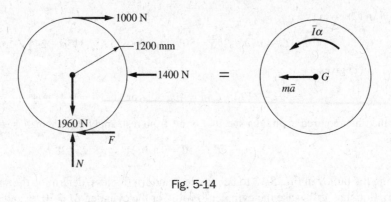

Fig. 5-14

SOLUTION

Assume that the sphere rolls to the left and that the friction force F is also to the left. By inspection, $N = 200 \times 9.8 = 1960$ N. The equations of motion are

$$\sum F = m\bar{a} \qquad \text{or} \qquad F + 1400 - 1000 = 200\bar{a}$$

$$\sum \bar{M} = \bar{I}\alpha \qquad \text{or} \qquad -F \times 1.2 - 1000 \times 1.2 = \frac{2}{5}(200)(1.2)^2\alpha$$

Using $\bar{a} = 1.2\alpha$, these equations yield $\bar{a} = -2.14$ m/s^2 and $F = -828$ N.

Thus, the sphere will roll to the right and the friction is to the right, instead of the direction originally assumed.

5.11. Assume that disk A in Fig. 5-15 rolls without slipping. Determine the tensions in the ropes and the acceleration of the mass center of disk A.

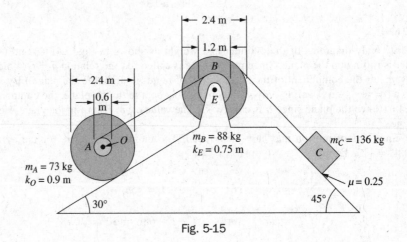

Fig. 5-15

SOLUTION

$$I_A = mk_O^2 = 73 \times 0.9^2 = 59.1 \text{ kg·m}^2 \qquad I_B = 88 \times 0.75^2 = 49.5 \text{ kg·m}^2$$

Draw free-body diagrams of A, B, and C. See Fig. 5-16. Assume in the diagrams that C moves down the plane and that A rolls up the plane. The following kinematic ideas are needed to solve the problem. First, $\alpha_A = \bar{a}/1.2$. Second, the acceleration of rope T_1 equals that component of the acceleration of point D (on disk A) that is parallel to the plane. This component has a magnitude that is the sum of the acceleration relative to the mass center ($0.3 \times \alpha_A$) and the acceleration $\bar{a}$ of the mass center. Therefore, the acceleration of T_1 is $1.25\bar{a}$. Since this is also the acceleration of a point 0.6 m from the center of B, $\alpha_B = 1.25\bar{a}/0.6 = 2.08\bar{a}$. The acceleration of T_2 is equal to that of $a_C = 2.08\bar{a} \times 1.2 = 2.5\bar{a}$.

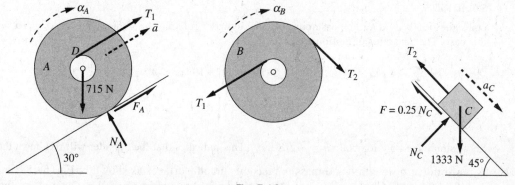

Fig. 5-16

The equations of motion for A are

$$\sum F = m\bar{a} \qquad \text{or} \qquad T_1 + F_A - 715 \sin 30° = 73\bar{a} \tag{1}$$

$$\sum \bar{M} = \bar{I}\alpha_A \qquad \text{or} \qquad T_1 \times 0.3 - F_A \times 1.2 = 59.1 \frac{\bar{a}}{1.2} \tag{2}$$

The equation of motion for B is

$$\sum \bar{M} = \bar{I}\alpha_B \qquad \text{or} \qquad T_2 \times 1.2 - T_1 \times 0.6 = 49.5(2.08\bar{a}) \tag{3}$$

The equations of motion for C parallel and perpendicular to the plane are

$$\sum F_\parallel = ma_C \qquad \text{or} \qquad 1333 \times 0.707 - 0.25N_C - T_2 = 136(2.5\bar{a}) \tag{4}$$

$$\sum F_\perp = 0 \qquad \text{or} \qquad N_C - 1333 \times 0.707 = 0 \tag{5}$$

Solve (5) for N_C; substitute this into (4). Solve (4) for T_2 in terms of $\bar{a}$. Eliminate F_A between (1) and (2) to find T_1 in terms of $\bar{a}$. Substitute these values of T_1 and T_2 into (3) to find $\bar{a}$. We find that $N_C = 942$ N and

$$T_2 = 706 - 340\bar{a}$$

$$T_1 = 286 + 91.2\bar{a}$$

Substitute into (3) to obtain $\bar{a} = 1.19$ m/s^2. Then $T_1 = 395$ N and $T_2 = 301$ N.

5.12. A solid homogeneous 293-kg cylinder rolls without slipping on the inclined rails shown in Fig. 5-17. Determine the mass center acceleration.

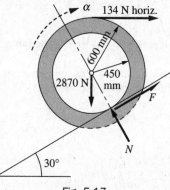

Fig. 5-17

SOLUTION

Draw the free-body diagram, assuming that the cylinder rolls up the plane. Note that $\bar{a} = r\alpha = 0.45\alpha$ in this case. The equations of motion are

$$\sum F = m\bar{a} \qquad \text{or} \qquad F - 2870\sin 30° + 134\cos 30° = 293\bar{a} \tag{1}$$

$$\sum \bar{M} = \bar{I}\alpha \qquad \text{or} \qquad -F \times 0.45 + 134 \times 0.6 = \frac{1}{2} \times 293 \times 0.6^2 \left(\frac{\bar{a}}{0.45}\right) \tag{2}$$

Solve simultaneously to obtain $\bar{a} = -2.78$ m/s². This indicates that the cylinder will roll down the plane.

5.13. A uniform bar of length ℓ and mass m rests on smooth surfaces as shown in Fig. 5-18. Find $\omega(\theta)$ if $\omega = 0$ at $\theta = 90°$. Choose the x axis along the horizontal surface and the y axis along the vertical surface.

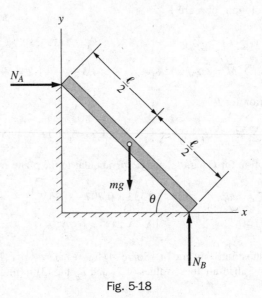

Fig. 5-18

SOLUTION

The free-body diagram shows only normal reactions of the surfaces on the bar, because friction is negligible (the surfaces are smooth). The equations of motion are

$$\sum F_x = N_A = m\bar{a}_x \tag{1}$$

$$\sum F_y = N_B - mg = m\bar{a}_y \tag{2}$$

$$\sum \bar{M} = N_A \left(\frac{1}{2}\ell\sin\theta\right) - N_B\left(\frac{1}{2}\ell\cos\theta\right) = \bar{I}\alpha \tag{3}$$

Note that $\bar{I}$ for the rod about its mass center is $\frac{1}{12}m\ell^2$ (see Table 5-1), and that α is the second derivative of θ with respect to time.

Since the three equations contain five unknowns, two more equations in the unknowns are required to obtain a solution. The geometry of the figure indicates that

$$\bar{x} = \frac{1}{2}\ell\cos\theta \qquad \text{and} \qquad \bar{y} = \frac{1}{2}\ell\sin\theta$$

Then,

$$\frac{d\bar{x}}{dt} = \frac{d\bar{x}}{d\theta}\frac{d\theta}{dt} = -\frac{1}{2}\ell\sin\theta\frac{d\theta}{dt} \qquad \text{and} \qquad \frac{d\bar{y}}{dt} = \frac{d\bar{y}}{d\theta}\frac{d\theta}{dt} = \frac{1}{2}\ell\cos\theta\frac{d\theta}{dt}$$

Differentiate again to obtain

$$\bar{a}_x = \frac{d^2\bar{x}}{dt^2} = -\frac{1}{2}\ell\left(\cos\theta\,\frac{d\theta}{dt}\right)\frac{d\theta}{dt} - \frac{1}{2}\ell\sin\theta\,\frac{d^2\theta}{dt^2}$$

$$\bar{a}_y = \frac{d^2\bar{y}}{dt^2} = \frac{1}{2}\ell\left(-\sin\theta\,\frac{d\theta}{dt}\right)\frac{d\theta}{dt} + \frac{1}{2}\ell\cos\theta\,\frac{d^2\theta}{dt^2}$$

These values will be substituted for $\bar{a}_x$ and $\bar{a}_y$ in the original equations. However, it is good to substitute first the values $N_A = m\bar{a}_x$ and $N_B = m\bar{a}_y + mg$ from (1) and (2), respectively, into (3). Thus, (3) becomes, after dividing all terms by $\frac{1}{2}mg\ell$,

$$\frac{\sin\theta}{g}\,\bar{a}_x - \frac{\cos\theta}{g}\,\bar{a}_y = \cos\theta + \frac{\ell}{6g}\frac{d^2\theta}{dt^2}$$

Next substitute the values of $\bar{a}_x$ and $\bar{a}_y$ just determined and simplify to obtain

$$\frac{d^2\theta}{dt^2} = -\frac{3g}{2\ell}\cos\theta$$

Recognizing that $d\theta/dt = \omega$, we can write $d^2\theta/dt^2 = d\omega/dt = (d\omega/d\theta)(d\theta/dt) = \omega\,d\omega/d\theta$. Then,

$$\int_0^\omega \omega\,d\omega = -\frac{3g}{2\ell}\int_{90°}^\theta \cos\theta\,d\theta$$

The angular speed ω is found to be

$$\omega = \sqrt{\frac{3g}{\ell}(1 - \sin\theta)}$$

5.14. Figure 5-19(*a*) shows a 3.6-m, 90-N ladder restrained in a 60° position. The floor and the wall are smooth. If the ladder is suddenly released and the top is given a speed of 0.9 m/s down, what will be the reactions of the floor and the wall on the ladder at the moment immediately following release?

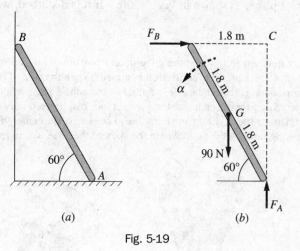

(*a*)   (*b*)

Fig. 5-19

SOLUTION

The free-body diagram in Fig. 5-19(*b*) shows the angular acceleration α assumed counterclockwise. Using the instant center C, the angular velocity is $\omega = v_B/BC = 0.9/1.8 = 0.5$ rad/s.

The acceleration of the mass center G will be written referring to the acceleration of both A and B:

$$a_G = (a_{G/A})_t + (a_{G/A})_n + a_A \tag{1}$$

$$a_G = (a_{G/B})_t + (a_{G/B})_n + a_B \tag{2}$$

The quantities on the right-hand sides of (1) and (2) have the following information:

(a) $(a_{G/A})_t$ is 1.8α down to left at 30° with horizontal.

(b) $(a_{G/A})_n$ is $1.8(0.5)^2 = 0.45$ down to right at 60° with horizontal.

(c) a_A is horizontal.

(d) $(a_{G/B})_t$ is 1.8α up to right at 30° with horizontal.

(e) $(a_{G/B})_n$ is $1.8(0.5)^2 = 0.45$ up to left at 60° with horizontal.

(f) a_B is vertical.

From (1), where a_A has no vertical component, we sum vertically to get

(g) $(a_G)_y$ with components $1.8\alpha \sin 30°$ down and $0.45 \cos 30°$ down or $(0.9\alpha + 0.39)$ down

From (2), where a_B has no horizontal component, we sum horizontally to get

(h) $(a_G)_x$ with components $1.8\alpha \cos 30°$ right and $0.45 \cos 60°$ left or $(1.56\alpha - 0.225)$ right.

The equations of motion are

$$\sum F_x = m(a_G)_x \qquad \text{or} \qquad F_B = \frac{90}{g}(1.56\alpha - 0.225) \tag{3}$$

$$\sum F_y = m(a_G)_y \qquad \text{or} \qquad 90 - F_A = \frac{90}{g}(0.9 + 0.39) \tag{4}$$

$$\sum M_G = I_G \alpha \qquad \text{or} \qquad F_A \times 0.9 - F_B \times 0.9 = \frac{1}{12}\left(\frac{90}{g}\right)(3.6)^2 \alpha \tag{5}$$

From these, we find

$$F_B = 14.33\alpha - 2.07 \qquad \text{and} \qquad F_A = 86.4 - 8.27\alpha$$

Substitute these values into (5) and solve for $\alpha = 2.64$ rad/s^2 counterclockwise.
  Hence, $F_A = 64.6$ N up and $F_B = 35.8$ N to the right.

5.15. A homogeneous right circular cone is precariously balanced on its apex in unstable equilibrium on a smooth horizontal plane, as shown in Fig. 5-20(a). If it is disturbed, what is the path of its center of mass G?

SOLUTION

Draw a free-body diagram to indicate the cone in any position while falling [see Fig. 5-20(b)]. Note that there is no horizontal force present (no friction assumed). Then the sum of the horizontal forces is zero. But the sum of the horizontal forces equals the product of the cone's mass m and the horizontal acceleration $\bar{a}_x$ of the mass center. Since there is a mass, the product $m\bar{a}_x$ can be zero only if $\bar{a}_x$ is zero.
  If the initial horizontal speed is zero and the value of $\bar{a}_x$ is zero, the center of mass G can only have motion in a vertical line passing through the point where the apex of the cone was in its original undisturbed position.

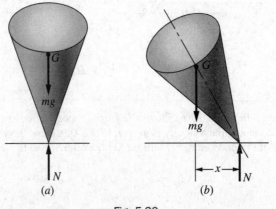

Fig. 5-20

5.16. Two homogeneous cylindrical disks are rigidly connected to an axle, as shown in Fig. 5-21. Each disk has a mass of 7.25 kg and is 900 mm in diameter. The axle is 200 mm in diameter and has a mass of 9 kg. A string wrapped as shown exerts a force of 45 N at the middle of the axle. Determine the acceleration of the mass center of the object.

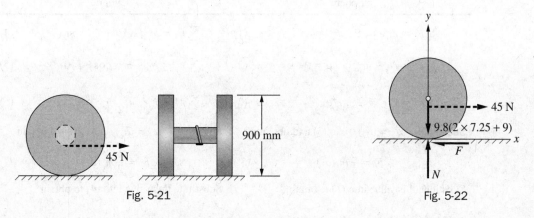

Fig. 5-21   Fig. 5-22

SOLUTION

A free-body diagram is shown indicating only a side view (see Fig. 5-22). Friction F is assumed acting to the left. The equations of motion are

$$\sum F_x = 45 - F = m\bar{a}_x \tag{1}$$

$$\sum F_y = N - 9.8 \times 23.5 = m\bar{a}_y = 0 \tag{2}$$

$$\sum \bar{M} = 0.45F - 45(0.1) = \bar{I}\alpha \tag{3}$$

The moment of inertia $\bar{I}$ is the sum of the moments of inertia of the two disks and the axle:

$$\bar{I} = 2\left(\frac{1}{2}m_{disk}r_{disk}^2\right) + \frac{1}{2}m_{axle}r_{axle}^2 = 7.25(0.45)^2 + \frac{1}{2}(9)(0.1)^2 = 1.513 \text{ kg·m}^2$$

Substituting these values, (1), (2), and (3) become

$$45 - F = 23.5\bar{a}_x \tag{4}$$

$$N - 230 = 0 \tag{5}$$

$$0.45F - 4.5 = 1.513\alpha \tag{6}$$

Since $\alpha = \bar{a}_x/0.45$ where the distance from the center of rolling to the surface on which the rolling occurs is 0.45 m, (6) may be written

$$0.45F - 4.5 = 1.513\left(\frac{\bar{a}_x}{0.45}\right) \quad \text{or} \quad F - 10 = 7.47\bar{a}_x$$

Add this equation to (4) to obtain $\bar{a}_x = 1.13$ m/s^2 to the right.

5.17. A homogeneous sphere and a homogeneous cylinder roll, without slipping, simultaneously from rest at the top of an inclined plane to the bottom. Which reaches the bottom first?

SOLUTION

It is likely from the wording of the problem that the radii of the solids have no influence. Let the subscripts s and c refer to the sphere and cylinder, respectively.

The free-body diagram shown in Fig. 5-23 refers to either solid. The equations of motion are listed below. The moments of inertia for the sphere and cylinder are, from Table 5-1, $\frac{2}{5}m_s r_s^2$ and $\frac{1}{2}m_c r_c^2$.

Sphere	Cylinder
$\sum F_x = m_s g \sin\theta - F_s = m_s(\bar{a}_x)_s$ (1)	$\sum F_x = m_c g \sin\theta - F_c = m_c(\bar{a}_x)_c$ (4)
$\sum F_y = N_s - m_s g \cos\theta = 0$ (2)	$\sum F_y = N_c - m_c g \cos\theta = 0$ (5)
$\sum \bar{M} = F_s r_s = \frac{2}{5}m_s r_s^2 \alpha_s$ (3)	$\sum \bar{M} = F_c r_c = \frac{1}{2}m_c r_c^2 \alpha_c$ (6)
Substitute, $r_s \alpha_s = (\bar{a}_s)_s$ into (3) and obtain	Substitute $r_c \alpha_c = (\bar{a}_s)_c$ into (6) and obtain
$F_s = \frac{2}{5}m_s(\bar{a}_x)_s$ (3′)	$F_c = \frac{1}{2}m_c(\bar{a}_x)_c$ (6′)
Substitute this value into (1) to obtain	Substitute this value into (4) to obtain
$(\bar{a}_x)_s = \frac{5}{7}g\sin\theta$ (1′)	$(\bar{a}_x)_c = \frac{2}{3}g\sin\theta$ (4′)

Thus, the sphere, having the larger acceleration, will reach the bottom first, or the object with the least mass moment of inertia.

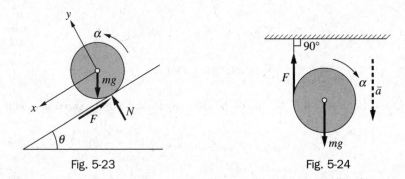

Fig. 5-23   Fig. 5-24

5.18. A cylinder 100 mm in diameter has a cord wrapped around it at the midsection as shown in Fig. 5-24. Attach the free end of the cord to a fixed support and allow the cylinder to fall. Find the acceleration of the mass center.

SOLUTION

Two equations—one summing the forces in the vertical direction assuming downward as positive and the other summing moments about the mass center assuming clockwise moments as positive—are useful here:

$$\sum F_v = mg - F = m\bar{a} \tag{1}$$

$$\sum \bar{M} = Fr = \bar{I}\alpha \tag{2}$$

Refer to Fig. 5-24. Force F is the tension in the cord. The moment of inertia $\bar{I} = \frac{1}{2}mr^2$, where $r = 0.05$ m. Also, $\alpha = \bar{a}/r = 20\bar{a}$, since the motion is equivalent to rolling the cylinder on the cord. Equation (2) becomes

$$0.05F = \frac{1}{2}m(0.05)^2(20\bar{a}) \qquad \text{or} \qquad F = \frac{1}{2}m\bar{a}$$

Substitute this value of F into (1) to obtain $\bar{a} = 2g/3 = 6.53$ m/s^2.

The yo-yo works on this principle. It is designed so that $\bar{a}$ is much smaller than g. This occurs if the cylinder is grooved so that F is applied at a smaller radius than that of the cylinder.

5.19. A solid sphere and a thin hoop of equal masses m and radii R are harnessed together by a rigging and are free to roll without slipping down the inclined plane shown in Fig. 5-25(a). Neglecting the mass of the rigging, determine the force in it. Assume frictionless bearings.

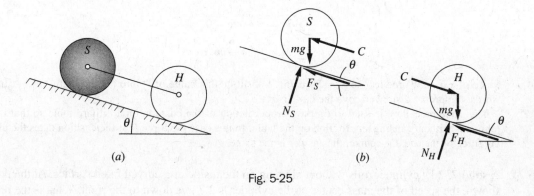

Fig. 5-25

SOLUTION

Draw free-body diagrams assuming that C is the compression in the rigging [see Fig. 5-25(b)]. If the sign should be negative, this merely indicates that C is tension. Let the subscripts S and H indicate sphere and hoop, respectively. The acceleration $\bar{a}$ is the same for either mass center; and since the radii are equal, the angular acceleration α for each is the same. Sum forces parallel to the plane to obtain the following two equations:

$$mg \sin\theta - F_S - C = m\bar{a} \tag{1}$$

$$mg \sin\theta - F_H + C = m\bar{a} \tag{2}$$

Since there are four unknowns (F_S, F_B, C, $\bar{a}$), two more equations are necessary. These are obtained by taking moments about the mass centers:

$$F_S \times r = I_S \alpha = \tfrac{2}{5} mr^2 \alpha \tag{3}$$

$$F_B \times r = I_H \alpha = mr^2 \alpha \tag{4}$$

The relation $\bar{a} = r\alpha$ holds for either equation. From (3) and (4), $F_S = \tfrac{2}{5} m\bar{a}$ and $F_H = m\bar{a}$. Substitute these values into Eqs. (1) and (2); then add the resulting equations to eliminate C and obtain $\bar{a} = \tfrac{10}{17} g \sin\theta$. Next solve for $C = \tfrac{3}{17} mg \sin\theta$ (compression).

5.20. At what height above the table should a billiard ball of radius r be hit with a cue held horizontally so that the ball will start moving with no friction between it and the table? Refer to Fig. 5-26.

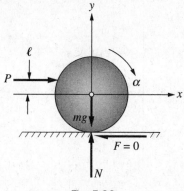

Fig. 5-26

SOLUTION

The equations of motion are

$$\sum F_x = P = m\bar{a}_x \tag{1}$$

$$\sum F_y = N - mg = 0 \tag{2}$$

$$\sum \bar{M} = P\ell = \frac{2}{5}mr^2\alpha \tag{3}$$

Since $\alpha = \bar{a}_x/r$, (3) reduces to $P = 2mr\bar{a}_x/5\ell$. Substitute this value of P into (1) to obtain $\ell = \frac{2}{5}r$. In other words, the proper distance above the table is $r + \frac{2}{5}r = \frac{7}{5}r$.

The same procedure is used to find the proper height of the cushion on a billiard table so that the ball rebounds without causing any friction on the table. In this case the force P of the cushion takes the place of the force of the cue. Of course, the answer is the same.

5.21. A solid 732-kg cylinder rolls without slipping on the inside of a curved fixed surface. At the position shown, the speed of the mass center of the cylinder is 2.7 m/s down to the right. What is the reaction of the surface on the cylinder?

SOLUTION

Draw a free-body diagram showing the reaction components F and N. See Fig. 5-27(a). Figure 5-27(b) illustrates the position of the cylinder needed to obtain the kinematic relationship necessary to solve the problem. Note that as the cylinder rolls up to the left, ϕ and θ increase as indicated. Hence, assume that the angular acceleration of the cylinder is positive in the counterclockwise direction and thus that $\bar{a}_t$ is positive to the upper left.

The equations of motion are

$$\sum F_n = m\bar{a}_n \quad \text{or} \quad N - 71.7\cos 30° = 7.32\bar{a}_n \tag{1}$$

$$\sum F_t = m\bar{a}_t \quad \text{or} \quad -71.7\sin 30 + F = 7.32\bar{a}_t \tag{2}$$

$$\sum \bar{M} = \bar{I}\alpha \quad \text{or} \quad -F \times 0.15 = \frac{1}{2}(7.32)(0.15)^2\alpha \tag{3}$$

where $\bar{a}_n$ and $\bar{a}_t$ are the magnitudes of the components of the acceleration of the mass center G and α is the magnitude of the angular acceleration of the cylinder about G. The mass center G can be considered as rotating about the center of curvature O. Hence, $\bar{a}_n = v^2/OG = 2.7^2/1.35 = 5.4$ m/s². The tangential magnitude $\bar{a}_t$ is as yet unknown.

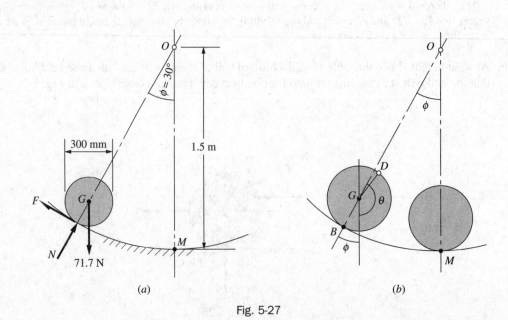

(a)   (b)

Fig. 5-27

To express α in terms of $\bar{a}_t$, refer to Fig. 5-27(*b*), where *D* is the point on the cylinder originally in contact with *M*. Since pure rolling is assumed, the arc *BM* in the curved surface must equal the arc *BD* on the cylinder (otherwise slip occurs). Let *r* = radius of the cylinder and *R* = radius of surface.

Then $r(\phi + \theta) = R\phi$ or $\theta = (R/r - 1)\phi$. This relationship also holds for the time derivatives of θ and ϕ. The second derivative of θ with respect to time is the magnitude of the angular acceleration α. The second derivative of ϕ with respect to time may be found in terms of $\bar{a}_t$, since center *G* is considered to rotate about *O*; that is, $\bar{a}_t = (R - r)(d^2\phi/dt^2)$. Hence,

$$\alpha = \frac{d^2\theta}{dt^2} = \frac{R - r}{r}\frac{d^2\phi}{dt^2} = \frac{R - r}{r}\frac{\bar{a}_t}{R - r} = \frac{\bar{a}_t}{r}$$

In this case, $R = 1.5$ m and $r = 0.15$ m; therefore $\alpha = 6.67\bar{a}_t$ and the equations of motion simplify to

$$N - 62.1 = (7.32)(5.4)$$
$$-35.8 + F = 7.32\bar{a}_t$$
$$-F = 3.66\bar{a}_t$$

from which $F = 11.9$ N and $N = 102$ N.

5.22. A disk of mass *m* and radius of gyration *k* has an angular speed ω_0 clockwise when set on a horizontal floor. See Fig. 5-28(*a*). If the coefficient of friction between the disk and the floor is μ, derive an expression for the time at which skidding stops and rolling occurs.

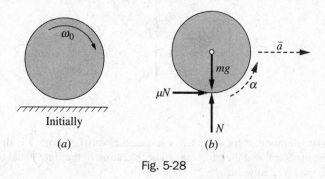

Fig. 5-28

SOLUTION

A free-body diagram shows the friction force μN, the normal force *N*, and the gravitational force *mg* acting on the disk. These will cause a mass center acceleration $\bar{a}$ to the right and an angular acceleration counterclockwise, as shown in Fig. 5-28(*b*).

The equations of motion are

$$\sum F_h = m\bar{a} \qquad \text{or} \qquad \mu N = m\bar{a} \tag{1}$$

$$\sum F_v = 0 \qquad \text{or} \qquad N = mg \tag{2}$$

$$\sum \bar{M} = \bar{I}\alpha \qquad \text{or} \qquad \mu Nr = mk^2\alpha \tag{3}$$

Substituting $N_A = mg$ into (1) and then into (3), we obtain

$$\bar{a} = \mu g \tag{4}$$

$$\alpha = \frac{\mu g r}{k^2} \tag{5}$$

The speed $\bar{v}$ at any time *t* can be found from (4) to be

$$\bar{v} = \bar{v}_0 + \mu g t = \mu g t \tag{6}$$

since $\bar{v}_0 = 0$. Also, the angular speed ω at any time t can be found from (5) to be

$$\omega = \omega_0 - \frac{\mu g r t}{k^2} \qquad (7)$$

where it is assumed that clockwise is positive.

When skidding stops and rolling occurs, $\bar{v} = r\omega$. Hence, we multiply (7) by r to obtain $\bar{v}$, which is also given by (6). This yields the required time t':

$$\mu g t' = r\omega_0 - \frac{\mu g r^2 t'}{k^2} \qquad \text{or} \qquad t' = \frac{r\omega_0}{\mu g(1 + r^2/k^2)}$$

5.23. A uniform cylinder of radius R is on a platform that is subjected to a constant horizontal acceleration of magnitude a. Assuming no slip, determine a_0, the magnitude of the acceleration of the mass center of the cylinder.

SOLUTION

The free-body diagram is shown in Fig. 5-29 with all forces acting on the cylinder.

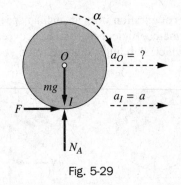

Fig. 5-29

The angular acceleration of the cylinder is assumed clockwise. Point I is the instant center between the cylinder and the platform, and therefore has an acceleration to the right equal to that of the platform. From kinematic concepts, we can write

$$\mathbf{a}_O = \mathbf{a}_{O/I} + \mathbf{a}_I$$

Summing in the horizontal direction, this equation yields (to the right is positive)

$$a_0 = R\alpha + a \qquad (1)$$

The equations of motion are next applied to the cylinder:

$$\sum F_h = ma_O \qquad \text{or} \qquad F = ma_O \qquad (2)$$

$$\sum M_O = I_O\alpha \qquad \text{or} \qquad -FR = \frac{1}{2}mR^2\alpha \qquad (3)$$

The sign on the moment of the frictional force F is negative because the moment is opposite to the assumed α.

Substituting from (1) into (2), we find $F = m(R\alpha + a)$. Now put this value of F into (3) to obtain $R\alpha = -\frac{2}{3}a$. Thus, by Eq. (1), the acceleration of the mass center is

$$a_0 = -\frac{2}{3}a + a = \frac{1}{3}a$$

5.24. The homogeneous bar ABC in Fig. 5-30(a) is 3 m long and has a mass of 20 kg. It is pinned to the ground at A and to the homogeneous bar CD, which is 2 m long with a mass of 10 kg. Determine the angular acceleration of each bar immediately after a restraining wire at C is cut. Assume no friction as D starts moving to the right on the rollers.

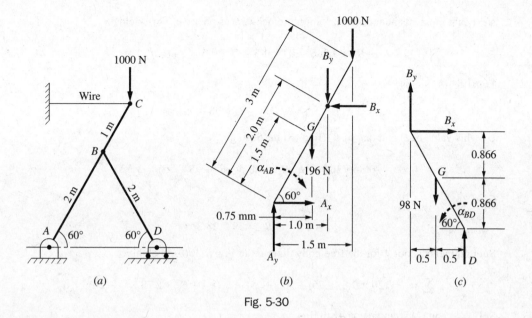

Fig. 5-30

SOLUTION

Figures 5-30(*b*) and (*c*) are free-body diagrams of the two bars at the instant of release. The acceleration of *B* is the same on either bar, and thus we write

$$a_B = (a_{B/D})_n + (a_{B/D})_t + a_D \tag{1}$$

The acceleration of *B* as a point on *ABC* is $2\alpha_{AB}$, and since it is perpendicular to the bar, it is directed to the right and down at an angle of 30° with the horizontal.

The acceleration $(a_{B/D})_n$ is zero because there is no angular velocity at the instant of release. The acceleration $(a_{B/D})_t$ is to the left and down at an angle of 30°, and equals $2\alpha_{BD}$. Finally, note that a_D can only be horizontal. In Eq. (1), we sum the vertical components of the accelerations with down positive and obtain

$$2\alpha_{AB} \sin 30° = 2\alpha_{BD} \sin 30° + 0$$

Since this shows $\alpha_{AB} = \alpha_{BD}$, we shall use α for both quantities.

Summing the horizontal components of the accelerations in (1), with positive to the right being used, we obtain

$$2\alpha_{AB} \cos 30° = -2\alpha_{BD} \cos 30° + a_D$$

Writing α for both angular accelerations, we now have $a_D = 3.46\alpha$ to the right.

For the bar *BD*, we can write for its mass center *G* the following acceleration:

$$a_G = (a_{G/D})_t + (a_{G/D})_n + a_D \tag{2}$$

As before, $(a_{G/D})_n$ is zero. The $(a_{G/D})_t = (1)\alpha$ is directed to the left and down at an angle of 30°. Of course, we have already shown $a_D = 3.46\alpha$ directed to the right. Sum components horizontally in (2) to get

$$(a_G)_x = 3.46\alpha - (1\alpha)(0.866) = 2.6\alpha \qquad \text{(to the right)}$$

Sum components vertically to obtain

$$(a_G)_y = (1\alpha)(0.5) = 0.5\alpha \qquad \text{(down)}$$

In Fig. 5-30(*c*), sum horizontally, using to the right as positive, and sum vertically, using down as positive. This yields

$$B_x = m(a_G)_x = 10(2.6\alpha) = 26\alpha \qquad \text{(to the right)} \tag{3}$$

$$-B_y + 10 \times 9.8 - D = 10(0.5\alpha) = 5\alpha \tag{4}$$

Next sum moments about G, using counterclockwise as positive. This yields

$$-B_x \times 0.866 - B_y \times 0.5 + D \times 0.5 = \frac{1}{12}(10)(2)^2\alpha \tag{5}$$

Equation (4) yields

$$D = -B_y + 98 - 5\alpha$$

Use this value of D and that of B_x from (3) in (5) to get

$$-0.866(26\alpha) - B_y(0.5) + 0.5(-B_y + 98 - 5\alpha) = 3.33\alpha \tag{6}$$

From (6),

$$B_y = 49 - 28.3\alpha$$

Sum moments about A for the free-body diagram in Fig. 5-30(b) to obtain

$$\sum M_A = (20 \times 9.8)(0.75) + 1000(1.5) + B_y \times 1 - B_x \times 1.732 = \frac{1}{3}(20)(3)^2\alpha \tag{7}$$

Solve (6) and (7) simultaneously to find

$$\alpha = 12.7 \text{ rad/s}^2$$

Translation

5.25. A 22.7-kg door is supported on rollers A and B resting on a horizontal track, as shown in Fig. 5-31. A constant force P of 45 N is applied. What will be the velocity of the door 5 s after starting from rest? What are the reactions of the rollers? Assume that roller friction is negligible.

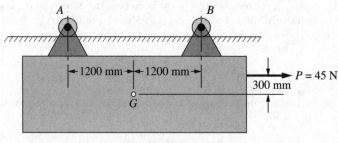

Fig. 5-31

SOLUTION

The acceleration a_x is assumed to be in the x direction in the free-body diagram shown in Fig. 5-32.
Apply the equations $\sum F_x = ma_x$, $\sum F_y = ma_y$, and $\sum \bar{M} = 0$ to obtain the following:

$$\sum F_x = 45 = 22.7 a_x \tag{1}$$

$$\sum F_y = F_A + B - 222 = 22.7 \times 0 \tag{2}$$

$$\sum \bar{M} = F_B \times 1.2 - A \times 1.2 - 45 \times 0.3 = 0 \tag{3}$$

From (1), $a_x = 1.98 \text{ m/s}^2$. Since a_x is constant, the formula involving the known quantities v_0, a, t, and the unknown v should be used:

$$v = v_0 + at = 0 + (1.98 \times 5) = 9.91 \text{ m/s}$$

Solve (2) and (3) simultaneously to obtain $F_A = 105$ N and $F_B = 117$ N.

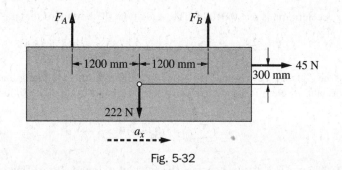

Fig. 5-32

5.26. Figure 5-33(*a*) shows a homogeneous bar *AB* that is 2 m long with a mass of 2 kg. The ends are constrained to slide in smooth horizontal guides. Determine the acceleration of the bar and the normal forces at *A* and *B* under the action of the horizontal 20-N force.

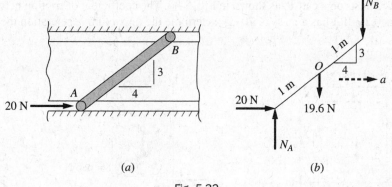

(*a*)   (*b*)

Fig. 5-33

SOLUTION

The free-body diagram in Fig. 5-33(*b*) shows the given force, the normal forces at *A* and *B*, and the gravitational force 19.6 N. The acceleration is shown acting to the right. (There is no angular acceleration.)

Assume that forces to the right are positive, forces up are positive, and counterclockwise moments are positive. The equations of motion are

$$\sum F_x = ma \qquad \text{or} \qquad 20 = 2a \tag{1}$$

$$\sum F_y = 0 \qquad \text{or} \qquad N_A - N_B = 19.6 = 0 \tag{2}$$

$$\sum \bar{M}_O = 0 \qquad \text{or} \qquad 20(1)(3/5) - N_A(1)(4/5) - N_B(1)(4/5) = 0 \tag{3}$$

From (1), $a = 10 \text{ m/s}^2$. Equations (2) and (3) solved simultaneously yield $N_A = 17.3$ N and $N_B = -2.3$ N. (The reactions are not in the directions assumed to be correct. A minus sign means that the reaction acts opposite to the direction shown, that is, N_B acts up.)

5.27. A freight car starts from rest at the top of a 1.5 percent grade 1600 m long. Assuming a resistance of 40 N/metric ton, how far along the level track at the bottom of the grade will it roll before coming to rest?

SOLUTION

On the incline the forces acting parallel to the track are the component of the weight $0.015W$ and the resistance of $(40/1000 \times 9.8)W = 0.0041W$ N. The first accelerates the car, but the latter decelerates it. The equation of motion is

$$\sum F_{\parallel} = 0.015W - 0.0041W = \left(\frac{W}{g}\right)a$$

The resulting acceleration is $a = 0.0109g$. The velocity at the bottom of the grade is found from

$$v^2 = v_0^2 + 2as = 0 + 2(0.0109g)(1600) = 34.9g$$

To determine the distance s along the level, apply the same formulas, noting that the only horizontal force acting is a constant decelerating one (because of train resistance), that is, $-0.0041W$. As before,

$$\sum F_\parallel = -0.0041W = \left(\frac{W}{g}\right)a$$

hence, $a = -0.0041g$. Also, v_0^2 now becomes the value of v^2 found above. Thus,

$$v^2 = v_0^2 + 2as \quad \text{or} \quad 0 = 34.9g - 2(0.0041gs)$$

from which $s = 4260$ m along the level.

5.28. A block A rests on a cart B, as shown in Fig. 5-34. The coefficient of friction between the cart and the block is 0.30. If A has a mass of 70 kg, determine the maximum acceleration the cart may have.

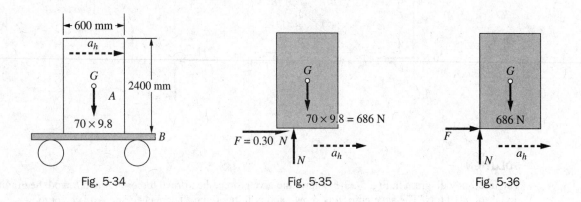

| Fig. 5-34 | Fig. 5-35 | Fig. 5-36 |

SOLUTION

The two possibilities to consider are (a) the acceleration that will cause the block to slide off the cart and (b) the acceleration that will cause tipping. In case (a), the friction force is a maximum and is equal to the product of the coefficient of friction and the normal force between the cart and the block. In case (b), the necessary friction may be (1) less than maximum, (2) maximum, or (3) greater than maximum. If the friction necessary to cause tipping is greater than maximum, naturally it cannot be obtained, and therefore the block will slide at the value of the acceleration calculated in (a).

The free-body diagram for case (a) is shown in Fig. 5-35. Note that the normal force N is applied at some unknown point in this representation. The friction is shown acting to the right, since the cart is being accelerated to the right and pulls the block with it because of friction. Only two equations are necessary:

$$\sum F_h = ma_h \qquad \text{that is,} \qquad F = ma_h \tag{1}$$

$$\sum F_v = ma_v = 0 \qquad \text{that is,} \qquad N - 686 = 0 \tag{2}$$

From Eq. (2), $N = 686$ N. Equation (1) becomes

$$0.30 \times 686 = 70a_h \qquad a_h = 2.94 \text{ m/s}^2$$

Acceleration above 2.94 m/s^2 will cause the block to slide.

The free-body diagram for the case of tipping is shown in Fig. 5-36. In this case tipping impends about the back edge of the block. Therefore the normal and frictional forces are shown there. Note that F is not placed equal to 0.30*N*. The limiting value of friction is used only if slipping impends, as in case (*a*). The equations are

$$\sum F_h = ma_h \tag{3}$$

$$\sum F_v = ma_v = 0 \tag{4}$$

$$\sum \bar{M} = 0 \tag{5}$$

These become

$$F = 70a_h \tag{6}$$

$$N - 686 = 0 \tag{7}$$

$$-N \times 0 \cdot 3 + F \times 1 \cdot 2 = 0 \tag{8}$$

From (7), $N = 686$ N. Substituting into (8), $F = 172$ N. Substituting $F = 172$ N into (6), $a_h = 2.46$ m/s^2.

Thus at an acceleration of 2.46 m/s^2 the block tends to tip. Since at this acceleration the force of friction is 172 N, which is less than the limiting value (0.30 × 686 N), the tipping will occur before sliding can exist.

Or, looking at the problem slightly differently, an acceleration of 2.46 m/s^2 will cause tipping, whereas an acceleration of 2.94 m/s^2 will cause sliding. Since the acceleration 2.46 m/s^2 is reached first, tipping will occur before sliding.

It can be deduced directly from Eq. (8) that tipping will occur first, since the ratio of *F* to *N* in this case is 0.25, which is less than the maximum ratio of 0.30.

5.29. A 60-kg operator is standing on a spring scale in a 600-kg elevator that is accelerating 6 m/s^2 up. What is the scale reading in kilograms and what is the tension in the cables?

SOLUTION

Figure 5-37 shows the free-body diagram of the 60-kg elevator operator, with the gravitational force acting down and the scale force acting up. Because the acceleration is up, we shall assume that direction is positive. The equation of motion for the operator is

$$\sum F_v = P - 60 \times 9.8 = ma = 60 \times 6$$

From this, $P = 948$ N and the scale reading will be $948/9.8 = 96.7$ kg. Thus, there is an apparent increase in the mass of the operator of $96.7 - 60 = 36.7$ kg.

To determine the tension *T* in the cable, draw the free-body diagram as shown in Fig. 5-38. If we call the up direction positive, the equation of motion for the elevator and the operator is

$$\sum F_v = ma \qquad \text{or} \qquad T - 660 \times 9.8 = 660 \times 6 \qquad \therefore T = 10\,400 \text{ N}$$

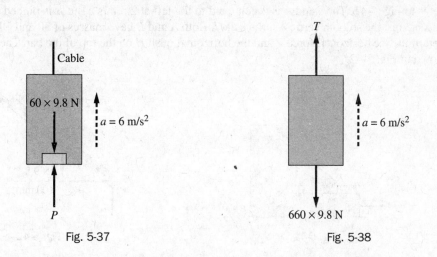

Fig. 5-37   Fig. 5-38

5.30. Block *B* is accelerated along the horizontal plane by means of a mass *A* attached to it by a flexible, inextensible, massless rope passing over a smooth pulley, as shown in Fig. 5-39. Assume that the coefficient of friction between *B*, which has a mass of 2 kg, and the plane is 0.20. Find the acceleration of *A* if it has a mass of 1.2 kg. Also, find the location of the normal force.

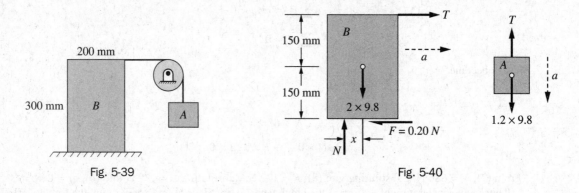

Fig. 5-39   Fig. 5-40

SOLUTION

Draw the free-body diagrams of *A* and *B* (see Fig. 5-40). Note that the force *N* is shown acting at a distance *x* from the vertical line through the mass center of block *B*. Assume the acceleration of block *B* is to the right as shown.

Sum the vertical forces acting on block *A* (assume down is positive to agree with the arrow on the acceleration vector) to obtain

$$1.2 \times 9.8 - T = 1.2a \qquad (1)$$

By inspection of the diagram for block *B*, $N = 2 \times 9.8 = 19.6$ N. Sum the horizontal forces on block *B* (to the right is positive) to obtain

$$T - 0.20N = 2a \qquad (2)$$

Equations (1) and (2) yield $a = 2.45$ m/s^2.

The sum of the moments of the force about the mass center of *B* is equal to zero since *B* is not rotating. Thus, with counterclockwise moments being positive, the equation is

$$-T \times 150 - N \times x - F \times 150 = 0 \qquad (3)$$

Substituting the values of *T*, *F*, and *N* into (3), the value of $x = -97.5$ mm. The minus sign indicates that the force *N* is acting to the right of the mass center instead of to the left as assumed.

5.31. Refer to Fig. 5-41. The stand *A* is accelerated to the left at 2.5 m/s^2. But *B* is pinned at *P*, and its top rests against the smooth vertical surface at *H*. Both *A* and *B* have masses of 30 and 7 kg, respectively. Determine the horizontal force *F* and the horizontal push *H* on the top of the bar. The dimensions are shown in Fig. 5-42.

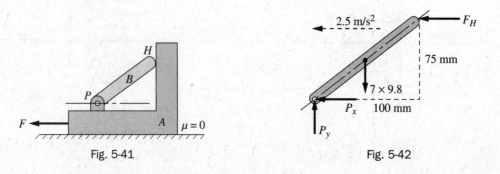

Fig. 5-41   Fig. 5-42

SOLUTION

Taking the entire body as a free body, sum the forces horizontally (to the left is positive) to obtain

$$F = Ma = (30 + 7)(2.5) = 92.5 \text{ N}$$

Next draw the free-body diagram of bar *B*, as shown in Fig. 5-42. The equations of motion are

$$\sum F_h = Ma_h \qquad \text{or} \qquad P_x + H = 7(2.5) \tag{1}$$

$$\sum F_v = Ma_v \qquad \text{or} \qquad P_y - 7 \times 9.8 = 0 \tag{2}$$

$$\sum \bar{M} = 0 \qquad \text{or} \qquad (0.0375)H - (0.0375)P_x - (0.05)P_y = 0 \tag{3}$$

From (1), $P_x = 17.5 - H$. From (2), $P_y = 68.6$ N. Substituting these in (3) gives

$$(0.0375)H - (0.0375)(17.5H) - (0.05)(68.6) = 0 \qquad \therefore H = 54.5 \text{ N}$$

5.32. Both bodies *A* and *B* in Fig. 5-43 have a mass of 20 kg. A small strip is nailed to *B* to prevent *A* from sliding while *P* accelerates the system to the left on the smooth plane. Determine the maximum acceleration without causing *A* to tip. If the system was moving initially with velocity 3 m/s is to the right, what will be its velocity after moving a total distance of 4 m?

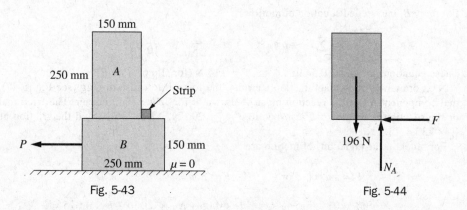

Fig. 5-43   Fig. 5-44

SOLUTION

Draw a free-body diagram of *A* (see Fig. 5-44). The normal force N_A is shown acting up at the extreme right, since the block is about to tip. The equations of motion are

$$\sum F_h = Ma_h \qquad \text{or} \qquad F = 20a \tag{1}$$

$$\sum F_v = Ma_v = 0 \qquad \text{or} \qquad N_A - 196 = 0 \tag{2}$$

$$\sum \bar{M} = 0 \qquad \text{or} \qquad N_A \times 0.075 - F \times 0.125 = 0 \tag{3}$$

Substitute $N_A = 196$ N from (2) into (3) to obtain $F = 118$ N. Substitute into (1) to find $a = 5.9$ m/s^2.

To determine the velocity after traversing 4 m, first consider the distance moved to the right before the velocity becomes zero. Applying the kinematics equation,

$$v^2 = v_0^2 + 2as \qquad 0 = (3)^2 - 2(5.9)s \qquad \therefore s = 0.76 \text{ m}$$

The remainder of the 4 m will be to the left, that is, 3.24 m. Using the same kinematics equation and assuming the positive direction to the left, $v^2 = 0 + 2 \times 5.9 \times 3.24$. Hence, the final velocity is to the left and equals 6.18 m/s.

5.33. Block A with mass 30 kg is located on top of a block B with mass 45 kg, as shown in Fig. 5-45. The coefficient of friction between the blocks is $\frac{1}{3}$. The coefficient of friction between block B and the horizontal plane is $\frac{1}{10}$. Determine the maximum value of P that will not cause block A to slide or tip.

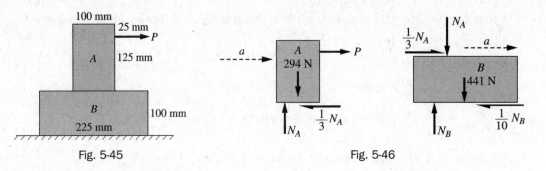

Fig. 5-45   Fig. 5-46

SOLUTION

First determine the force P to cause A to slip on B. Draw the free-body using friction as $\frac{1}{3}N_A$ between blocks. See Fig. 5-46. The location of the reaction N_A is unknown, but is also immaterial in this solution. By inspection, its value is 294 N. Likewise by inspection, the reaction $N_B = 294 + 441 = 735$ N. The other equation of motion for body A is

$$\sum F_h = ma_h \qquad \text{or} \qquad P - \frac{1}{3}N_A = 30a \tag{1}$$

For body B, the needed equation of motion is

$$\sum F_h = ma_h \qquad \text{or} \qquad \frac{1}{3}N_A - \frac{1}{10}N_B = 45a \tag{2}$$

These equations yield $a = 0.544$ m/s^2 and $P = 114$ N (for slip of A on B).

Next draw free-body diagrams to determine the force P to cause tipping (see Fig. 5-47). Now the normal component N_A of the reaction must be drawn at the lower right corner. The frictional component F of the reaction is unknown. By inspection, $N_A = 294$ N, $N_B = 735$ N, and the friction at the bottom is $\frac{1}{10} \times 735 = 73.5$ N.

For block A, the equations of motion are

$$\sum F = ma_h \qquad \text{or} \qquad P - F = 30a \tag{1}$$

$$\sum \bar{M} = 0 \qquad \text{or} \qquad -P \times 0.05 + N_A \times 0.050 - F \times 0.075 = 0 \tag{2}$$

For block B, the needed equation is

$$\sum F_h = ma_h \qquad \text{or} \qquad F - 73.5 = 45a \tag{3}$$

Solve (1), (2), and (3) simultaneously to obtain $P = 132$ N.

The smaller of the two values of P, that is, 114 N, is the maximum value that will not cause block A to slide or tip.

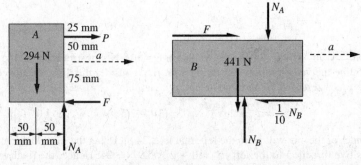

Fig. 5-47

5.34. Find the minimum time for the lift truck shown in Fig. 5-48 to attain its rated speed of 1.5 m/s without causing the six cartons, 1500 mm high, to tip. Each carton is 250 mm high with a base 400 mm on each side. Assume sufficient friction present so that sliding will not occur.

SOLUTION

The free-body diagram of the cartons on the truck indicates motion to the right. Thus, the reversed effective force $m\bar{a}$ is applied to the left through the mass center to hold the system in "equilibrium for study purposes." Tipping will tend to occur about the left lower edge. Moments about that edge yield

$$m \times 9.8 \times 0.2 = m\bar{a} \times 0.75$$

The maximum acceleration is thus $\bar{a} = 2.61$ m/s^2. Since $v = v_0 + at$,

$$1.5 = 0 + 2.61t \qquad \therefore t = 0.575 \text{ s}$$

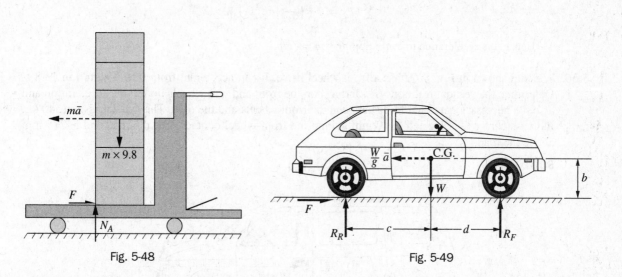

Fig. 5-48   Fig. 5-49

5.35. The car shown in Fig. 5-49 has a rear-wheel drive.

(a) What force must the rear wheels exert on the ground to cause an acceleration $\bar{a}$ to the right?

(b) Neglecting the rotational inertia effect of the wheels, what is the maximum acceleration possible for a given coefficient of friction μ between the tires and the pavement?

(c) Assuming sufficient friction to be available, what acceleration is required to cause the car to start to tip backward?

SOLUTION

(a) Apply the reversed effective force $(W/g)\bar{a}$ to hold the car in "equilibrium for study purposes." Summing forces horizontally,

$$\sum F_h = F - (W/g)\bar{a} = 0$$

From this equation, $F = (W/g)\bar{a}$.

(b) To find the maximum acceleration possible, solve for R_R. The greatest friction may be determined from $F = \mu R_R$. Take moments about R_F to obtain

$$-R_R \times (c + d) + W \times d + \frac{W}{g}\bar{a} \times b = 0$$

From this equation,

$$R_R = \frac{Wd}{c + d} + \frac{Wb\bar{a}}{g(c + d)}$$

Hence,
$$F = \mu R_R = \mu \frac{Wd}{c+d} + \mu \frac{Wb\bar{a}}{g(c+d)}$$

Substitute this value of F into $F = (W/g)\bar{a}$ and then solve for $\bar{a}$.

$$\bar{a} = \frac{\mu dg}{c + d - \mu b}$$

Notice that with this type of solution (using the equations of motion as shown), moments may be taken about a point other than the mass center since the car is not rotating.

(c) In this case the reaction R_F on the front wheels is zero. Sum moments about the line of contact of the rear wheels with the ground to obtain

$$\sum M_{R_R} = 0 = \left(\frac{W}{g}\right)\bar{a} \times b - W \times c$$

Hence, the acceleration to cause tipping is $\bar{a} = cg/b$.

5.36. The van shown in Fig. 5-50 has a front-wheel drive. It can accelerate from 0 to 95 km/h in 13.8 s. Determine the vertical reactions (R_F and R_R) on the front and rear wheels under this acceleration and also the necessary frictional force between the front wheels and the road. The van's mass is 1540 kg. Its mass center G is located 1150 mm behind the front-wheel contact with the road and is 775 mm above the road.

SOLUTION

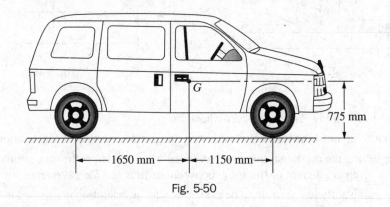

1650 mm   1150 mm

775 mm

Fig. 5-50

To find $\bar{a}$, use the equation $v = v_0 + \bar{a}t$ or

$$95 \times \frac{1000}{3600} = 0 + \bar{a} \times 13.8$$

Hence, $\bar{a} = 1.912$ m/s^2.
Summation of the horizontal forces yields the friction force:

$$\sum F_h = F = 1540 \times 1.912 = 2940 \text{ N}$$

Summation of the vertical forces and summation of the moments about G yield simultaneous equations in R_F and R_R:

$$\sum F_v = R_R + R_F - 1540 \times 9.8 = 0$$

$$\sum \bar{M} = -1.05R_R + 1.15R_F + 2940 \times 0.775 = 0$$

$$R_F = 8082 \text{ N} \qquad R_R = 7010 \text{ N}$$

Rotation

5.37. The bar in Fig. 5-51 is on a smooth horizontal table. It is pivoted at its left end about a vertical axis through O. A horizontal force F is applied perpendicular to the bar at its free end. Note that the mass of the bar is m, its length is ℓ, and its angular speed is zero when the force is applied. Determine the acceleration α and the reaction at the axis through O.

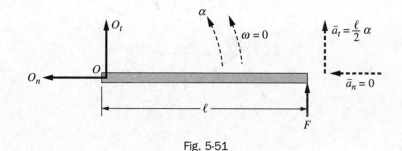

Fig. 5-51

SOLUTION

The equations of motion are

$$\sum F_n = m\bar{r}\omega^2 = 0 \qquad \text{or} \qquad O_n = 0$$

$$\sum F_t = m\bar{r}\alpha \qquad \text{or} \qquad O_t + F = m\frac{\ell}{2}\alpha$$

$$\sum M_O = I_O\alpha \qquad \text{or} \qquad F\ell = \frac{1}{3}m\ell^2\alpha$$

It can readily be determined that

$$\alpha = \frac{3F}{m\ell} \qquad \text{and} \qquad O_t = \frac{1}{2}F$$

5.38. Solve Problem 5.37, but assume that the horizontal force F is applied through the center of percussion.

SOLUTION

The arm for F is the distance q to the center of percussion. Thus,

$$q = \frac{k_O^2}{\bar{r}} = \frac{I_O}{m\bar{r}} = \frac{\frac{1}{3}m\ell^2}{\frac{1}{2}m\ell} = \frac{2}{3}\ell$$

The equations of motion then become

$$\sum F_t = m\bar{r}\alpha \qquad \text{or} \qquad O_n + F = \frac{1}{2}m\ell\alpha$$

$$\sum M_O = I_O\alpha \qquad \text{or} \qquad F\left(\frac{2}{3}\ell\right) = \left(\frac{1}{3}m\ell^2\right)\alpha$$

The solution becomes $\alpha = 2F/m\ell$ and $O_t = 0$.

This means that the tangential component of the reaction is zero if the force passes through the center of percussion. Batters know that a ball that hits their bat about two-thirds of its length from their hands does not sting. This is so because O_t equals zero or nearly so.

5.39. A bar pivoted about a horizontal axis through its lower end is allowed to fall from a vertical position. Determine the angular velocity and angular acceleration.

SOLUTION

Assume that the bar is of length ℓ and mass m. The free-body diagram in Fig. 5-52 shows the gravitational force mg acting down and the reactions R_t and R_n at the point of support. The bar is assumed to be at a position θ from the vertical rest position.

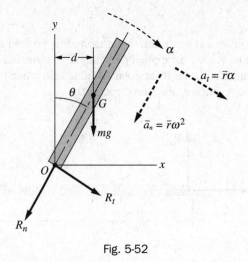

Fig. 5-52

The bar is rotating about a fixed axis under the action of an unbalanced force system, the equations of motion being $\sum F_n = m\bar{r}\omega^2$, $\sum F_t = m\bar{r}\alpha$, and $\sum M_O = I_O\alpha$. However, in describing the motion, it will be sufficient to determine the magnitudes of the angular velocity and acceleration as functions of displacement θ. Hence, the only equation needed is $\sum M_O = I_O\alpha$.

The only external force with a moment about O is the gravitational force mg. Hence, taking clockwise moments as positive, $mgd = I_O\alpha$. But $d = \frac{1}{2}\ell\sin\theta$, $I_O = \frac{1}{3}m\ell^2$, and $\alpha = d^2\theta/dt^2$. Hence,

$$mg\left(\frac{1}{2}\ell\sin\theta\right) = \frac{1}{3}m\ell^2\frac{d^2\theta}{dt^2} \qquad \text{or} \qquad \frac{d^2\theta}{dt^2} = \frac{3g}{2\ell}\sin\theta \qquad (1)$$

This differential equation may be solved by noting that

$$\frac{d^2\theta}{dt^2} = \frac{d}{dt}\left(\frac{d\theta}{dt}\right) = \frac{d\omega}{dt} = \frac{d\omega}{d\theta}\frac{d\theta}{dt} = \frac{d\omega}{d\theta}\omega$$

The original equations $d^2\theta/dt^2 = (3g/2\ell)\sin\theta$ is now written as

$$\omega\,d\omega = \frac{3g}{2\ell}\sin\theta\,d\theta$$

Integration yields $\frac{1}{2}\omega^2 = -(3g/\ell)\cos\theta + C$, where C is the constant of integration. To evaluate C, note that $\omega = 0$ when $\theta = 0$. Hence, $0 = -(3g/2\ell)(1) + C$ or $C = 3g/2\ell$. Then

$$\frac{1}{2}\omega^2 = -\left(\frac{3g}{2\ell}\right)\cos\theta + \frac{3g}{2\ell} \qquad \text{and} \qquad \omega = \sqrt{\frac{3g}{\ell}(1 - \cos\theta)}$$

This indicates a method of finding ω in terms of θ and, of course, from (1)

$$\alpha = \frac{3g}{2\ell}\sin\theta$$

5.40. A uniform bar of length ℓ and mass m is rotating at a constant angular velocity ω about a vertical axis through a point at a distance a from one end. For the position shown in Fig. 5-53, when the bar is passing through the plane of the paper, determine the horizontal and vertical components of the reaction of the support on the bar.

SOLUTION

Assume that a is less than one-half the length. The gravitational force is mg, and the horizontal and vertical components of the bearing reactions are shown acting on the rod. Since the angular speed ω is constant, the magnitude α of the angular acceleration (about the vertical axis) is zero. Hence, there are no forces acting that have moments about the vertical axis. This conclusion should be evident from the scalar equation $\sum M_v = I_v\alpha = I_v(0) = 0$.

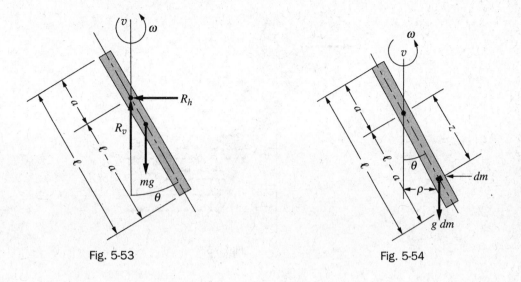

Fig. 5-53   Fig. 5-54

Also the equations of motion ($\sum F_n = m\bar{r}\omega^2$, $\sum F_t = m\bar{r}\alpha$, and $\sum M_O = I_O\alpha$) do *not* apply to the body as a whole, since the rod is rotating about an axis that is *not* perpendicular to the plane of symmetry of the rod.

Consider the bar to be composed of a series of thin pieces, each moving with an angular velocity ω about the vertical axis and at a distance ρ from the axis.

From Fig. 5-54,

$$\rho = z\sin\theta \qquad \text{and} \qquad dm = \frac{dz}{\ell}m$$

Each differential mass is moving on a circular path (radius ρ). Hence, a normal acceleration a_n exists toward the axis along ρ. A normal force dF accompanies this acceleration a_n ($dF = dm \times a_n$). The sum of all such normal forces (all horizontal) must be the horizontal bearing reaction R_h on the bar. Since in the position shown the longer length of the bar is to the right of the axis, the normal forces directed to the left will be greater than those acting to the right. Hence, the reaction R_h that supplies these normal forces acts to the left.

The normal force on the particle dm shown in the figure acts toward the axis, i.e., to the left, which means in a negative direction. Hence,

$$dF = -dm\,\rho\omega^2 = -\frac{dz}{\ell}m\,(z\sin\theta)\omega^2$$

and the total force is

$$F = \int dF = \int_{-a}^{\ell-a} -\frac{dz}{\ell}m(z\sin\theta)\omega^2$$

Since θ and ω^2 may be removed from within the integral sign:

$$F = R_h = -\frac{m\omega^2\sin\theta}{\ell}\int_{-a}^{\ell-a} z\,dz = -m\omega^2\left(\frac{1}{2}\ell - a\right)\sin\theta$$

To determine R_v, use the fact that the sum of the vertical forces must equal zero:

$$\sum F_v = 0 = R_v - mg \qquad \text{or} \qquad R_v = mg$$

For an explanation of the relation between ω and θ, refer to Problem 5.42.

5.41. In Problem 5.40, determine the location of the reversed effective forces to hold the bar in "equilibrium."

SOLUTION

Since the rod is rotating about an axis that is not perpendicular to a plane of symmetry, we take elements of mass of the rod that may be considered as rotating in a plane perpendicular to the axis of rotation.

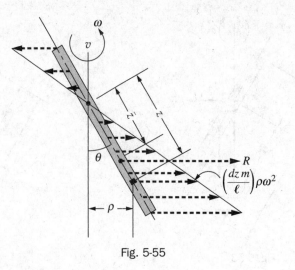

Fig. 5-55

In Fig. 5-55, the reversed effective normal forces are

$$dm\,\rho\omega^2 = \frac{dz}{\ell}m\omega^2\rho$$

for the mass of rod element of length dz shown acting away from the vertical axis. Also shown is the single force R at a distance $\bar{z}$ from the support. This force must equal the sum of the individual forces for the elements and must also have the same moment as they have about the pivot point. Hence,

$$R = \int_{-a}^{\ell-a} \frac{dz}{\ell}m\omega^2\rho = \int_{-a}^{\ell-a} \frac{dz}{\ell}m\omega^2 z\sin\theta = \frac{m\omega^2\sin\theta}{2}(\ell - 2a)$$

This result is equivalent in magnitude to that derived in Problem 5.40.

Next take moments of the individual inertia forces about the pivot point and equate them to the moment of R about the pivot point. In all cases, the forces are horizontal and the moment arms are vertical (therefore equal to the product of the z distance and $\cos\theta$):

$$R\bar{z}\cos\theta = \int_{-a}^{\ell-a} z\cos\theta\frac{dz}{\ell}m\omega^2\rho$$

Substituting for R its value just derived and for ρ its value $z\sin\theta$, the equation becomes

$$\frac{m\omega^2\sin\theta}{2}(\ell - 2a)\bar{z}\cos\theta = \frac{\cos\theta\,m\omega^2\sin\theta}{\ell}\int_{-a}^{\ell-a} z^2\,dz$$

When $\theta \ne 0$ or $90°$,

$$\frac{\bar{z}}{2}(\ell - 2a) = \frac{1}{3\ell}[(\ell-a)^3 - (-a)^3]   \text{ or }   \bar{z} = \frac{2}{3}\left(\frac{\ell^2 - 3\ell a + 3a^2}{\ell - 2a}\right)$$

If the bar is turning about an axis through its end, the above equation for $\bar{z}$ reduces to $\frac{2}{3}\ell$ (the center of percussion), since a will then be zero. Under this condition, the force R is

$$R = \frac{1}{2}m\omega^2\ell\sin\theta$$

5.42. Rework Problem 5.40 using the inertia force method.

SOLUTION

In Problem 5.41 it was indicated that the reversed effective force ($\frac{1}{2}m\ell\omega^2\sin\theta$) for a bar of length ℓ rotating about its end should be applied at a distance two-thirds of the length of the bar. Apply reversed effective forces to each of the two parts of the bar as shown in Fig. 5-56 and solve as a problem in statics.

Note that a couple M is applied to the system at the pivot point because it is not evident by inspection that the sum of moments about the pivot point of the two reversed effective forces and the weight will be zero.

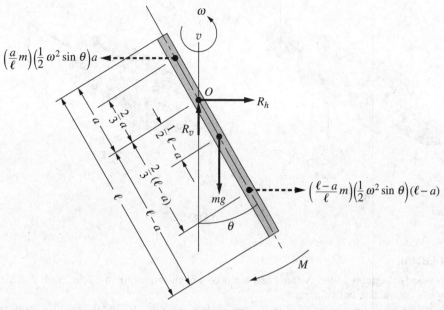

Fig. 5-56

The equations of equilibrium are

$$\sum F_h = 0 = R_h - \left(\frac{a}{\ell}m\right)\left(\frac{1}{2}\omega^2\sin\theta\right)a + \left(\frac{\ell-a}{\ell}m\right)\left(\frac{1}{2}\omega^2\sin\theta\right)(\ell-a) \tag{1}$$

$$\sum F_v = 0 = -mg + R_v \tag{2}$$

$$\sum M_O = 0 = \left[\left(\frac{\ell-a}{\ell}m\right)\left(\frac{1}{2}\omega^2\sin\theta\right)(\ell-a)\right]\left[\frac{2}{3}(\ell-a)\cos\theta\right]$$

$$+ \left[\left(\frac{a}{\ell}m\right)\left(\frac{1}{2}\omega^2\sin\theta\right)a\right]\left(\frac{2}{3}a\cos\theta\right) - mg\left(\frac{1}{2}\ell-a\right)\sin\theta + M \tag{3}$$

From (1), $R_h = -\frac{1}{2}\omega^2 m(\ell-2a)\sin\theta$. This checks, of course, with the result in Problem 5.40. The minus sign indicates that R_h was assumed in the wrong direction, and therefore actually acts to the left.

From (3),

$$M = m\sin\theta\left[g\left(\frac{1}{2}\ell-a\right) - \frac{1}{3}\omega^2(\ell^2 - 3\ell a + 3a^2)\cos\theta\right]$$

The significance of the couple M may now be appreciated. To hold the bar at a desired angle θ when it is rotating with a given angular speed ω requires a couple M of the magnitude just derived. However, if the couple M is not available at the pivot, then the rod will seek and maintain a definite angle θ for a given angular speed ω. This angle θ may be determined by setting the expression by M equal to zero. Then, since $\sin\theta \neq 0$,

$$\cos\theta = \frac{3g(\ell-2a)}{2(\ell^2 - 3\ell a + a^2)\omega^2}$$

Once the value of θ has been determined, the value of R_h may be found by substituting the value of $\sin\theta$ in its equation.

5.43. A 75-kg cylinder rotates from rest in frictionless bearings under the action of a mass of 7.5 kg carried by a rope wrapped around the cylinder, as shown in Fig. 5-57(a). If the diameter is 900 mm, what will be the angular velocity of the cylinder 2 s after motion starts?

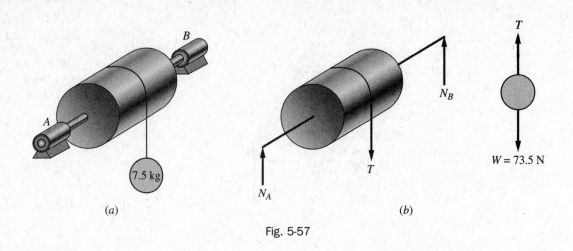

(a)   (b)

Fig. 5-57

SOLUTION

Free-body diagrams of the cylinder and the mass are shown in Fig. 5-57(b). Note that the tension T in the rope is common to both diagrams.

To determine the angular velocity ω after 2 s, it is necessary to find the angular acceleration α. The only equation necessary for the cylinder is the moment equation:

$$\sum M_{AB} = \bar{I}_{AB}\alpha$$

The subscript AB means *with respect to the axis AB*. Then,

$$T \times r = \frac{1}{2}mr^2\alpha \qquad \text{or} \qquad T \times 0.45 = \frac{1}{2} \times 75 \times 0.45^2\alpha \tag{1}$$

Since both tension T and angular acceleration α are unknown, another equation is required. Write the equation of motion of the mass by summing forces in the vertical direction:

$$\sum F_v = ma_v \qquad \text{or} \qquad 73.5 - T = 7.5a_v \tag{2}$$

Substituting $a_v = r\alpha = 0.45\alpha$ into (2),

$$73.5 - T = 7.5 \times 0.45\alpha \tag{3}$$

Solve (1) for $T = 16.88\alpha$ and substitute into (3) to obtain $\alpha = 3.63$ rad/s^2.
To find ω after 2 s:

$$\omega = \omega_0 + at = 0 + (3.63)(2) = 7.26 \text{ rad/s}$$

5.44. Relate the period of the compound pendulum shown in Fig. 5-58 to that of a simple pendulum.

SOLUTION

The compound pendulum differs from the simple pendulum in which only one small object is considered. Here we are concerned with a large mass with various linear velocities and accelerations. The system rotates about an axis perpendicular to the plane of the paper but not a centroidal axis.

Assume the pendulum to be moving counterclockwise, that is, θ is positive in that direction. Then its weight has a moment about the axis of rotation that tends to retard motion. Hence, the equation obtained by taking moments about the axis of rotation is

$$\sum M_O = I_O\alpha$$

and since the horizontal moment arm for the weight (force) mg is $\bar{r}\sin\theta$, this is written

$$\sum M_O = -(mg)(\bar{r}\sin\theta) = I_O\frac{d^2\theta}{dt^2} \qquad \text{or} \qquad \frac{d^2\theta}{dt^2} = -\frac{g}{I_O/m\bar{r}}\sin\theta$$

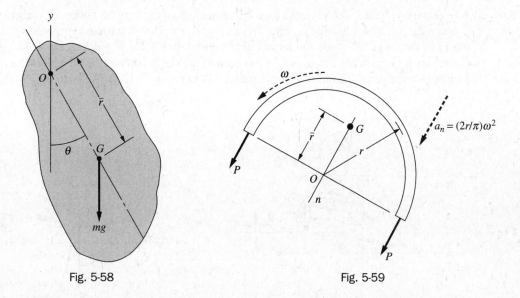

Fig. 5-58   Fig. 5-59

This is the same type of equation as derived for the simple pendulum of length ℓ, that is,

$$\frac{d^2\theta}{dt^2} = -\frac{g}{\ell}\sin\theta$$

Thus, it is seen that the compound pendulum has the same period as a simple pendulum whose length ℓ is equal to $I_O/m\bar{r}$.

This may be stated somewhat differently if instead of I_O there is substituted its value in terms of the radius of gyration k_O of the compound pendulum about the axis of rotation. Since $I_O = mk_O^2$, the length ℓ of the equivalent simple pendulum may be written

$$l = \frac{I_O}{m\bar{r}} = \frac{mk_O^2}{m\bar{r}} = \frac{k_O^2}{\bar{r}}$$

The compound pendulum behaves as a simple pendulum with its mass concentrated at a point $k_O^2\bar{r}$ from the axis of rotation. This value $k_O^2/\bar{r}$ occurs frequently in problems of rotation.

5.45. Find the stress in the rim of a flywheel rotating with constant angular speed ω.

SOLUTION

Draw a free-body diagram of one-half the rim, as shown in Fig. 5-59. The n axis is shown with sense from G to O, where G is the mass center of the thin semicircular rim. Note that $OG = 2r/\pi$.

The tensile forces P represent the pull of the other half of the rim on the free body shown. Since the wheel is rotating with constant angular speed, there is no angular acceleration; thus, no tangential effective force $(m\bar{r}\alpha)$ need be shown.

Only one equation of motion is necessary:

$$\sum F_n = m\bar{a}_n \qquad \text{or} \qquad 2P = m\frac{2r}{\pi}\omega^2$$

But the unit tensile stress is the force P (on one side) divided by the cross-sectional area A on which or through which it acts, or $\sigma = P/A = mr\omega^2/\pi A$.

Next express the mass m of the half rim in terms of the rim cross section A, half rim length πr, and mass density δ (kg/m^3):

$$m = A\pi r\delta$$

This is based on the assumption that the rim thickness is small compared with the rim diameter. Substituting this expression for m in the equation for stress, we have

$$\sigma = \delta r^2\omega^2$$

Since the rim speed v is equal to $r\omega$, the formula for stress may also be written

$$\sigma = \delta v^2$$

5.46. It is necessary to bank railroad tracks on a curve in such a way that the outer rail is above the inner rail. Since the roadbed is usually at some elevation above sea level, it is customary to call the vertical distance of the outer rail above the inner rail not elevation but the *superelevation e*. The greater the superelevation *e*, the faster a train may travel around the curve. Determine the value of *e* in terms of the speed v of the train, the radius r of the curve, and the angle of bank. Refer to Fig. 5-60(a).

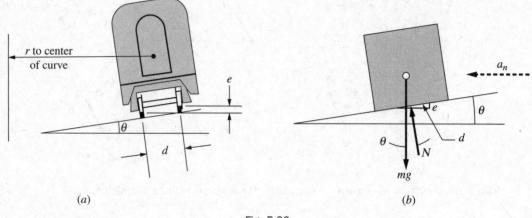

(a)   (b)

Fig. 5-60

SOLUTION

This is an example of rotation of a mass that can be considered a particle turning about the center of the curve.

Consider the curve to be properly banked for a specified speed v. This means that the train has no tendency to slide in or out along the radius. Hence, the side pressure of either rail against the flange of the wheel is zero. The only reaction on the car is the push, perpendicular to the roadbed, of the tracks on the wheels.

The free-body diagram in Fig. 5-60(b) shows the gravitational force mg and a single reaction N of the tracks. Note that the acceleration a_n is directed to the left toward the center of the curve. Its magnitude is $r\omega^2$. The equations of motion are

$$\sum F_n = mr\omega^2 \qquad \text{or} \qquad N\sin\theta = mr\omega^2$$

$$\sum F_v = 0 \qquad \text{or} \qquad N\cos\theta = mg$$

Dividing, $\tan\theta = r\omega^2/g$.

Since θ is usually a small angle, $\tan\theta$ is approximately equal to $\sin\theta$ (up to about 10°); therefore the above equation may be written

$$\sin\theta \approx \tan\theta = \frac{r\omega^2}{g} = \frac{r^2\omega^2}{gr}$$

But from the figure, $\sin\theta = e/d$. Hence,

$$\frac{e}{d} = \frac{r^2\omega^2}{gr} = \frac{v^2}{gr} \qquad \text{or} \qquad e = \frac{v^2 d}{gr}$$

5.47. In Problem 5.46, calculate the superelevation for a 600-m curve for a speed of 190 km/h. Assume that the gage $d = 1412$ mm.

SOLUTION

$$C = \frac{v^2 d}{gr} = \frac{(190 \times 1000/3600)^2 \times 1.412}{9.8 \times 600} = 0.669 \text{ m}$$

5.48. In Fig. 5-61(a) mass A is accelerating down at 1.5 m/s². It is connected by a weightless rope passing over a smooth drum to a cylinder B of mass 74 kg. The cylinder is acted upon by a moment $M = 66$ N·m counterclockwise. Determine the mass of A and the components of the reaction at O on cylinder B.

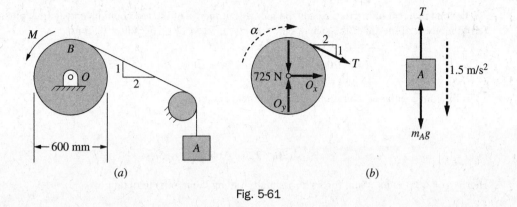

Fig. 5-61

SOLUTION

Draw free-body diagrams of the cylinder and the mass A, as shown in Fig. 5-61(b). The magnitude a_t of the tangential acceleration of a point on the rim is 1.5 m/s². Hence, the magnitude α of the angular acceleration is $a_t/r = 1.5/0.3 = 5$ rad/s². Since the cylinder rotates about an axis of symmetry, the three equations of motion for the cylinder are

$$\sum \bar{M} = \bar{I}\alpha \qquad \text{or} \qquad T \times 0.3 - 66 = \frac{1}{2} \times 74 \times 0.3^2 \times 5 \tag{1}$$

$$\sum F_x = 0 \qquad \text{or} \qquad O_x + T\left(\frac{2}{\sqrt{5}}\right) = 0 \tag{2}$$

$$\sum F_y = 0 \qquad \text{or} \qquad O_y - 725 - T\left(\frac{1}{\sqrt{5}}\right) = 0 \tag{3}$$

The equation of motion for weight A is

$$m_A g - T = m_A \times 1.5 \tag{4}$$

Equation (1) yields $T = 275$ N. Hence, from (4), $m_A = 33$ kg. Solve (2) and (3) to obtain $O_x = -246$ N and $O_y = 848$ N.

5.49. A homogeneous sphere having a mass of 100 kg is attached rigidly to a slender rod having a mass of 20 kg. In the horizontal position shown in Fig. 5-62(a), the angular velocity of the system is 8 rad/s. Determine the magnitude of the angular acceleration of the system and the reaction at O on the rod.

SOLUTION

Locate the centroid of the system with reference to the pin O:

$$\bar{r} = \frac{20(300) + 100(750)}{120} = 675 \text{ mm}$$

Figure 5-62(b) shows the three external forces O_n, O_t, and the gravitational force through G acting on the system. Note that O_n is shown to the left to agree with the sense of $\bar{a}_n$, and O_t is shown down to agree with the sense of $\bar{a}_t$.

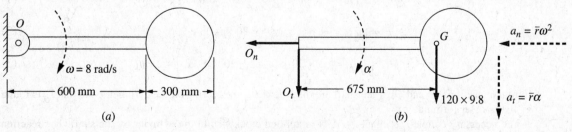

Fig. 5-62

The total moment of inertia I equals the moment of inertia of the rod about its end ($\frac{1}{3} m\ell^2$) plus the transferred moment of inertia of the sphere ($\frac{2}{5} mr^2 + md^2$), where $d = 750$ mm $= 0.75$ m:

$$I = \frac{1}{3} 20(0.6)^2 + \frac{2}{5} 100(0.15)^2 + 100(0.75)^2 = 59.55 \text{ kg·m}^2$$

To find the angular acceleration, use the moment equation

$$\sum M_O = I\alpha$$

or

$$120 \times 9.8 \times 0.675 = 59.55\alpha$$

Hence, $\alpha = 13.3$ rad/s^2. Sum forces horizontally (along the n axis) to obtain

$$O_n = m\bar{r}\omega^2 = 120(0.675)(8)^2 = 5180 \text{ N} \qquad \text{(to the left)}$$

Sum forces vertically (along the t axis) to obtain

$$O_t + 120 \times 9.8 = m\bar{r}\alpha = 120 \times 0.675 \times 13.3 \qquad \therefore O_t = -98.7 \text{ N} \qquad \text{(therefore up)}$$

5.50. An eccentric 20-kg cylinder used in a vibrator rotates about an axis 50 mm from its geometric center and perpendicular to the top view, shown in Fig. 5-63. If the magnitudes ω and α of its angular velocity and angular acceleration are, respectively, 10 rad/s and 2 rad/s^2 in the position shown, determine the reaction of the vertical shaft on the cylinder.

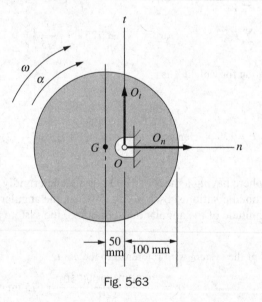

Fig. 5-63

SOLUTION

Choose n and t axes as shown. The distance from O to the mass center G is $\bar{r} = 50$ mm. The equations of motion are

$$\sum F_n = m\bar{r}\omega^2 \qquad \text{or} \qquad O_n = 20 \times 0.05 \times 10^2 \tag{1}$$

$$\sum F_t = m\bar{r}\alpha \qquad \text{or} \qquad O_t = 20 \times 0.05 \times 2 \tag{2}$$

$$\sum M_O = I_O\alpha = \left(\frac{1}{2} \times 20 \times 0.15^2 + 20 \times 0.05^2 \right) \times 2 \tag{3}$$

Here the reactions O_n and O_t of the shaft are the external forces acting *on* the cylinder. Note that I_O is found by the transfer formula.

The necessary couple $I_O\alpha$ equals 0.55 N·m applied clockwise to the cylinder by the shaft. The reaction components are $O_n = 100$ N to the right and $O_t = 2.0$ N up.

5.51. The 2.7-kg sphere in Fig. 5-64(*a*) moves in a circular and horizontal path under the actions of the 2.4-m weightless bar *AB* and the very light cord *BC*. When the speed of *B* is 3 m/s, determine the forces in the two supporting members.

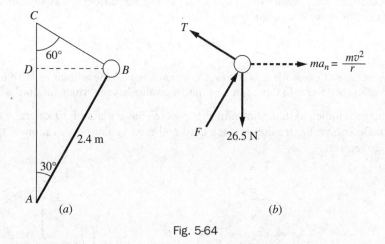

Fig. 5-64

SOLUTION

The free-body diagram in Fig. 5-64(*b*) shows the reversed effective force applied to hold the sphere in "equilibrium for study purposes." From trigonometry, $DB = \bar{r} = 1.2$ m. Hence, the reversed effective force is

$$m\frac{v^2}{r} = 2.7\left(\frac{3^2}{1.2}\right) = 20.2 \text{ N}$$

The equations become

$$\sum F_h = 0 = F\cos 60° - T\cos 30° + 20.2$$

$$\sum F_v = 0 = F\cos 30° + T\cos 60° - 26.5$$

The solutions are $F = 12.86$ N and $T = 30.7$ N.

In this particular problem, with *BC* and *AB* at right angles, a neat solution would be found by setting the sum of the forces along *AB* and the sum along *BC* equal to zero. In each equation, only one unknown is involved.

5.52. In Fig. 5-65(*a*) the bar *AB* is held in a vertical position by the weightless cord *BC* as the system rotates about the vertical *y*–*y* axis. The pin at *A* is smooth and the bar *AB* has a mass of 14.6 kg. If the breaking strength of the cord is 540 N, how fast can the system rotate without breaking the cord?

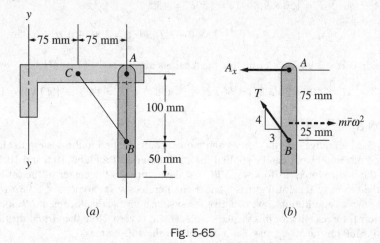

Fig. 5-65

SOLUTION

Draw a free-body diagram of the bar AB, omitting the weight and the vertical component A_y for simplicity [see Fig. 5-65(b)]. The reversed effective force is placed at the center of the bar. This is true only because each horizontal element or slice of the bar is the same distance from y–y as any other element.

Take moments about A to obtain

$$m\bar{r}\omega^2 \times 3 - T \times \frac{3}{5} \times 4 = 0$$

Only the horizontal component of T has a moment about A. Substitute $m = 14.6$ kg, $T = 540$ N, and $\bar{r} = 0.15$ m to obtain $\omega = 14.0$ rad/s or 134 rpm, the value beyond which the cord will break.

5.53. The compound pulley system shown in Fig. 5-66(a) has a mass of 30 kg and a radius of gyration of 450 mm. Determine the tension in each cord and the angular acceleration of the pulleys when the masses are released.

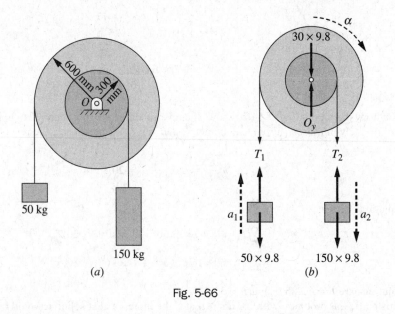

(a)   (b)

Fig. 5-66

SOLUTION A

Draw free-body diagrams of the three components of the system as in Fig. 5-66(b). Note that $a_1 = 0.6\alpha$ and $a_2 = 0.3\alpha$. The equations of motion are

$$\sum F = T_1 - 50 \times 9.8 = 50a_1 = 30\alpha \tag{1}$$

$$\sum F = 150 \times 9.8 - T_2 = 150a_2 = 45\alpha \tag{2}$$

$$\sum \bar{M} = \bar{I}\alpha \qquad \text{or} \qquad T_2 \times 0.3 - T_1 \times 0.6 = 30(0.45)^2\alpha = 6.08\alpha \tag{3}$$

The solution is $\alpha = 3.9$ rad/s^2. From this the tensions are

$$T_1 = 490 + 30(3.91) = 607 \text{ N} \qquad \text{and} \qquad T_2 = 1470 - 45(3.91) = 1290 \text{ N}$$

SOLUTION B

If the reversed effective (inertia) forces and moments are applied to the system to "hold it in equilibrium," the method of virtual work can be applied for a solution (see Problems 10.6 and 10.7).

Apply the inertia forces as follows: $50(0.6\alpha)$ down through the center of the 50-kg mass, $150(0.3\alpha)$ up through the center of the 150-kg mass, and a counterclockwise moment of 6.08α to the pulley. Since the system is now in "equilibrium," we can give the system a virtual displacement $\delta\theta$ clockwise. The work done by the external forces *only* for this virtual displacement is zero. Note the virtual displacement are $\delta\theta$ for the pulleys, $(0.6)\delta\theta$ for the 50-kg mass, and $(0.3)\delta\theta$ for the 150-kg mass.

Hence,

$$\delta U = [-50 \times 9.8 - 50(0.6)\alpha](0.6)\delta\theta + [150 \times 9.8 - 150(0.3)\alpha](0.3)\,\delta\theta + (-6.08\alpha)\,\delta\theta = 0$$

The solution after dividing each term of $\delta\theta$ is $\alpha = 3.9$ rad/s^2. The convenience in finding the angular acceleration is noted, but to find the tension in either cord, a free-body diagram must be drawn.

5.54. In Fig. 5-67(a), the mass C of 12 kg is moving down with a velocity of 4.8 m/s. The moment of inertia of drum B is 16 kg $\cdot$ s^2, and it rotates in frictionless bearings. If the coefficient of friction between the brake A and the drum is 0.40, what force P is necessary to stop the system in 2 s? What is the reaction at D on the rod?

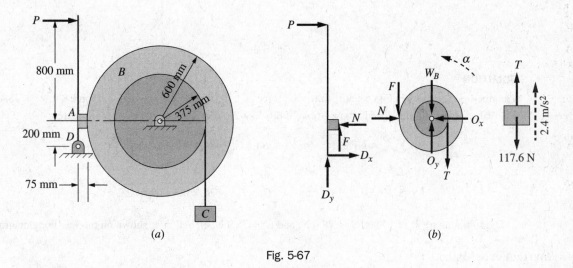

(a)   (b)

Fig. 5-67

SOLUTION

Draw the free-body diagrams of the bar, drum, and weight as shown in Fig. 5-67(b). Next determine the magnitude of the acceleration of the weight given the initial speed is 4.8 m/s down, the final speed is zero, and the time is 2 s. The value of acceleration is $a = 4.8/2 = 2.4$ m/s^2 up.

A vertical summation of forces acting on mass C gives

$$T - 117.6 = 12 \times 2.4 \qquad \therefore T = 146.4 \text{ N}$$

To find the friction F, sum moments of forces about O of the drum. The angular acceleration of the drum is counterclockwise (the system is slowing down). Its magnitude may be determined because the linear acceleration of a point 375 mm from O is 2.4 m/s^2. Hence, $\alpha = 2.4/0.375 = 6.40$ rad/s^2. The moment equation is

$$F \times 0.6 - T \times 0.375 + I\alpha \qquad \text{or} \qquad F \times 0.6 - 146.4 \times 0.375 = 16 \times 6.4 \qquad \therefore F = 262 \text{ N}$$

But $F = \mu N$. Hence the necessary normal force $N = 262/0.4 = 655$ N. Summing moments about D of all forces acting on the bar,

$$-P \times 1.0 + N \times 0.2 + F \times 0.075 = 0 \qquad \therefore P = 151 \text{ N}$$

Sum forces horizontally on the bar to obtain

$$151 - 655 + D_x = 0 \qquad \therefore D_x = 504 \text{ N to the right}$$

Summing forces vertically,

$$D_y + 262 = 0 \qquad \therefore D_y = 262 \text{ N down}$$

5.55. An 8-kg ball A is mounted on a horizontal bar attached to a vertical shaft, as shown in Fig. 5-68(a). Neglecting the mass of the bar and shaft, what are the reactions at B and C when the system is rotating at a constant speed of 90 rpm?

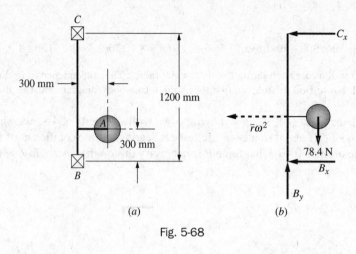

Fig. 5-68

SOLUTION

The free-body diagram in Fig. 5-68(b) shows the forces C_x, B_x, B_y, and the gravitational force $8 \times 9.8 = 78.4$ N. The value of the angular speed is $\omega = 90 \times 2\pi/60 = 9.42$ rad/s. The equations of motion are

$$\sum F_x = m\bar{r}\omega^2 \qquad \text{or} \qquad B_x + C_x = 8(0.3)(9.42)^2 = 213$$

$$\sum F_y = 0 \qquad \text{or} \qquad B_y - 78.4 = 0$$

$$\sum \bar{M} = 0 \qquad \text{or} \qquad C_x \times 0.9 - B_x \times 0.3 - B_y \times 0.3 = 0$$

The solutions are $B_x = 140$ N, $B_y = 78.4$ N, and $C_x = 72.9$ N, all acting as shown on the free-body diagram.

Inertia Force Method

5.56. Solve Problem 5.55 using the inertia force method.

SOLUTION

Figure 5-69 shows the free-body diagram, with the reversed effective force indicated. The equivalent "equilibrium" equations are then

$$m\bar{r}\omega^2 = 8(0.3)(9.42)^2 = 213 \text{ N}$$

$$\sum M_B = 1.2C_x - 78.4(0.3) - 213(0.3) = 0 \qquad \therefore C_x = 72.9 \text{ N}$$

$$\sum F_x = -B_x - C_x + 213 = 0 \qquad \therefore B_x = 140 \text{ N}$$

$$\sum F_y = B_y - 78.4 = 0 \qquad \therefore B_y = 78.4 \text{ N}$$

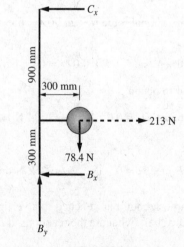

Fig. 5-69

5.57. Solve Problem 5.33 using the inertia force method.

SOLUTION

Figure 5-70 shows the free-body diagrams of A and B, with the reversed effective forces indicated. Writing the equivalent "equilibrium" equations, we have

$$\sum F_h = P - 30a - \frac{1}{3}N_A = 0 \qquad \text{(for } A\text{)} \tag{1}$$

$$\sum F_h = -45a + \frac{1}{3}N_A - \frac{1}{10}N_B = 0 \qquad \text{(for } B\text{)} \tag{2}$$

But $N_A = 294$ N and $N_B = 735$ N. From (2), $a = 0.544$ m/s^2. Substituting in (1), $P = 114$ N.

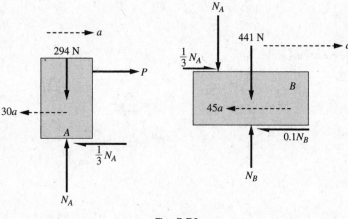

Fig. 5-70

5.58. Find the reactions on the rollers in Problem 5.25 using the inertia force method. The height of the door is 1.5 m.

SOLUTION

Figure 5-71 shows the free-body diagram of the door, with the inertia force indicated. The equations for the equivalent equilibrium problem can be written as

$$\sum F_x = 45 - 22.7a = 0 \tag{1}$$

$$\sum M_A = 2.4F_B + (0.45)45 - (0.75)(22.7a) - (1.2)222 = 0 \tag{2}$$

$$\sum F_y = F_A + F_B - 222 = 0 \tag{3}$$

Hence, from (1), $a = 1.98$ m/s^2; from (2), $F_B = 117$ N; and from (3), $F_A = 105$ N.

It can be seen that, without D'Alembert's principle, it would require a solution of simultaneous equations to determine F_A and F_B. This is generally the advantage to be gained from the inertia force method.

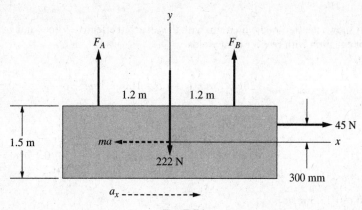

Fig. 5-71

5.59. Solve Problem 5.37 using the inertia force method.

SOLUTION

Figure 5-72 shows the free-body diagram, with the reversed effective force and reversed effective couple indicated. The equivalent equilibrium equations are

$$\sum F_n = O_n = 0$$

$$\sum M_O = F\ell - m\left(\frac{1}{2}\ell\right)\alpha\left(\frac{1}{2}\ell\right) - \frac{1}{12}m\ell^2\alpha = 0$$

$$\sum F_t = O_t + F - m\left(\frac{1}{2}\ell\right)\alpha = 0$$

Solving these equations yields

$$\alpha = \frac{3F}{m\ell}$$

$$O_t = \frac{1}{2}F$$

$$O_n = 0$$

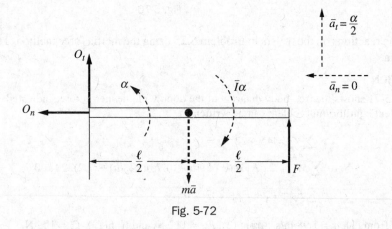

Fig. 5-72

5.60. Find the angular acceleration in Problem 5.49 using the inertia force method.

SOLUTION

Figure 5-73 shows the free-body diagram, with the reversed effective forces and couples indicated. Writing the moment equation with respect to O yields

$$\sum M_O = -20g\bar{r}_1 + m_1\bar{r}_1^2\alpha + \bar{I}_1\alpha - 100g\bar{r}_2 + m_2\bar{r}_2^2\alpha + \bar{I}_2\alpha = 0$$

$$\sum M_O = -20(9.8)(0.3) + 20(0.3)^2\alpha + (20/12)(0.6)^2\alpha$$

$$-100(9.8)(0.75) + 100(0.75)^2\alpha + \frac{2}{5}(100)(0.15)^2\alpha = 0$$

from which $\alpha = 13.3$ rad/s^2.

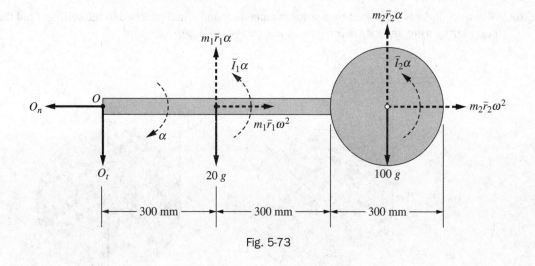

Fig. 5-73

5.61. Solve Problem 5.2 using the inertia force method.

SOLUTION

Figure 5-74 shows the free-body diagram, with the reversed effective force and couple indicated. Now, $\bar{I} = \frac{1}{2}mr^2 = \frac{1}{2}(300)(0.375)^2 = 21.1 \text{ kg·m}^2$. The equations for the equivalent "equilibrium" problem become

$$\sum M_O = r(W \sin 25°) - rm\bar{a}_x - \bar{I}\alpha = 0$$

$$\sum F_x = W \sin 25° - F - m\bar{a}_x = 0$$

Or

$$\sum M_O = (0.375)300 \times 9.8 \sin 25° - (0.375)300\bar{a}_x - 21.1\alpha = 0$$

where $\bar{a}_x = 0.375\alpha$. Solving yields $\bar{a}_x = 2.76 \text{ m/s}^2$ and

$$\sum F_x = 300 \times 9.8 \sin 25° - F - 300(2.76) = 0 \qquad \therefore F = 415 \text{ N}$$

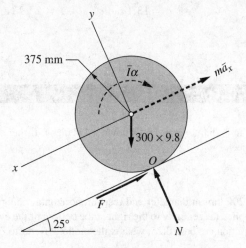

Fig. 5-74

5.62. A 7-kg cylinder is supported by a string wrapped around it and attached to the ceiling. Find the force in the string using the inertia force method [see Fig. 5-75(a)].

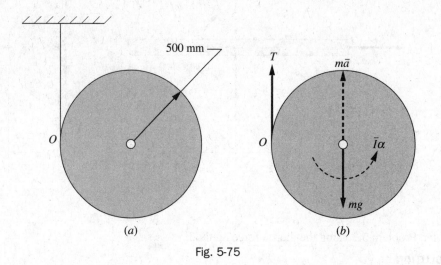

Fig. 5-75

SOLUTION

Figure 5-75(b) shows the free-body diagram with the reversed effective force and couple indicated. Now, $m = 7 \text{ kg}, \bar{I} = \frac{1}{2}mr^2 = \left(\frac{1}{2}\right)7(0.5)^2 = 0.875 \text{ kg·m}^2$, and $\bar{a} = 0.5\alpha$. The equations of the equivalent "equilibrium" problem become

$$\sum M_O = -mgr + m\bar{a}r + \bar{I}\alpha = 0$$

$$\sum \bar{M} = -rT + \bar{I}\alpha = 0$$

Or

$$\sum M_O = -(7)(9.8)(0.5) + 7\bar{a}(0.5) + 0.875\alpha = 0$$

$$\sum \bar{M} = -0.5T + 0.875\alpha = 0$$

Solving for α and T gives

$$\alpha = 13.1 \text{ rad/s}^2 \qquad T = 22.9 \text{ N}$$

SUPPLEMENTARY PROBLEMS

5.63. A 68-kg cylinder is 3 m in diameter and rolls down a 45° plane. If there is no slippage, determine the angular acceleration of the wheel.

Ans. 3.08 rad/s^2

5.64. A 180-kg cylinder is 200 mm in diameter and rests on horizontal rails perpendicular to its geometric axis. A force P of 550 N is applied tangentially to the right at the bottom of the cylinder. Assuming that the coefficients of static and kinetic friction are both 0.25, what is the motion? Refer to Problem 5.4.

Ans. Cylinder slides to right and rotates counterclockwise; $\bar{a} = 0.6 \text{ m/s}^2$; $\alpha = 12.1 \text{ rad/s}^2$

5.65. In Fig. 5-76, a cylinder of weight W and radius of gyration k has a rope wrapped around a groove of radius r. Determine the acceleration of the mass center if pure rolling is assumed. Force P is horizontal, as is the plane.

Ans. $\bar{a} = PgR(R - r)/W(R^2 + k^2)$

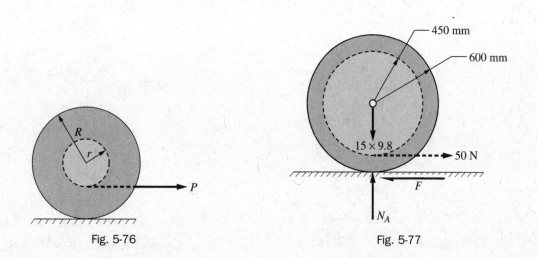

Fig. 5-76   Fig. 5-77

5.66. A homogeneous cylinder having a mass of 15 kg has a narrow peripheral slot cut in it as shown in Fig. 5-77. A force of 50 N is exerted on a string wrapped in the slot. If the cylinder rolls without slipping, determine the acceleration of its mass center and the frictional force F. Neglect the effect of the slot.

Ans. $a = 0.556$ m/s^2, $F = 41.7$ N to the left

5.67. A 50-kg cylindrical wheel 1200 mm in diameter is pulled up a 20° plane by a cord wrapped around its circumference. If the cord passes over a smooth pulley at the top of the plane and supports a hanging mass of 90 kg, determine the angular acceleration of the wheel. Assume that the cord pulls at the top of the wheel and is parallel to the plane.

Ans. 6.1 rad/s^2

5.68. What is the tension in the cable in the system shown in Fig. 5-78? Assume that the pulleys are weightless and frictionless. The plane is smooth.

Ans. $T = 45$ N

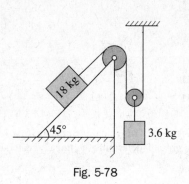

Fig. 5-78

5.69. In Problem 5.68 assume that the lower pulley is a homogeneous cylinder 300 mm in diameter and weighing 17.8 N. What is the tension in the cable parallel to the plane?

Ans. $T = 26$ N

5.70. A constant force of 225 N is applied as shown in Fig. 5-79. How long will it take the 25-kg block to reach a speed of 1.2 m/s starting from rest? Assume that the pulleys are massless and frictionless.

     Ans. $t = 0.146$ s

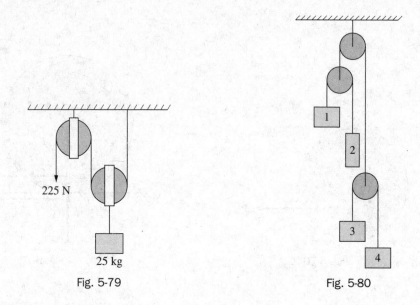

Fig. 5-79   Fig. 5-80

5.71. In the system shown in Fig. 5-80, the pulleys are to be considered weightless and frictionless. The masses are 1, 2, 3, and 4 kg. Determine the acceleration of each mass and the tension in the top cord.

     Ans. $a_1 = 9.02$ m/s^2 up, $a_2 = 0.39$ m/s^2 down, $a_3 = 3.53$ m/s^2 down, $a_4 = 5.10$ m/s^2 down, $T = 75.3$ N

5.72. In Fig. 5-81 assume that the weight W_2 is moving down. If the pulleys are weightless and frictionless, determine the tension in the rope when the weights are released from rest.

     Ans. $I = 3W_1 W_2/(4W_1 + W_2)$

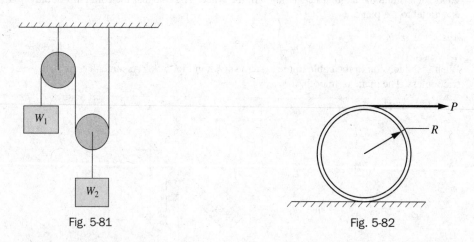

Fig. 5-81   Fig. 5-82

5.73. A horizontal force of 180 N is applied tangentially at the top of a 45-kg cylinder having a diameter of 1500 mm. If there is no slippage at the ground, determine the linear acceleration of the center of the cylinder.

     Ans.   5.33 m/s^2 horizontally

5.74. A thin hoop of weight W rolls horizontally under the action of a horizontal force P applied at the top as shown in Fig. 5-82. Express the mass center acceleration $\bar{a}$ in terms of P, W, and radius R. Also show that the frictional force of the plane on the hoop is zero.

     Ans. $\bar{a} = Pg/W$

5.75. In Problem 5.11, what must be the mass of C so that the acceleration of the disk A is 1.8 m/s² up the plane?

Ans. 610 kg

5.76. A 2-kg sphere of radius 80 mm has a rope wrapped around it with one end attached to the ceiling as shown in Fig. 5-83. If the sphere is released from rest, determine the tension in the rope, the angular acceleration of the sphere, and the speed of the mass center 2 s after release.

Ans. $T = 5.6$ N, $\alpha = 87.5$ rad/s² clockwise, $\bar{v} = 14.0$ m/s down

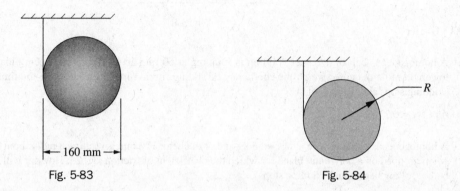

Fig. 5-83   Fig. 5-84

5.77. A cylinder of weight W and radius of gyration k is released from rest in the position shown in Fig. 5-84. If the radius of the cylinder is R, determine the tension in the cord and the acceleration of the mass center.

Ans. $T = W/(1 + R^2/k^2)$, $\bar{a} = g/(1 + k^2/R^2)$

5.78. Figure 5-85 shows an 18.2-kg mass attached to a cord that wraps around a 13.6-kg cylinder. The cylinder rolls without slipping up the 30° plane. Determine the angular acceleration of the cylinder.

Ans. 1.2 rad/s²

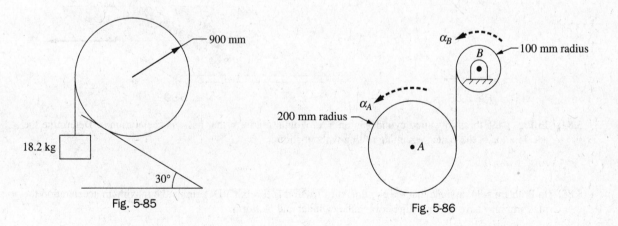

Fig. 5-85   Fig. 5-86

5.79. In Fig. 5-86, a 2-kg cylinder B with a radius of 100 mm is free to turn in frictionless bearings. A rope is wrapped around B and is then wrapped around the 4-kg cylinder A with radius 200 mm. Determine the angular accelerations of A and B immediately after the cylinders are released with zero velocity.

Ans. $\alpha_A = 14$ rad/s², $\alpha_B = 56$ rad/s²

5.80. In Fig. 5-87, a homogeneous cylinder and a homogeneous sphere of equal masses $m = 8$ kg and equal radii R are harnessed together by a light frame and are free to roll without slipping down the plane inclined 28° with the horizontal. Determine the force in the frame. Assume that the bearings are frictionless.

Ans. 1.27 N

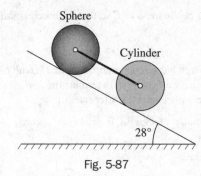

Fig. 5-87

5.81. A homogeneous ball of diameter 900 mm is spinning at 60 rpm about a horizontal centroidal axis when it is lowered onto a plane for which the coefficient of sliding friction is 0.30. Determine the time it takes before skidding stops and rolling begins.

Ans. $t = 0.275$ s

5.82. A homogeneous cylinder with a mass of 3 kg and of diameter 120 mm is rotating 8 rad/s about a horizontal axis when dropped on a horizontal plane for which the coefficient of friction is 0.25. How far will the center travel before rolling begins and skidding stops?

Ans. $d = 5.2$ mm

5.83. A homogeneous cylinder of mass m and radius R is at rest on a horizontal plane when a couple M is applied as shown in Fig. 5-88. Determine the magnitude of the coefficient of friction between the wheel and the plane so that rolling will occur.

Ans. $\mu \geq \frac{2}{3} M/mgR$

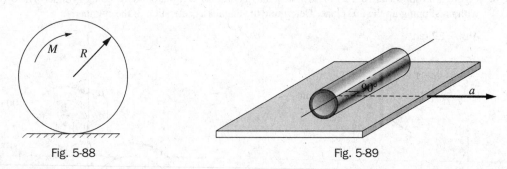

Fig. 5-88   Fig. 5-89

5.84. In Fig. 5-89, the thin-walled cylinder is on a horizontal platform that has an acceleration a. Determine the acceleration of the center, assuming rolling without sliding.

Ans. $\bar{a} = 0.5a$

5.85. In Problem 5.84, assume that the coefficient of friction is $\mu = 0.35$. Determine the maximum acceleration the platform may have without slip between the cylinder and platform.

Ans. $a = 6.86$ m/s^2

5.86. The 30-kg semicircular homogeneous plate of 900-mm radius is released from rest in the position shown in Fig. 5-90. What are the forces in the two cords for this position? The plate and cords are in the vertical plane.

Ans. $T_{AB} = 148$ N, $T_{CD} = 59.8$ N

5.87. In a loop-the-loop 6 m in diameter, a car of mass 227 kg leaves the platform 9 m above the bottom of the loop. What is the normal force of the track on the car at the top of the loop? Assume that the center of gravity of the car is 3 m from the center of the loop.

Ans. 2200 N

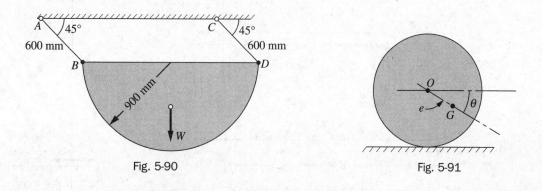

Fig. 5-90

Fig. 5-91

5.88. A disk of weight W has its center of mass G at a distance e from the geometric center O. The disk is rolling on the horizontal plane with constant angular velocity ω. Determine the normal force and frictional force of the floor on the disk when it is in the position shown in Fig. 5-91.

Ans. $F = (W/g)e\omega^2 \cos\theta$ to the left, $N = W + (W/g)e\omega^2 \sin\theta$ up

5.89. The disk in Fig. 5-92 has a mass of 50 kg, a radius of 600 mm, and a radius of gyration relative to its mass center of 450 mm. The uniform slender bar is 1200 mm long and has a mass of 18 kg. Its left end is pinned to the disk. It is initially held in the horizontal position. Determine the angular accelerations of the disk and bar immediately after release from rest.

Ans. $\alpha_{\text{disk}} = 2.25 \text{ rad/s}^2$ clockwise, $\alpha_{\text{bar}} = 10.5 \text{ rad/s}^2$ clockwise

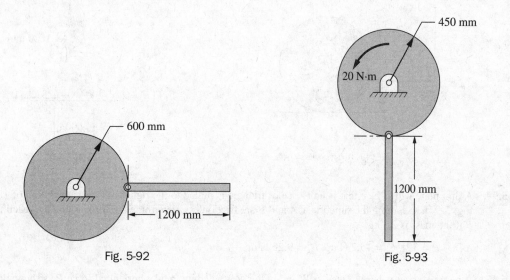

Fig. 5-92

Fig. 5-93

5.90. A 20-kg drum is homogeneous and has a radius of 450 mm. A 10-kg bar is 1200 mm long. Its upper end is pinned to the bottom of the drum and is at rest as shown in Fig. 5-93. Assuming that a couple with moment of 20 N·m is applied to the drum, determine the angular accelerations of the drum and the bar.

Ans. $\alpha_{\text{drum}} = 7.9 \text{ rad/s}^2$ counterclockwise, $\alpha_{\text{bar}} = 4.44 \text{ rad/s}^2$ clockwise

Translation

5.91. A 91-kg door is free to roll on a horizontal track on frictionless wheels at A and B, as shown in Fig. 5-94. What horizontal force P will reduce the vertical reaction at A to zero? What will be the acceleration of the door under the action of this force P? What will be the reaction at B?

Ans. $P = 2000 \text{ N}$, $a = 22 \text{ m/s}^2$, $B = 892 \text{ N}$

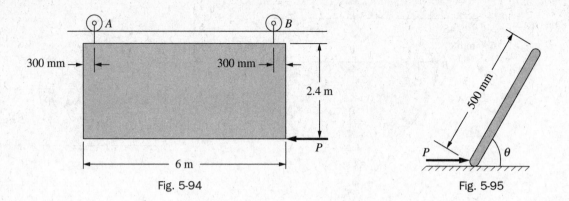

Fig. 5-94   Fig. 5-95

5.92. Figure 5-95 shows a homogeneous slender bar with a mass of 4 kg and a length of 500 mm being pushed along the smooth horizontal surface by a horizontal force $P = 60$ N. Determine the angle θ for translation. What is the accompanying acceleration?

Ans. $\theta = 33.2°$, $a = 15$ m/s^2

5.93. A cylinder having a mass of 10 kg with a diameter of 1.2 m is pushed to the right without rotation and with acceleration 2 m/s^2 (see Fig. 5-96). Determine the magnitude and location of the force P if the coefficient of friction is 0.20.

Ans. $P = 39.6$ N, $h = 0.3$ m

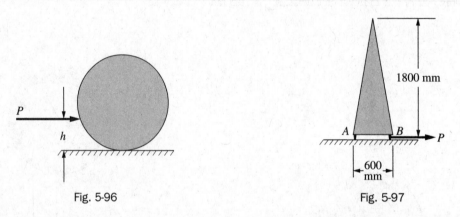

Fig. 5-96   Fig. 5-97

5.94. A thin plate with a face that is an isosceles triangle is pulled to the right with an acceleration of 1.2 m/s^2 (see Fig. 5-97). Assuming the supports at A and B are frictionless, determine the force P and the reactions at A and B if the mass is 2.5 kg.

Ans. $P = 3$ N, $F_A = 15.2$ N up, $F_B = 9.25$ N up

5.95. A homogeneous sphere of weight W and radius R is acted upon by a horizontal force P as shown in Fig. 5-98. The coefficient of sliding friction between the plane and sphere is μ. Where must the force P be applied in order that the sphere skid without rotating? What is the acceleration?

Ans. $d = R(1 - \mu W/P)$, $a = g(P/W - \mu)$ to the right

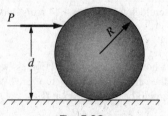

Fig. 5-98

5.96. Refer to Fig. 5-99. What horizontal force is needed to give the 50-kg block an acceleration of 3 m/s² up the 20° plane? Assume a coefficient of friction between the block and the plane of 0.25.

Ans. $P = 507$ N

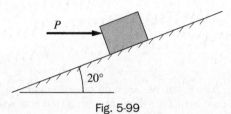

Fig. 5-99

5.97. A body is projected up a 30° plane with an initial velocity of 12 m/s. If the coefficient of friction between the body and the plane is 0.20, how far will the body move up the plane and how long will it take to reach this point?

Ans.   10.9 m, 1.82 s

5.98. A water slide is inclined 35° with the horizontal. Assuming no friction, how long will it take a child starting from a rest position to slide 4.5 m?

Ans. $t = 1.27$ s

5.99. An object starts from rest at the top of a plane that has a slope of 50° with the horizontal. After traveling 6 m down the plane, the object slides 9 m across the horizontal floor and comes to rest. Determine the coefficient of friction between the object and the surfaces.

Ans. $\mu = 0.357$

5.100. A dish slides 500 mm on a level table before coming to rest. If the coefficient of friction between the dish and the table is 0.12, what was the time of travel?

Ans. $t = 0.92$ s

5.101. A 4.23-kg block is prevented from slipping by means of a strip as shown in Fig. 5-100. What is the greatest acceleration the cart may have without tipping the block?

Ans. $a = 3.68$ m/s² to the left

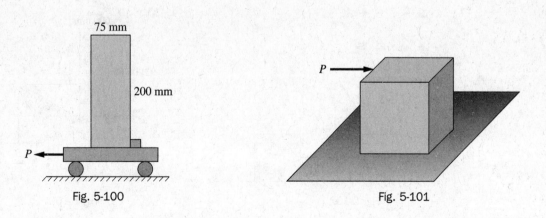

Fig. 5-100   Fig. 5-101

5.102. The cube in Fig. 5-101 is moved along the horizontal plane by a horizontal force *P*. If the coefficient of friction is 0.2, what is the maximum acceleration the block can have if it slides but is on the verge of tipping?

Ans. $a = 5.88$ m/s² to the right

5.103. A cylindrical shaft 1800 mm in diameter and 875 mm high stands on end in a flat car that has a constant velocity of 56 km/h. If the car is brought to rest with uniform deceleration in a distance of 12 m, will the cylinder tip?

Ans.   No

5.104. An 8-kg mass is being lowered by a rope that can safely support 60 N. What is the minimum acceleration the mass can have under these conditions?

Ans.   2.3 m/s^2 down

5.105. An elevator requires 2 s from rest to acquire a downward velocity of 3 m/s. Assuming uniform acceleration, what is the force of the floor during this time on an operator whose normal mass is 68.2 kg?

Ans.   566 N (57.8 kg)

5.106. An elevator having a mass of 500 kg is ascending with an acceleration of 4 m/s^2. The mass of the operator is 65 kg. Assuming that the operator is standing on spring scales during the ascension, what is the scale reading? What is the tension in the cables?

Ans.   897 N (91.5 kg), 7800 N

5.107. A body with a mass of 12 kg at rest is on a spring scale in an elevator. What should be the acceleration (magnitude and direction) in order for the scale to indicate a mass of 10 kg (98 N)?

Ans. $a = 1.63$ m/s^2 down

5.108. A 4.55-kg mass hangs from a spring balance in an elevator that is accelerating upward at 2.42 m/s^2. Determine the spring balance reading.

Ans.   5.67 kg (55.6 N)

5.109. An elevator having a mass of 700 kg is designed for a maximum acceleration of 2 m/s^2. What is the maximum load that may be placed in the elevator if the allowable load in the supporting cable is 19 kN?

Ans. $m = 910$ kg

5.110. A horizontal cord connects the 10- and 20-kg masses shown in Fig. 5-102. There is no friction between the masses and the horizontal plane. If a 4-N force is applied horizontally, determine the acceleration of the masses and the tension in the cord.

Ans. $a = 0.133$ m/s^2, $T = 2.67$ N

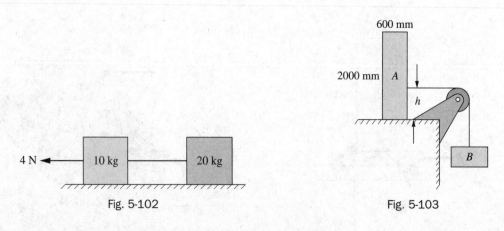

Fig. 5-102   Fig. 5-103

5.111. A 5-kg block A is pulled to the right under the action of a falling 5-kg mass B, as shown in Fig. 5-103. If the coefficient of friction between the block A and the plane is 0.5, determine the acceleration of B. Also determine the maximum value of h if block A is not to tip.

Ans. $a = 2.45$ m/s^2 down, $h = 730$ mm above plane

5.112. In Fig. 5-104, the 13.6-kg mass is on a horizontal floor. The coefficient of friction between the weight and the floor is 0.2. The pulleys shown are assumed to be massless and frictionless. Determine the acceleration of the system and the tensions in the cords.

Ans. $a = 1.02$ m/s^2, $T_{AB} = 39.4$ N, $T_{BC} = 79.9$ N

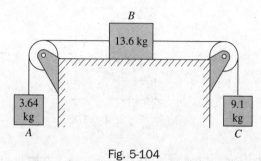

Fig. 5-104

5.113. A 455-kg block stands on a truck, as shown in Fig. 5-105. The mass m falls, accelerating the block and the 136-kg truck. A strip nailed to the platform prevents sliding. What is the maximum acceleration the system may have without tipping the block? Find m. Assume rolling of the truck and neglect the inertia effects of the wheels and pulley.

Ans. $a = 1.54$ m/s^2 to the right, $m = 110$ kg

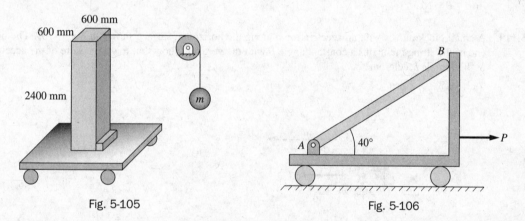

Fig. 5-105   Fig. 5-106

5.114. In Fig. 5-106, the homogeneous bar AB has a mass of 5 kg and is fastened by a frictionless pin at A and rests against the smooth vertical part of the cart. The cart has a mass of 50 kg and is pulled to the right by a horizontal force P. At what value of P will the bar exert no force on the cart at B?

Ans. $P = 642$ N

5.115. The platform A accelerates to the right on a horizontal plane as shown in Fig. 5-107. The bar B is held in the vertical position by the horizontal cord CD. Determine the tension in the cord and the pin reactions at E. The bar is uniform and has a mass of 8.18 kg.

Ans. $T = 30$ N, $E_y = 80.2$ N up, $E_x = 10$ N to the right

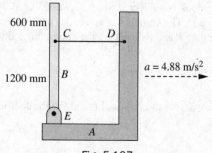

Fig. 5-107

5.116. In Fig. 5-108, a block is moving to the right with an acceleration of 1.5 m/s². A bar 1 m long with a mass of 1 kg is suspended as shown by means of a frictionless pin at A. What is the angle θ?

Ans. $\theta = 8.7°$

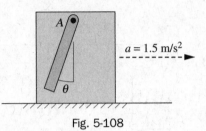

Fig. 5-108

5.117. A chain hangs freely from the rear of a truck. If the truck accelerates to the left 3 m/s², what will be the approximate angle between the chain and the horizontal?

Ans. $\theta = 73°$

5.118. A helicopter carrying a 24-m transmission-line tower by means of a sling accelerates at 1.83 m/s² horizontally. What angle does the centerline of the tower make with the vertical?

Ans. $\theta = 10.5°$

5.119. A small block moves with an acceleration a along the horizontal surface shown in Fig. 5-109. The bob at the end of the string maintains a constant angle θ with the vertical. Show that θ is a measure of the acceleration a in this simple accelerometer.

Ans. $a = g \tan \theta$

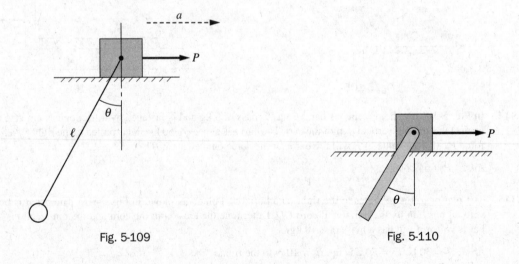

Fig. 5-109   Fig. 5-110

5.120. In Fig. 5-110, the 50-kg block is subjected to an acceleration of 3 m/s² horizontally to the right under the action of a horizontal force P. The homogeneous slender bar has a mass of 8 kg and is 1 m long. Assuming a frictionless surface, determine P and θ. What are the horizontal and vertical components of the pin reaction on the bar?

Ans. $P = 174$ N, $\theta = 17°$, $V = 78.4$ N, $H = 24.0$ N

5.121. In the preceding problem, what are the values of P and θ if the coefficient of friction between the block and the plane is 0.25?

Ans. $P = 316$ N, $\theta = 17°$

5.122. A 455-kg box is shown on a truck in Fig. 5-111. The coefficient of friction between the block and the truck is 0.30. Determine the forward acceleration at which the box tips or slides. Assume homogeneity.

Ans. Tips at $a = 2.45$ m/s^2 to the right

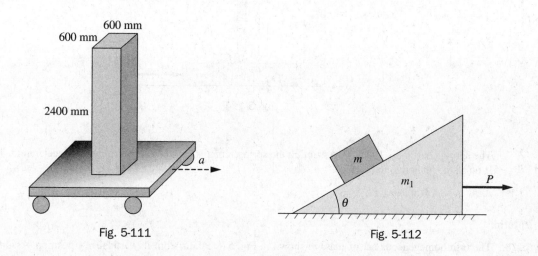

Fig. 5-111   Fig. 5-112

5.123. A block of mass m is on a surface (the coefficient of friction is μ) that is inclined at angle θ with the horizontal, as shown in Fig. 5-112. This surface is part of a triangular block of mass m_1. A horizontal force P causes the system to have an acceleration a to the right. What value of P will cause the top block to move relative to the surface? Assume no friction on the bottom surface. What is the acceleration?

Ans. $\quad a = \dfrac{g(\mu - \tan\theta)}{1 + \mu\tan\theta}, P = \dfrac{[(m + m_1)g](\mu - \tan\theta)}{1 + \mu\tan\theta}$

5.124. A truck is traveling along a level road at a constant speed. Its body has a mass m_1 and contains a box of mass m_2. Show that if the box drops off the truck, the truck body will experience an upward acceleration of $m_2 g/m_1$.

5.125. The 1140-kg automobile in Fig. 5-113 is brought to rest on a horizontal road from a speed of 96 km/h. If the car is equipped with four-wheel-drive brakes, and the coefficient of friction between the tires and the road is 0.6, what will be the time required to bring the car to rest? What will be the distance covered in the time?

Ans. 4.54 s, 60.6 m

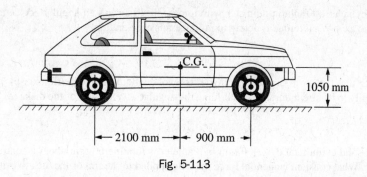

Fig. 5-113

5.126. In Problem 5.125, determine the vertical reaction on the two front tires and on the two rear tires.

Ans. $R_F = 10\,160$ N, $R_R = 1050$ N

5.127. The bar BC in Fig. 5-114 is vertical when the system is at rest. The bar is 3 m long and has a mass of 10 kg. It is noted that the bar is rotated 5° from the vertical when the cart to which it is fastened is accelerating.

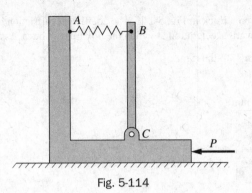

Fig. 5-114

The spring constant is 100 N/m. Assuming the spring remains horizontal, determine the acceleration. The cart is on a smooth horizontal surface.

Ans. $a = 4.37$ m/s^2

Rotation

5.128. The 6-m homogeneous bar of mass m shown in Fig. 5-115 falls from its vertical rest position. Assuming no friction at the pivot, determine the angular velocity of the bar at the time when the tangential component of the acceleration of the unpivoted end is equal to g, the acceleration due to gravity.

Ans. $\theta = 0.73$ rad (41.8°), $\dot\theta = 1.12$ rad/s

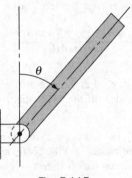

Fig. 5-115

5.129. A solid cylinder 600 mm in diameter with a mass of 40 kg has an angular acceleration of 2 rad/s^2 about its geometric axis. What torque is acting to produce this acceleration?

Ans.   3.6 N·m

5.130. A 364-kg disk 1.8 m in diameter is made to revolve about its geometric axis by a force of 356 N applied tangentially to its circumference. Determine the angular acceleration of the disk.

Ans.   2.17 rad/s^2

5.131. A 41-kg solid cylindrical disk 900 mm in diameter is rotating 60 rpm about a central axis perpendicular to a diameter. What constant tangential force must be applied to the rim of the disk to bring it to rest in 2 min?

Ans.   0.482 N

5.132. A 200-kg mass hangs vertically downward from the end of a massless cord that is wrapped around a cylinder 900 mm in diameter held by frictionless bearings. The mass descends from rest 8 m in 4 s. What is the mass of the cylinder?

Ans.   3920 kg

5.133. A 13.6-kg bar 3 m long is suspended vertically from a pivot point located at the end. The bar is struck a horizontal blow of 400 N at a point 600 mm below the pivot point. Determine the horizontal reaction at the pivot. What is the angular acceleration of the bar due to the blow?

Ans. $R_h = 280$ N to the left, $\alpha = 5.89$ rad/s^2

5.134. A slender bar 1200 mm long has a mass of 3.6 kg. It is hanging vertically at rest when struck by a horizontal force $P = 12$ N at the lower end, as shown in Fig. 5-116. Determine (*a*) the angular acceleration of the bar and (*b*) the horizontal and vertical components of the pin reaction O on the bar.

Ans. $\alpha = 8.33$ rad/s^2 counterclockwise, $O_y = 35.3$ N up, $O_x = 6$ N to right

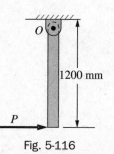

1200 mm

P

Fig. 5-116

5.135. In the preceding problem, where should the 12-N force be applied so that the horizontal component of the pin reaction will be zero?

Ans. 800 mm below O

5.136. At what point should the horizontal force P be applied to the bar, the solid cylinder, and the solid sphere so that the horizontal component of the pin reaction at O is zero? See Fig. 5-117.

Ans. $d_b = \frac{2}{3}d$, $d_s = \frac{7}{10}d$, $d_c = \frac{3}{4}d$

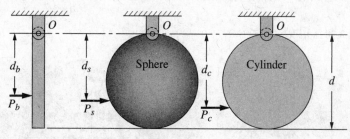

Fig. 5-117

5.137. The coefficient of friction between the horizontal floor and a runner's shoes is 0.5. Find the radius of the smallest circular path around which the runner can travel at a constant speed of 4.9 m/s without slipping.

Ans. $r = 4.9$ m

5.138. A train weighing 440 kN has its center of gravity 1650 mm above the rails. At a speed of 48 km/h, it rounds an unbanked curve having a radius of 760 m. If the centerlines of the rails are 1412 mm in apart, find the vertical force on the outer rail.

Ans. 227 kN

5.139. Determine the angle of banking for a roadway to allow an automobile speed of 100 km/h on a curve of 90-m radius so that there will be no side thrust on the wheels.

Ans. 41.2°

5.140. A highway curve with a radius of 600 m is to be banked for a speed of 80 km/h. What should be the angle of bank so that at the design speed there will be no frictional force necessary on the tire in a direction perpendicular to the direction of motion?

Ans. $\theta = 0.084$ rad or 4.8°

5.141. A 0.027-kg bead slides from rest at A down a frictionless wire, bent as shown in Fig. 5-118. Determine the normal force of the wire on the bead at point A. Next find the normal forces at points B and C with the bead on the arc of the circle.

Ans. $N_A = 0.159$ N, $N_B = 1.22$ N, $N_C = 1.54$ N

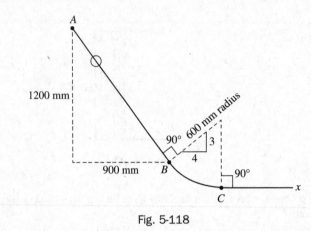

Fig. 5-118

5.142. A flywheel with radius of gyration k and mass m is acted upon by a constant torque M. What will be the angular speed after the flywheel has rotated θ rad from rest?

Ans. $\omega = (1/k)\sqrt{2M\theta/m}$

5.143. A 15.5-kg disk of diameter 1.8 m is released from rest when O and G are on a horizontal line, as shown in Fig. 5-119. What will be the angular velocity when G is vertically below O? Determine the reactions at O at that time.

Ans. $\omega = 3.78$ rad/s, $O_n = 353$ N, $O_t = 0$

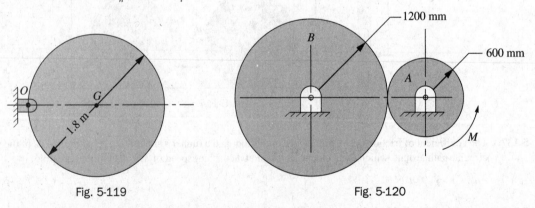

Fig. 5-119   Fig. 5-120

5.144. If the complete object of Problem 5.49 is used as a pendulum, determine the frequency of the motion.

Ans.   0.58 Hz

5.145. Refer to Fig. 5-120. A moment M of 61 N·m is applied to the uniform disk A, which then drives uniform disk B without slip occurring between the two disks. What is the angular acceleration of each disk? Disk A has a mass of 29.3 kg and disk B has a mass of 58.6 kg.

Ans. $\alpha_A = 3.75$ rad/s^2 counterclockwise, $\alpha_B = 1.88$ rad/s^2 clockwise

5.146. A massless bar 4 m long rotates in a horizontal plane about its center. From each end of the rod is suspended a 4-kg mass on a 600-mm cord. When the system is revolving at $\frac{2}{3}$ rev/s, determine the angle between the cords and the vertical.

Ans. $(2 + 0.6 \sin \theta)/\tan \theta = 0.559$ and $\theta = 78°$ by trial and error

5.147. What is the angular acceleration of the pulley shown in Fig. 5-121 turning under the action of the two masses?

Ans. $\alpha = 0.666$ rad/s^2 clockwise

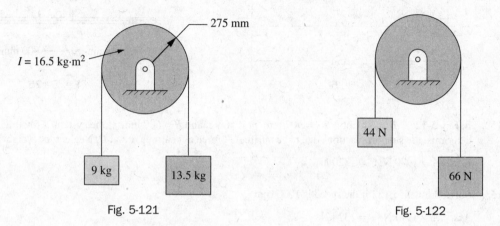

<table>
<tr><td>Fig. 5-121</td><td>Fig. 5-122</td></tr>
</table>

5.148. In Fig. 5-122, the simple Atwood machine has weights of 44 and 66 N measured on earth with $g = 9.8$ m/s^2. If the machine is moved to the moon, where the acceleration of gravity is 0.16 of the earth's g, what will be the tension in the cord when the masses are released from rest?

Ans. $T = 8.45$ N

5.149. Two masses are connected by a rope that passes over a pulley rotating in frictionless bearings. See Fig. 5-123. The masses of A, B, and the pulley are 14, 9, and 5 kg, respectively. The radius of gyration of the pulley is 400 mm. Determine the acceleration of the masses if the system is released from rest at a place where $g = 4.9$ m/s^2.

Ans. $a = 0.935$ m/s^2

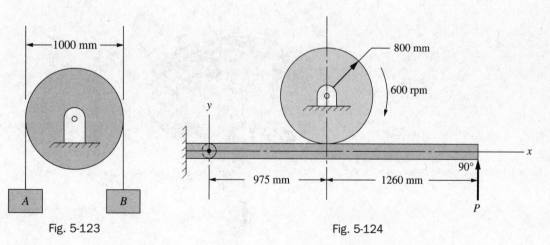

Fig. 5-123   Fig. 5-124

5.150. In Fig. 5-124, the disk of mass 34.5 kg and radius of gyration 700 mm is rotating at 600 rpm. What force P must be applied to the braking mechanism to stop the disk in 25 s? Use a coefficient of friction between the disk and the horizontal member equal to 0.30.

Ans. $P = 154$ N

5.151. A thin triangular 9.1-kg lamina is rotating at 30 rpm about a horizontal axis in two frictionless bearings at *A* and *B*, as shown in Fig. 5-125. When the lamina is vertical as shown, determine the bearing reactions at *A* and *B*.

Ans.   *A* = 0.65 N up, *B* = 1.15 N up

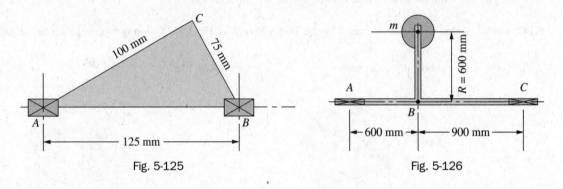

Fig. 5-125   Fig. 5-126

5.152. In Fig. 5-126, *AB* = 600 mm, *BC* = 900 mm, *m* = 50 kg, and *R* = 600 mm. If the system is rotating at 100 rpm, determine the support reactions due to "centrifugal" force at bearings *A* and *C*. Neglect the weight of the arm.

Ans.   *A* = 1970 N, *C* = 1320 N

5.153. Solve Problem 5.152 if the speed is 1000 rpm.

Ans.   *A* = 197 kN, *C* = 132 kN

5.154. A disk rotates uniformly at 15 rpm in a horizontal plane. A 40-kg mass is placed on the disk at a point 2 m distant from the axis of rotation. If the mass is just about to slip in this position, what is the coefficient of friction between the mass and the disk?

Ans. $\mu = 0.5$

5.155. A 2-kg block is held stationary on a stop 450 mm from the center of a horizontal turntable by a cord that passes down through a hole in the shaft, as shown in Fig. 5-127. What tension in the cord is necessary to keep the block 450 mm from the center as the block and turntable rotate around the center at 30 rad/s? Assume that there is no friction.

Ans.   *T* = 810 N

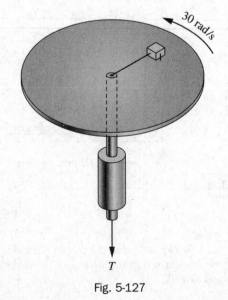

Fig. 5-127

5.156. The horizontal member in Fig. 5-128 is 1200 mm long and is rotating at 20 rpm about a vertical axis through its midpoint. Each 2-kg ball is suspended by a cord that is 300 mm long. What angle θ will each cord make with the vertical? A trial-and-error solution may be useful.

Ans. $\theta = 24.5°$

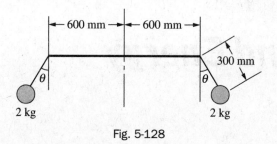

Fig. 5-128

Do the following problems using the inertia force method.

5.157. Solve Problem 5.78.

5.158. Solve Problem 5.79.

5.159. Solve Problem 5.90.

5.160. Solve Problem 5.93.

5.161. Solve Problem 5.101.

5.162. Solve Problem 5.111.

5.163. Solve Problem 5.115.

5.164. Solve Problem 5.145.

5.165. Solve Problem 5.150.

CHAPTER 6

Work and Energy

6.1 Work

Work U done by a force $\mathbf{F}$ acting on a particle moving along any path is defined as the line integral from position P_1 at time t_1 to position P_2 at time t_2:

$$U = \int_{r_1}^{r_2} \mathbf{F} \cdot d\mathbf{r} \tag{1}$$

where $d\mathbf{r}$ is the infinitesimal change in the position vector $\mathbf{r}$ that locates P_1 and P_2. The expression for work may also be written (see Fig. 6-1) as

$$U = \int_{s_1}^{s_2} F_t\, ds \tag{2}$$

where s_1, s_2 = respective distances of particle at the beginning and end of motion from reference point P_0
$\qquad F_t$ = magnitude of tangential component of force $\mathbf{F}$, as indicated in Fig. 6-1
$\qquad ds$ = infinitesimal change in position of particle along path

Then, since

$$\mathbf{F} = F_x\mathbf{i} + F_y\mathbf{j} + F_z\mathbf{k} \qquad \text{and} \qquad d\mathbf{r} = dx\mathbf{i} + dy\mathbf{j} + dz\mathbf{k}$$

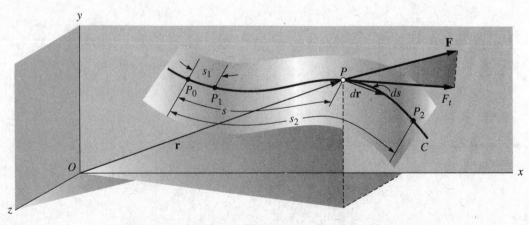

Fig. 6-1   The path of a particle as it moves from P_1 to P_2.

the line integral of Eq. (1) can be written as

$$U = \int_{P_1}^{P_2} (F_x \mathbf{i} + F_y \mathbf{j} + F_z \mathbf{k}) \cdot (dx\,\mathbf{i} + dy\,\mathbf{j} + dz\,\mathbf{k}) = \int_{P_1}^{P_2} (F_x dx + F_y dy + F_z dz) \tag{3}$$

This can be converted to a time integral as

$$U = \int_{t_1}^{t_2} \left(F_x \frac{dx}{dt} + F_y \frac{dy}{dt} + F_z \frac{dz}{dt} \right) dt \tag{4}$$

6.2   Special Cases

Special cases arise as follows:

1. Force constant in magnitude and with its direction along a straight line:

$$U = Fs \tag{5}$$

where U = work done
  F = constant force
  s = displacement during motion along straight line

2. Force constant in magnitude but at a constant angle with a straight-line displacement:

$$U = Fs \cos\theta \tag{6}$$

where U = work done
  F = constant force
  s = displacement during motion along straight line
  θ = angle between action line of force and displacement

3. Forces constituting a couple (rotation occurs):

$$U = \int_{\theta_1}^{\theta_2} M\,d\theta \tag{7}$$

where M = couple
  $d\theta$ = differential angular displacement
  $\theta_1,\ \theta_2$ = initial and final angular displacements

Work is positive if the force acts in the direction of motion. Work is negative if the force acts opposite to the direction of motion.

The work done on a rigid body by two or more concurrent forces during a displacement is equivalent to the work done by the resultant of the force system during the same displacement.

Work is a scalar quantity. In SI, the unit is the newton-meter (N·m), which is called the joule (J). (1 N·m = 1 J.)

6.3   Power

Power, which is the rate of doing work, equals $dU/dt = \mathbf{F} \cdot d\mathbf{r}/dt = \mathbf{F} \cdot \mathbf{v}$. In the case of a couple, this would be power = $dU/dt = \mathbf{M} \cdot \boldsymbol{\omega}$.

The unit of power is the joule per second (J/s). This is, of course, the watt (1 J/s = 1 W). This unit is small; thus, a larger unit, the horsepower, which equals 746 W, is often used.

The formulas used for horsepower (hp) are

$$\text{hp} = \frac{\mathbf{F} \cdot \mathbf{v}}{746} \quad \text{and} \quad \text{hp} = \frac{\mathbf{M} \cdot \boldsymbol{\omega}}{746} \tag{8}$$

where $\mathbf{F}$ = force, N
$\qquad \mathbf{v}$ = velocity, m/s
$\qquad \mathbf{M}$ = moment of couple, N·m
$\qquad \boldsymbol{\omega}$ = angular velocity, rad/s

6.4   Efficiency

Efficiency is equal to work output divided by work input for the same period of time. It is also expressed as power output divided by power input. The work output is less than the work input by the work dissipated (usually in overcoming friction).

6.5   Kinetic Energy of a Particle

Kinetic energy T of a particle with mass m and moving with speed v is defined as $\frac{1}{2}mv^2$. The unit is the joule (J).

6.6   Work-Energy Relations for a Particle

Work done on a particle by all forces is equal to the change in the kinetic energy T of the particle. Equation (1) is manipulated in the following way to show this to be true:

$$U = \int_{r_1}^{r_2} \mathbf{F} \cdot d\mathbf{r} = \int_{r_1}^{r_2} m\ddot{\mathbf{r}} \cdot d\mathbf{r} = \int_{t_1}^{t_2} m\ddot{\mathbf{r}} \cdot \frac{d\mathbf{r}}{dt} dt$$

$$= m \int_{t_1}^{t_2} (\ddot{\mathbf{r}} \cdot \dot{\mathbf{r}}) dt = \frac{1}{2} m \int_{t_1}^{t_2} \frac{d}{dt} (\dot{\mathbf{r}})^2 dt = \frac{1}{2} m \int_{v_1}^{v_2} d(v^2)$$

$$= \frac{1}{2} m [v^2]_{v_1}^{v_2} = \frac{1}{2} mv_2^2 - \frac{1}{2} mv_1^2 = T_2 - T_1 \tag{9}$$

where U = work done
$\qquad m$ = mass of particle
$\qquad v_1, v_2$ = initial and final speeds, respectively, at P_1 and P_2
$\qquad T_1, T_2$ = initial and final kinetic energy, respectively, at P_1 and P_2

6.7   Kinetic Energy of a Rigid Body in Translation

All particles of a rigid body in translation have the same velocity $\mathbf{v}$. Then if m_i = mass of the ith particle, the kinetic energy T of the translating body is

$$T = \sum_{i=1}^{n} \frac{1}{2} m_i v^2 = \frac{1}{2} v^2 \sum_{i=1}^{n} m_i = \frac{1}{2} mv^2 \tag{10}$$

where m = mass of entire body
$\qquad v$ = speed of body

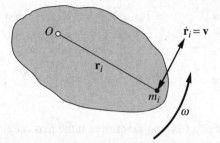

Fig. 6-2   A rigid body rotating about O.

6.8   Kinetic Energy of a Rigid Body in Rotation

If $\mathbf{r}_i$ is the radius vector of the ith particle of a body that is rotating and $\dot{\mathbf{r}}_i$ is its velocity, as shown in Fig. 6-2, then the kinetic energy is

$$T = \sum_{i=1}^{n} \frac{1}{2} m_i (\dot{\mathbf{r}}_i)^2 = \sum_{i=1}^{n} \frac{1}{2} m_i (\boldsymbol{\omega} \times \mathbf{r}_i)^2$$

$$= \sum_{i=1}^{n} \frac{1}{2} m_i \omega^2 r_i^2 = \frac{1}{2} \omega^2 \sum_{i=1}^{n} m_i r_i^2 = \frac{1}{2} I_O \omega^2 \qquad (11)$$

where I_O = mass moment of inertia of entire body about axis of rotation
  ω = angular velocity ($\boldsymbol{\omega} \perp \mathbf{v}$)

6.9   Kinetic Energy of a Body in Plane Motion

Consider the motion of the ith particle of the body in Fig. 6-3 that is rotating and translating in the xy plane, choosing the mass center C as the base point (Fig. 6-3). Then we have

$$\mathbf{r}_i = \mathbf{R} + \boldsymbol{\rho}_i \qquad \text{and} \qquad \dot{\mathbf{r}}_i = \dot{\mathbf{R}} + \dot{\boldsymbol{\rho}}_i$$

$$\dot{\mathbf{r}}_i^2 = (\dot{\mathbf{R}} + \dot{\boldsymbol{\rho}}_i) \cdot (\dot{\mathbf{R}} + \dot{\boldsymbol{\rho}}_i) = \dot{\mathbf{R}}^2 + 2\dot{\mathbf{R}} \cdot \dot{\boldsymbol{\rho}}_i + \dot{\boldsymbol{\rho}}_i^2$$

The kinetic energy is

$$T = \sum_{i=1}^{n} \frac{1}{2} m_i \dot{\mathbf{r}}_i^2 = \frac{1}{2} \sum_{i=1}^{n} m_i \dot{\mathbf{R}}^2 + \sum_{i=1}^{n} m_i \dot{\mathbf{R}} \cdot \dot{\boldsymbol{\rho}}_i + \frac{1}{2} \sum_{i=1}^{n} m_i \dot{\boldsymbol{\rho}}_i^2$$

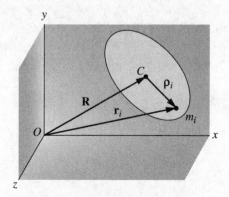

Fig. 6-3   A rigid body in plane motion.

But $\dot{\mathbf{R}}^2 = \bar{v}^2$, $\dot{\boldsymbol{\rho}}_i = \boldsymbol{\omega} \times \boldsymbol{\rho}_i$, and $m_i \dot{\mathbf{R}} \cdot \dot{\boldsymbol{\rho}}_i = m_i \dot{\mathbf{R}} \cdot (\boldsymbol{\omega} \times \boldsymbol{\rho}_i) = \dot{\mathbf{R}} \cdot (\boldsymbol{\omega} \times m_i \boldsymbol{\rho}_i)$. Thus,

$$T = \frac{1}{2}\bar{v}^2 \sum_{i=1}^{n} m_i + \dot{\mathbf{R}} \cdot \left(\boldsymbol{\omega} \times \sum_{i=1}^{n} m_i \boldsymbol{\rho}_i \right) + \frac{1}{2} \sum_{i=1}^{n} m_i \omega^2 \rho_i^2 \tag{12}$$

Since $\sum_{t=1}^{n} m_i \boldsymbol{\rho}_i = 0$ when $\boldsymbol{\rho}_i$ is drawn from an axis through the mass center, Eq. (12) is written as

$$T = \frac{1}{2} m \bar{v}^2 + \frac{1}{2} \bar{I} \omega^2 \tag{13}$$

where $\bar{v}$ = velocity of mass center
       ω = angular velocity
       $\bar{I}$ = moment of inertia about an axis through mass center parallel to z axis

6.10 · Potential Energy

A conservative force is one that does the same amount of work regardless of the path taken by its point of application (see Problem 6.1).

    The potential energy V of a body is measured by the work done against conservative forces (usually gravitational forces) acting on the body in bringing the body from some reference or datum position to the position in question. The *potential energy* may be defined as the negative of the work done by the conservative force acting on the body in bringing it from the datum position to the position in question. The selection of the datum position is arbitrary—usually for convenience.

6.11 Work-Energy Relations for a Rigid Body

The principle of work and energy states that the work done by the external forces acting on a rigid body during a displacement is equal to the change in kinetic energy of the body during the same displacement.

6.12 Conservation of Energy

The conservation of energy states that if a particle (or body) is acted upon by a conservative force system, the sum of the kinetic energy and potential energy is a constant. To show this to be true, let P and Q be, respectively, the initial and final positions of the particle. The work done by a conservative force $\mathbf{F}$ as the particle moves from P to Q has been defined as $U = \int_P^Q \mathbf{F} \cdot d\mathbf{r}$. But in terms of a general position S with radius vector $\mathbf{r}_s$, we can write

$$U = \int_P^S \mathbf{F} \cdot d\mathbf{r} + \int_S^Q \mathbf{F} \cdot d\mathbf{r} = \int_P^S \mathbf{F} \cdot d\mathbf{r} - \int_Q^S \mathbf{F} \cdot d\mathbf{r} \tag{14}$$

But potential energy $V_P = \int_P^S \mathbf{F} \cdot d\mathbf{r}$ and $V_Q = \int_Q^S \mathbf{F} \cdot d\mathbf{r}$. Hence,

$$U = V_P - V_Q$$

We have also shown in terms of kinetic energy that

$$U = T_Q - T_P$$

Thus,

$$V_P - V_Q = T_Q - T_P \qquad \text{or} \qquad V_P + T_P = V_Q + T_Q \tag{15}$$

SOLVED PROBLEMS

6.1. Determine the potential energy V of a body of mass m at a height h above a datum plane. Assume that h is small enough that the gravitational force on m does not vary. ($g = 9.77$ m/s^2 at the top of Mt. Everest.)

SOLUTION

The potential energy V is the negative of the work done against the only force acting on the body as it traverses any smooth path from S (datum) to P. This force is the gravitational force mg, which is assumed constant throughout this travel. Refer to Fig. 6-4(a).

Since **F** is constant, we can write

$$V = \int_C -\mathbf{F} \cdot d\mathbf{r} = -\mathbf{F} \cdot \int_{\mathbf{r}_S}^{\mathbf{r}_P} d\mathbf{r} = -\mathbf{F} \cdot (\mathbf{r}_P - \mathbf{r}_S)$$

But Fig. 6-4(b) indicates that

$$-\mathbf{F} \cdot (\mathbf{r}_P - \mathbf{r}_S) = \mathbf{F} \cdot (\mathbf{r}_S - \mathbf{r}_P) = mg \left| \mathbf{r}_S - \mathbf{r}_P \right| \cos\theta = mgh$$

Thus, the potential energy V of a body of mass m at a height h above an arbitrary datum plane is

$$V = mgh$$

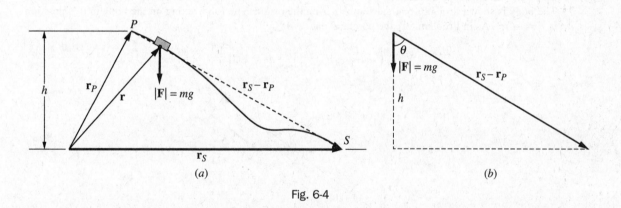

Fig. 6-4

6.2. A body on a frictionless horizontal plane is attached to a spring with modulus k N/m. Refer to Fig. 6-5(a). When the body is at S, a distance ℓ from the wall, there is no force in the spring ($k\ell = 0$). When the body is at P, a distance $\ell + x$ from the wall, determine the potential energy V_P of the system (actually of the spring, because there is no change in the potential energy of the body, which is always on the horizontal plane).

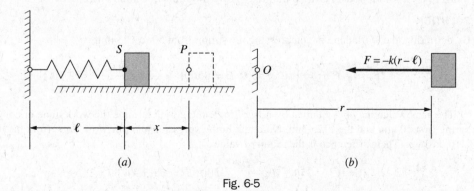

Fig. 6-5

SOLUTION

Figure 6-5(*b*) shows the body at a distance *r* from the wall as the origin. The force acting on the body in this position is proportional to the deformation of the spring from the unstressed length; thus $F = -k(r - \ell)$. The work done by this force for a differential change *dr* is $-k(r - \ell)\,dr$. The potential energy is

$$V_P = -\int_{\ell}^{\ell+x} [-k(r - \ell)]\,dr = \frac{1}{2}kx^2$$

The body possesses this potential energy because of the pull of the spring, which varies with the deformation. The pull of the spring (its internal force) always acts along the spring. Thus, the potential energy for a given deformation remains the same even if the spring is moved sideways on the smooth horizontal surface, provided the distance from *O* does not change. This sideways motion could be conceived as the sum of infinitesimal displacements perpendicular to the spring during which forces act only along the spring (zero work is done perpendicular to the spring).

In Problem 6.1, the given body possesses potential energy because the attraction of the earth is considered constant. Problem 6.3 introduces an attractive force that varies inversely with the square of the distance.

6.3. Determine the potential energy of a body attracted by a force that is inversely proportional to the square of the distance *r* from the source *O* of the force.

SOLUTION

The initial position *S* of the body will be chosen at infinity. (*Note:* Some students might have tried the source *O* as the initial position, but *V* is defined in terms of the work done in moving from the initial position; and since the force tends to become infinite as the body approaches *O*, the choice of *O* is seen to be unwise.) The body is shown in Fig. 6-6 at a distance *r* from the origin. The force acting on the body in this position is $F = -C/r^2$. As in Problem 6.2, the potential energy is

$$V_P = \int_{\infty}^{r} -\frac{C}{r^2}(-dr) = \int_{\infty}^{r} \frac{C}{r^2}\,dr = -C\left[\frac{1}{r}\right]_{\infty}^{r} = -\frac{C}{r}$$

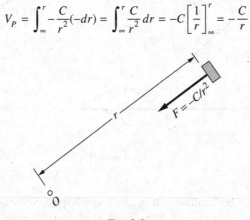

Fig. 6-6

6.4. A spring is initially compressed from a free length of 200 mm to a length of 150 mm, making a net initial compression of 50 mm. What additional work is done in compressing it 75 mm more (125 mm total) to a 75-mm length? Assume that the spring modulus $k = 3500$ N/m.

SOLUTION

By definition, the work done in compressing the spring from 50 to 125 mm is

$$U = \int_{S_1}^{S_2} F_t\,ds = \int_{0.05}^{0.125} 3500s\,ds = 3500\left(\frac{0.125^2}{2} - \frac{0.05^2}{2}\right) = 23.0 \text{ N·m}$$

The same value may be obtained by using Problem 6.2 to determine the work done in compressing the spring first 50 mm and then 125 mm. Note that both expressions are for work done from the unstressed or zero position. Their difference is the required value:

$$U = \int_{0}^{0.125} 3500s\,ds - \int_{0}^{0.05} 3500s\,ds = 27.3 - 4.37 = 22.9 \text{ N·m}$$

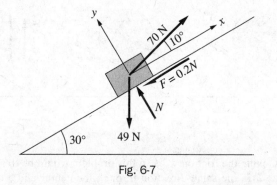

Fig. 6-7

6.5. Determine the total work done on a 5-kg body that is pulled 6 m up a plane inclined 30° with the horizontal, as shown in Fig. 6-7. Assume that the coefficient of friction $\mu = 0.2$.

SOLUTION

From the free-body diagram, N may be found by summing forces perpendicular to the plane:

$$\sum F_y = 0 = -5 \times 9.8\cos 30° + N + 70\sin 10° \qquad \therefore N = 30.3 \text{ N}$$

The work done by the forces is now determined. The sign is positive if the force acts in the direction of motion of the body:

$$U = (70\cos 10°) \times 6 - (49\sin 30°) \times 6 - 0.2 \times 30.3 \times 6 = 230 \text{ N·m}$$

Alternatively, find the resultant of all the forces and then determine the work done by the resultant. Since only the x component of the resultant will do work, determine only that

$$R_x = \sum F_x = 70\cos 10° - 0.20(30.3) - 49\sin 30° = 38.4 \text{ N}$$

Since 38.4 N is a constant force, the total work done is

$$R_x(6) = 38.4(6) = 230 \text{ N·m}$$

6.6. Find the work done in rolling a 20-kg wheel a distance 1.5 m up a plane inclined 30° with the horizontal, as shown in Fig. 6-8. Assume a coefficient of friction of 0.25.

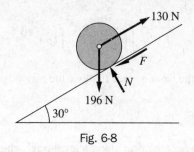

Fig. 6-8

SOLUTION

The normal force N does no work because it has no component in the direction of motion. The force of friction F also does no work, but for a different reason. Its point of contact moves with the wheel, and since there is no relative motion (no slipping involved) between the frictional force and the wheel, the work done by the force of friction is zero. This is not the case in the preceding problem, where the frictional force did negative work, but of course the body slipped along the plane.

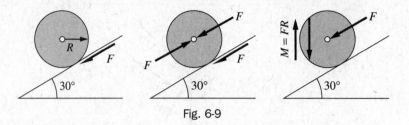

Fig. 6-9

Another way of attacking the friction work in this problem would be to replace the single force F with an equal parallel force F in the same direction through the center and a couple, as shown in Fig. 6-9. (Neglect all other forces for the time being.) The work done by the equivalent system is equal to the sum of the work done by the force F and the couple M (magnitude FR). Let us determine the work done for one revolution, i.e., for 2π rad. The center moves a distance $2\pi R$ during one revolution. Hence, the work done by F is $-F(2\pi R)$. The work done by the couple aids the wheel in moving up the hill and equals $+M\theta$ or $2\pi FR$. The sum of these two values is zero work.

Therefore, of the given system, the only forces doing work are the 130-N force and the component of the gravitational force along the plane (refer to Fig. 6-8). The component of the gravitational force is 196 sin 30°.

$$\text{Work done} = 130(1.5) - (196 \sin 30°)(1.5) = 48 \text{ N·m}$$

6.7. Work is expressed as $U = \int(\Sigma F_x)\,dx$, where ΣF_x is the sum of the forces in the x direction. Show that the work of the expanding gas in the simple engine shown in Fig. 6-10 is $U = \int p\,dV$, where p is the gas pressure in N/m^2 and V is the volume in m^3.

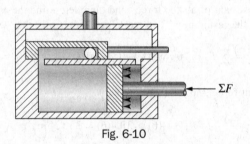

Fig. 6-10

SOLUTION

The force F acting on the piston is pA, where A is the area of the piston. Also, the change in volume $dV = A\,dx$. Hence,

$$U = \int\left(\Sigma F_x\right)dx = \int (pA)\frac{dV}{A} = \int p\,dV$$

If the pressure is constant, the work needed to move the piston is

$$U = p(V_2 - V_1)$$

6.8. At a certain instant during acceleration along a level track, the drawbar pull of an electric locomotive is 100 kN. What power is being developed if the speed of the train is 90 km/h?

SOLUTION

$$\text{Power} = (100\,000 \text{ N})\left(\frac{90\,000 \text{ m}}{3600 \text{ s}}\right) = 2\,500\,000 \text{ N·m/s} = 2.5 \times 10^6 \text{ J/s}$$

$$= 2.5 \text{ MJ/s or } 2.5 \text{ MW}$$

6.9. A belt wrapped around a pulley 600 mm in diameter has a tension of 800 N on the tight side and a tension of 180 N on the slack side. What power is being transmitted if the pulley is rotating at 200 rpm?

SOLUTION

The torque M is found by taking the algebraic sum of the moments of the tensile forces in the belt about the center of the pulley:

$$\text{Torque } M = 800 \times 0.3 - 180 \times 0.3 = 186 \text{ N·m}$$

The power is then

$$\text{Power} = M \cdot \omega = (186 \text{ N·m})(200 \times 2\pi/60 \text{ rad/s}) = 3900 \text{ W or 3.9 kW}$$

6.10. The force P acting on the Prony brake shown in Fig. 6-11 is 114 N. The moment M transmitted by the shaft of the engine to the brake drum gives it a counterclockwise angular velocity of 600 rpm. Determine the power dissipated by the Prony brake. Neglect the weight of the brake.

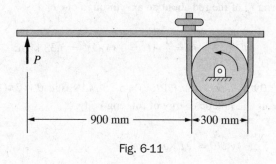

$\uparrow P$

$\leftarrow$ 900 mm $\rightarrow$ $\leftarrow$ 300 mm $\rightarrow$

Fig. 6-11

SOLUTION

The moment M being transmitted by the shaft equals the moment of the force P about the drum center; that is, $114 \times (0.9 + 0.15) = 119.7$ N·m. The horse power is

$$\text{Power} = \frac{M\omega}{746} = \frac{119.7 \times (600 \times 2\pi/60) \text{ rad/s}}{746 \text{ W/hp}} = 10.1 \text{ hp}$$

Thus, the power being transmitted by the shaft is 10.1 hp.

6.11. The power measured by a Prony brake attached to an engine flywheel is 3.8 brake horsepower (bhp). The indicated horsepower (ihp), as measured by means of indicator cards, is 4.1. What is the efficiency of the engine?

SOLUTION

$$\text{Efficiency} = \frac{\text{power output}}{\text{power input}} = \frac{3.8 \text{ bhp}}{4.1 \text{ ihp}} \times 100 = 93\%$$

The Prony brake measures the output, and an indicator card measures the power generated by the pistons.

6.12. A block of mass m is projected with initial speed v_0 along a horizontal plane. If it covers a distance s before coming to rest, what is the coefficient of friction, assuming that the force of friction is proportional to the normal force?

SOLUTION

The initial kinetic energy is $T_1 = \frac{1}{2}mv_0^2$. Its final kinetic energy is zero. The normal force is mg. Hence, the friction is μmg and does an amount of work equal to $-\mu mgs$. Hence:

$$U = T_2 - T_1 \quad \text{or} \quad -\mu mgs = -\frac{1}{2}mv_0^2 \quad \therefore \mu = \frac{v_0^2}{2gs}$$

6.13. The strength of a magnetic field is given by $F = -40/x$, where F is in newtons and x is the distance from the magnet in meters. A 0.1-kg disk is placed 1.8 m from the magnet on a smooth horizontal plane. What will be the speed of the disk when it is 900 mm from the magnet?

SOLUTION

Using the work-energy relation $U = \Delta T$, we have

$$U = \int F\, dx = \int_{1.8}^{0.9} -\frac{40}{x}\, dx = \left[-40 \ln x\right]_{1.8}^{0.9} = 27.7 \text{ N·m}$$

$$\Delta T = \frac{1}{2}(0.1)v^2 - 0 = 27.7 \qquad \therefore v = 23.5 \text{ m/s}$$

6.14. A slender rod 2 m long and having a mass of 4 kg increases its speed about a vertical axis through one end from 20 to 50 rpm in 10 rev. Find the constant moment M required to do this.

SOLUTION

The moment of inertia I_O of the rod about an axis through one end is

$$I_O = \frac{1}{3} m\ell^2 = \frac{1}{3}(4)(2)^2 = 5.33 \text{ kg·m}^2$$

$\omega_1 = 2\pi(20/60) = 2.09$ rad/s, $\omega_2 = 2\pi(50/60) = 5.23$ rad/s, and $\theta = 2\pi(10) = 62.8$ rad.

Work done = change in kinetic energy of rotating body:

$$M\theta = \frac{1}{2} I_O(\omega_2^2 - \omega_1^2)$$

$$M(62.8) = \frac{1}{2}(5.33)(5.23^2 - 2.09^2) \qquad \therefore M = 0.975 \text{ N·m}$$

6.15. A slender rod having a mass m and length ℓ is pinned at one end to a horizontal plane. The rod, initially in a vertical position, is allowed to fall (see Fig. 6-12). What will be its angular speed when it strikes the floor?

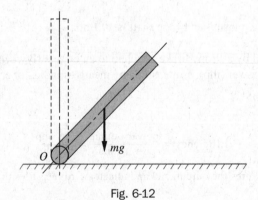

Fig. 6-12

SOLUTION

The only force doing work is the gravitational force mg, assumed concentrated at the center of gravity, which falls a total vertical distance $\frac{1}{2}\ell$. The work done by gravity is thus $\frac{1}{2} mg\ell$. The kinetic energy is changed from $T_1 = 0$ to $T_2 = \frac{1}{2} I_O \omega^2 = \frac{1}{2}(\frac{1}{3} m\ell^2)\omega^2$. Then

$$U = T_2 - T_1 \qquad \text{or} \qquad \frac{1}{2} mg\ell = \frac{1}{6} m\ell^2 \omega^2 \qquad \therefore \omega = \sqrt{\frac{3g}{\ell}}$$

6.16. A car and four wheels have a mass of 818 kg. Each 91-kg wheel is 750 mm in diameter. The car, moving at 24 km/h, coasts to rest on a level track in 3.2 km. What is the rolling resistance *F*?

SOLUTION

The initial kinetic energy T_1 is the kinetic energy (of translation) of the car and the kinetic energy (both rotation and translation) of the four wheels. The final kinetic energy $T_2 = 0$. The work done by the rolling resistance *F* equals $F(3200)$ N·m. Also, 24 km/h = 6.67 m/s. Then

$$U = T_2 - T_1$$

$$3200F = 0 - \frac{1}{2}(454)(6.67)^2 - 4\left(\frac{1}{2}\right)(91)\left[(6.67)^2 + \frac{1}{2}r^2\omega^2\right]$$

Substitute $\omega^2 r^2 = v^2 = 6.67^2$ into the expression to obtain $F = -6.95$ N.

6.17. A 45-kg slender rod 1.2 m long falls from rest from the horizontal line to the 45° angle shown in Fig. 6-13. In that position, what are the bearing reactions at *O* on the rod?

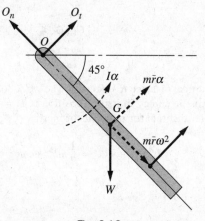

Fig. 6-13

SOLUTION

Choose *n* and *t* axes along and perpendicular to the bar. Apply reversed effective forces through the center of gravity to hold the bar in "equilibrium." To determine the angular acceleration α in the position shown, use the equation $\sum M_O = 0$ or

$$\sum M_O = 0 = I\alpha + m\bar{r}\alpha(\bar{r}) + m\bar{r}\omega^2(0) - mgF\cos 45°$$

$$0 = \frac{1}{12}m(1.2)^2\alpha + m(0.6)^2\alpha - mg(0.6)(0.707) \qquad \therefore \alpha = 8.66 \text{ rad/s}^2$$

Summing forces along the *t* axis,

$$\sum F_t = 0 = m\bar{r}\alpha - mg\cos 45° + O_t \qquad \therefore O_t = 45 \times 9.8 \times 0.707 - 45 \times 0.6 \times 8.66 = 78.0 \text{ N}$$

To determine O_n, it is first necessary to find ω. Use the work-energy method. The only force doing work is the weight, whose center *G* has fallen 0.6(0.707) = 0.4242 m. The kinetic energy change is $\frac{1}{2}\bar{I}\omega^2 + \frac{1}{2}m\bar{v}^2$. Hence,

$$U = \frac{1}{2}\left(\frac{1}{12}m\ell^2\right)\omega^2 + \frac{1}{2}m(\bar{r}\omega)^2$$

$$0.4242 \times 45g = \frac{1}{24} \times 45 \times 1.2^2\omega^2 + \frac{1}{2} \times 45 \times 0.6^2\omega^2 \qquad \therefore \omega = 4.16 \text{ rad/s}$$

Summing the forces along the bar,

$$-O_n + 45 \times 9.8(0.707) + 45 \times 0.6 \times 17.32 \qquad \therefore O_n = 180 \text{ N}$$

6.18. A solid homogeneous cylinder of radius R and mass m rotates freely from its initial rest position (G vertically above O) about a horizontal axis perpendicular to the plane of the paper. What is the value of its angular speed in the position θ? Refer to Fig. 6-14.

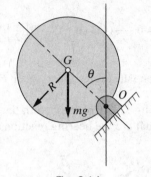

Fig. 6-14

SOLUTION

By inspection, G falls a vertical distance $R - R\cos\theta$. The work done on the cylinder by gravity is thus equal to $mgR(1 - \cos\theta)$. The kinetic energy in its rest position is $T_1 = 0$. The kinetic energy in the θ position is $T_2 = \frac{1}{2}I_0\omega^2$. By the transfer theorem for moments of inertia,

$$I_O = \bar{I} + mR^2 = \frac{1}{2}mR^2 + mR^2 = \frac{3}{2}mR^2$$

Then

$$U = T_2 - T_1 \qquad mgR(1 - \cos\theta) = \frac{3}{4}mR^2\omega^2 \qquad \therefore \omega = \sqrt{\frac{4g(1 - \cos\theta)}{3R}}$$

6.19. A sphere, rolling with an initial velocity of 9 m/s, starts up a plane inclined 30° with the horizontal, as shown in Fig. 6-15. How far will it roll up the plane?

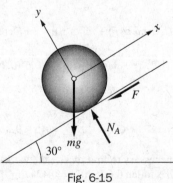

Fig. 6-15

SOLUTION

The initial kinetic energy T_1 decreases to $T_2 = 0$ at the top of the travel. The only force that does work is the component of the weight W along the plane. This gives, x being the required distance,

$$U = -(mg\sin 30°)x$$

The initial kinetic energy T_1 for the body in plane motion is $T_1 = \frac{1}{2}m\bar{v}_1^2 + \frac{1}{2}\bar{I}\omega_1^2$. Since $\bar{I} = \frac{2}{5}mR^2$ and $v_1 = -\omega_1 R$, we have

$$T_1 = \frac{1}{2}m\bar{v}_1^2 + \frac{1}{5}m\bar{v}_1^2 = \frac{7}{10}m(9)^2$$

Then

$$U = T_2 - T_1 \qquad -(mg\sin 30°)x = 0 - \frac{7}{10}m(9)^2 \qquad \therefore x = 11.42 \text{ m}$$

6.20. Figure 6-16(*a*) shows a 146-kg solid cylinder that rolls from rest without slipping on a horizontal plane under the action of the horizontal 54-N force. Determine the angular velocity of the cylinder after it has rotated 90°. The diameter is 960 mm.

SOLUTION

In the free-body diagram, Fig. 6-16(*b*), the only force doing work is the 54-N force. Using an (x, y) set of axes as shown, the radius vector for the top point is $\mathbf{r} = \mathbf{R} + 0.48\mathbf{j}$.

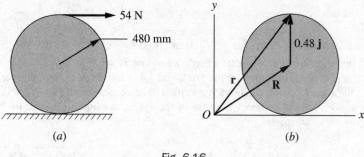

Fig. 6-16

When the cylinder rotates through a differential angular displacement $d\phi$, we can write

$$d\mathbf{r} = d\mathbf{R} + 0.48(d\phi)\mathbf{i}$$

Also note that $d\mathbf{R} = 0.48(d\phi)\mathbf{i}$. The differential work done by the 54-N force is now written

$$dU = 54\mathbf{i} \cdot d\mathbf{r} = 54\mathbf{i} \cdot (0.48\,d\phi\,\mathbf{i} + 0.48\,d\phi\,\mathbf{i})$$

$$= 51.84\,d\phi$$

and

$$U = \int_0^{\pi/2} 51.84\,d\phi = 81.4 \text{ N·m}$$

Alternatively, this result can be obtained by replacing the 54-N force at the top with a horizontal 54-N force through the center *and* a clockwise couple equal to $54 \times 0.48 = 25.92$ N·m.

The work done by the force through the center and the couple is

$$U = 54 \times 0.48 \times \frac{\pi}{2} + 25.92 \times \frac{\pi}{2} = 81.4 \text{ N·m}$$

Next determine the kinetic energy of the cylinder as

$$\frac{1}{2}m\bar{v}^2 + \frac{1}{2}\bar{I}\omega^2 = \frac{1}{2}(146)(0.48\omega)^2 + \frac{1}{2}\left[\frac{1}{2}(146)(0.48)^2\right]\omega^2$$

Equating this to the work done, we find

$$\omega = 1.797 \text{ rad/s}$$

6.21. A solid homogeneous cylinder rolls without slipping on the horizontal plane. The cylinder has a mass of 90 kg and is at rest in the position shown in Fig. 6-17. The modulus of the spring is 450 N/m, and its unstretched length is 600 mm. What will be the angular speed of the cylinder when the spring has moved the center 500 mm to the right?

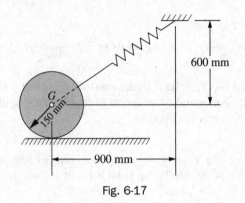

Fig. 6-17

SOLUTION

The cylinder has initial kinetic energy $T_1 = 0$ and final $T_2 = \frac{1}{2}m\bar{v}_2^2 + \frac{1}{2}\bar{I}\omega_2^2$. Substituting $\bar{v}_2 = r\omega_2 = 0.15\omega_2$ and $\bar{I} = \frac{1}{2}mr^2 = \frac{1}{2}(90)(0.15)^2 = 1.01$ kg·m^2, we obtain $T_2 = 1.52\omega_2^2$.

Since the point of application of the frictional force moves with the cylinder (and hence friction does no work), the forces form a conservative system. The conservation of energy law will be used here. The system possesses potential energy because of the spring configuration (see Problem 6.2). The initial and final lengths of the spring are

$$s_1 = \sqrt{(0.6)^2 + (0.9)^2} = 1.08 \text{ m} \qquad \text{and} \qquad s_2 = \sqrt{(0.6)^2 + (0.4)^2} = 0.72 \text{ m}$$

Then

$$V_1 = \frac{1}{2}k(1.08 - 0.6)^2 = \frac{1}{2}(450)(0.48)^2 = 51.8 \text{ N·m} \qquad V_2 = \frac{1}{2}k(0.72 - 0.6)^2 = 3.24 \text{ N·m}$$

Finally, from the conservation of energy law, we can write

$$T_1 + V_1 = T_2 + V_2 \qquad 0 + 51.8 = 1.52\omega_2^2 + 3.24 \qquad \therefore \omega_2 = 5.65 \text{ rad/s}$$

6.22. In the position shown in Fig. 6-18, block A is moving down 1.5 m/s. Cylinder B is considered solid and moves in frictionless bearings. The spring is originally compressed 150 mm and has a modulus of 80 N/m. What will be the speed v of A after dropping 1.2 m^2.

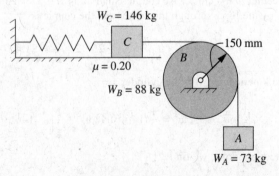

Fig. 6-18

SOLUTION

Consider the system consisting of bodies A, B, and C as a unit in determining kinetic energy and work done. The initial kinetic energy T_1 is the sum

$$T_1 = \frac{1}{2} m_C v_1^2 + \frac{1}{2} \bar{I}_B \omega_1^2 + \frac{1}{2} m_A v_1^2 = \frac{1}{2} \times 146 \times 1.5^2 + \frac{1}{2}\left(\frac{1}{2} \times 88 \times 0.15^2\right)\left(\frac{1.5}{0.15}\right)^2$$

$$+ \frac{1}{2} \times 73 \times 1.5^2 = 296 \text{ N·m}$$

where $\bar{I}_B = \frac{1}{2} m_B (0.15)^2$, and $\omega_1 = v_1/r = 1.5/0.15 = 10$ rad/s, since no slip of rope on cylinder is assumed. The final kinetic energy T_2 in terms of the required speed v is

$$T_2 = \frac{1}{2} m_C v_2^2 + \frac{1}{2} \bar{I}_B \omega_2^2 + \frac{1}{2} m_A v_2^2$$

$$= \frac{1}{2} \times 146 v_2^2 + \frac{1}{2}\left(\frac{1}{2} \times 88 \times 0.15^2\right)\left(\frac{v_2}{0.15}\right)^2 + \frac{1}{2} \times 73 v_2^2 = 131.5 v_2^2$$

Work is done on the system as follows: W_A does positive work; friction on C ($0.20 \times 146 \times 9.8 = 263$ N) does negative work; for the first 0.15 m, the spring does positive work, but after that it is being stretched from the neutral position and does negative work. Thus,

$$\text{Work } U = 73 \times 9.8 \times 1.2 - 286 \times 1.2 + \int_0^{0.15} 80s\, ds - \int_0^{1.05} 80s\, ds = 472 \text{ N·m}$$

Then

$$U = T_2 - T_1 \qquad 472 = 131.5 v_2^2 - 296 \qquad \therefore v_2 = 2.42 \text{ m/s}$$

6.23. A cylinder is pulled up a plane by the tension in a rope that passes over a frictionless pulley, assumed weightless, and is attached to a 70-kg mass, as shown in Fig. 6-19. The 45-kg cylinder has radius 600 mm. The cylinder moves from rest up a distance of 5 m. What will be its speed?

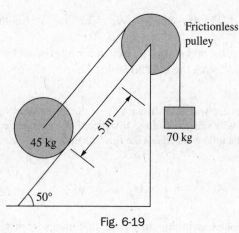

Fig. 6-19

SOLUTION

The initial kinetic energy of the cylinder and mass is $T_1 = 0$. The final kinetic energy of the system is

$$T_2 = T_c + T_m = \frac{1}{2} m_c \bar{v}^2 + \frac{1}{2} \bar{I}_c \omega^2 + \frac{1}{2} m_m \bar{v}^2 = 68.75\bar{v}^2$$

after substituting $m_c = 45$, $m_m = 70$, $\bar{I}_c = \frac{1}{2} m_c R^2$, and $R^2 \omega^2 = \bar{v}^2$. The initial potential energy for the cylinder will be assumed zero, and it will gain potential energy $(9.8 \times 45)(5 \sin 50°)$. The initial potential energy of the 70-kg mass will be assumed zero, and it will lose potential energy $(9.8 \times 70)(5)$. Thus,

$$T_1 + V_1 = T_2 + V_2 \qquad 0 + 0 = 68.75\bar{v}^2 + (9.8 \times 45)(5 \sin 50°) - (9.8 \times 70)(5) \qquad \therefore \bar{v} = 5.03 \text{ m/s}$$

6.24. The 44-kg block shown in Fig. 6-20 rests on a smooth plane. It is connected by a cord that passes around weightless, frictionless pulleys to a support. The 60-kg mass is attached as shown. After the system is released from rest, in what distance will the block on the plane attain a speed of 3 m/s?

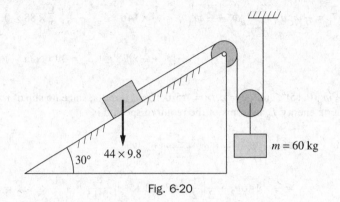

Fig. 6-20

SOLUTION

The 60-kg block travels one-half the distance that the 44-kg block does, and its speed is one-half the speed of the 44-kg block. The initial kinetic energy of the system is $T_1 = 0$. Its final kinetic energy is

$$T_2 = \frac{1}{2} \times 44 \times 3^2 + \frac{1}{2} \times 60 \times 1.5^2 = 266 \text{ N·m}$$

Work is done by the component of the weight along the plane and by the 60-kg mass. Assuming motion up the plane, the work done is

$$U = -(44 \times 9.8 \sin 30°)s + 60 \times 9.8 \frac{s}{2} = 78.4s$$

Using $U = T_2 - T_1$ provides

$$78.4s = 266 \qquad \therefore s = 3.39 \text{ m} \qquad \text{(up the plane)}$$

6.25. A flexible chain of length ℓ and mass a kg/m is released when it has a free overhang of c meters. What will be its speed when leaving the smooth table?

SOLUTION

The free overhang of mass ac will fall a distance $\ell - c$; the work done by gravity on this part will be $gac(\ell - c)$. The part on the table will fall an average distance $\frac{1}{2}(\ell - c)$; the work done by gravity on this part is $ga(\ell - c) \times \frac{1}{2}(\ell - c)$. The total work equals the final kinetic energy $\frac{1}{2}mv^2 = \frac{1}{2}a\ell v^2$. Then

$$gac(\ell - c) + \frac{1}{2}ga(\ell - c)^2 = \frac{1}{2}a\ell v^2 \qquad \therefore v = \sqrt{\frac{g(\ell^2 - c^2)}{\ell}}$$

6.26. Determine the work done in winding up a cable that hangs from a horizontal drum if its free length is 6 m and it has a mass of 50 kg.

SOLUTION

Figure 6-21 shows the cable in its original position. In analyzing the problem, note that any element dx is acted upon by a gravitational force $9.8(50/6)\,dx = 81.7\,dx$. Assume that the element dx is at a distance x from the free end. This element is then raised a distance $(6 - x)$ m. The work done on it is the product of the gravitational force and the distance raised.

The total work is the integral of the work done on a differential element. Note that x varies from 0 to 6 m:

$$\text{Work} = \int_0^6 81.7(6 - x)\,dx = 1470 \text{ N·m}$$

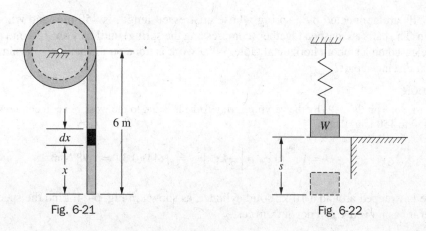

Fig. 6-21   Fig. 6-22

6.27. Refer to Fig. 6-22. If weight W hangs freely, it stretches the spring a distance c. Show that if the weight (held so that the spring is unstressed) is suddenly released, the spring stretches a distance $2c$ before the weight starts to return upward.

SOLUTION

The kinetic energy of the weight at the top (initial position) and the bottom is zero. Hence, the total work done on the weight must be zero. But the total work done is the positive work of gravity (Ws) offset by the negative work of the spring ($\frac{1}{2}ks^2$) the weight. Therefore,

$$Ws - \frac{1}{2}ks^2 = 0$$

But in the freely hanging position, W is balanced by the spring force kc. Putting $W = kc$ in the above equation, we obtain $s = 2c$.

The maximum tension in the spring for this type of loading is double the weight.

6.28. In the spring gun shown in Fig. 6-23, ball W rests against the compressed spring of constant k. Its initial compression is x_0. What will be the speed of the ball leaving the gun? Assume the spring is unstressed when the bearing plate is at the end of the gun.

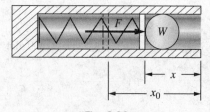

Fig. 6-23

SOLUTION

The ball is shown in any position x from the end of the gun. The force doing work is the spring force $F = kx$. The work done is

$$U = \int_0^{x_0} kx\, dx = \frac{1}{2}kx_0^2$$

This is equal to the gain in kinetic energy of the ball, which is

$$T_2 = \frac{1}{2}\left(\frac{W}{g}\right)v^2$$

Hence,

$$\frac{1}{2}kx_0^2 = \frac{1}{2}\frac{W}{g}v^2 \qquad \therefore\ v = x_0\sqrt{\frac{kg}{W}}$$

6.29. Two balls are connected by a spring whose unstressed length is 450 mm and whose modulus is 44 N/m. The balls are pushed together (compressing the spring) until they are 150 mm apart. They are then released on a smooth horizontal table. What work is done on the balls in returning them to their original distance apart?

SOLUTION

The work done on the balls by the varying spring force is equal to the work done in compressing the spring from 450 to 150 mm; that is,

$$U = \int_0^{300} kx\,dx = \left[\frac{1}{2}kx^2\right]_0^{300} = \frac{1}{2}(44)(0.3)^2 = 1.98 \text{ N·m}$$

6.30. A rope is wrapped around a 10-kg solid cylinder, as shown in Fig. 6-24. Find the speed of its center G after it has moved 1.2 m down from rest.

SOLUTION

The only force doing work is the gravitational force $10 \times 9.8 = 98$ N. Its initial kinetic energy is zero. Its final kinetic energy is

$$T_2 = \frac{1}{2}m\bar{v}^2 + \frac{1}{2}\bar{I}\omega^2 = \frac{1}{2}m\bar{v}^2 + \frac{1}{2}\left(\frac{1}{2}mr^2\right)\left(\frac{\bar{v}^2}{r^2}\right) = \frac{3}{4}m\bar{v}^2 = 7.5\bar{v}^2$$

Then

$$U = 98 \times 1.2 = 7.5\bar{v}_1^2 \qquad \therefore \bar{v} = 3.96 \text{ m/s}$$

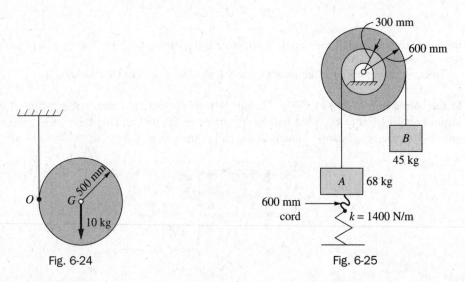

Fig. 6-24   Fig. 6-25

6.31. In Fig. 6-25, block A initially rests on the spring, to which it is connected by a 600-mm inextensible cord, which becomes taut after the system is released. What will be the stretch of the spring to bring the system to rest? The 74-kg cylinder may be considered homogeneous; it rotates in frictionless bearings.

SOLUTION

First, determine the kinetic energy of the system before the spring comes into play; i.e., while the mass A rises 600 mm, the work done equals $(45s_B - 68s_A)9.8$.

The mass B drops 1200 mm while A rises 600 mm. Hence, the kinetic energy when the spring action starts is the work done, or $T_1 = (45 \times 1.2 - 68 \times 0.6)9.8 = 129.4$ N·m. The final kinetic energy of the system is $T_2 = 0$.

The work done on the system (A and B) by gravity and by the spring as it stretches a distance x is

$$U = 45 \times 9.8\,(2x) - 68 \times 9.8\,(x) - \frac{1}{2} \times 1400(x^2) = 216x - 700x^2$$

Since $U = T_2 - T_1$,

$$216x - 700x^2 = 129.4 \qquad x^2 - 0.309x - 0.1849 = 0 \qquad \therefore x = 0.611 \text{ m}$$

6.32. In Fig. 6-26, what mass of B will cause the cylinder having an $I = 136$ kg·m^2 to attain an angular velocity of 4 rad/s after rotating counterclockwise 6 rad from rest?

SOLUTION

The work done in the counterclockwise direction is, with $\theta = 6$ rad,

$$U = W_B \times 0.6\theta - W_C \times 0.3\theta = 3.6W_B - 529$$

The system has initial kinetic energy $T_1 = 0$ and final kinetic energy

$$T_2 = \frac{1}{2} m_B v_B^2 + \frac{1}{2} I\omega^2 + \frac{1}{2} m_C v_C^2$$

But $v_B = 0.6\omega$, $v_C = 0.3\omega$, and $\omega = 4$ rad/s; hence,

$$T_2 = \frac{1}{2} m_B (2.4)^2 + \frac{1}{2} \times 136(4)^2 + \frac{1}{2} \times 30(1.2)^2 = 2.88\, m_B + 1100$$

Then using $U = T_2 - T_1$,

$$3.6 \times 9.8 m_B - 529 = 2.88 m_B + 1100 \qquad \therefore m_B = 50.3 \text{ kg}$$

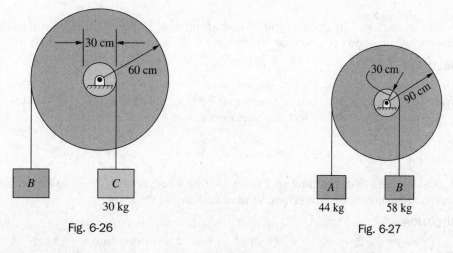

Fig. 6-26   Fig. 6-27

6.33. In Fig. 6-27, the mass of block A is 44 kg and that of block B is 58 kg. The drum has a moment of inertia $\bar{I} = 16$ kg·m^2. Through what distance will A fall before it reaches a speed of 2 m/s?

SOLUTION

The work done with $s_B = s_A/3$ is

$$U = (44 s_A - 58 s_B)(9.8) = 242 s_A$$

The system has initial kinetic energy $T_1 = 0$ and final kinetic energy

$$T_2 = \frac{1}{2} \times 44 \times 2^2 + \frac{1}{2} \times 16 \times 2.22^2 + \frac{1}{2} \times 58 \times 0.667^2 = 140.3$$

since $v_A = 2$ m/s, $v_B = 0.667$ m/s, and $\omega = v_A/r = 2/0.9 = 2.22$ rad/s.
Then

$$242 s_A = 140.3 \qquad \therefore s_A = 0.580 \text{ m}$$

6.34. The 7-kg block shown in Fig. 6-28 is released from rest and slides a distance s down the inclined plane. It strikes the spring, which it compresses 75 mm before motion impends up the plane. Assuming that the coefficient of friction is 0.25 and that the spring constant $k = 2800$ N/m, determine the value of s.

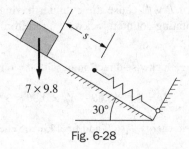

7×9.8

$30°$

Fig. 6-28

SOLUTION

The initial kinetic energy and the final kinetic energy (when the block has moved $s + 0.075$ m) are zero. Hence, the work done by friction, gravity, and the spring must be zero. The normal reaction, $9.8 \times 7 \cos 30° = 59.4$ N, does no work. The friction $= 0.25 \times 59.4 = 14.85$ N. The component of the gravitational force along the plane $= 9.8 \times 7 \sin 30° = 34.3$ N. Each of these forces does work for $s + 0.075$ m; frictional work is negative, the other is positive.

The work of the spring is negative and equals $\frac{1}{2} k(0.075)^2 = 7.88$ N·m. Hence,

$$U = (34.3 - 14.85)(s + 0.075) - 7.88 = 0 \qquad \therefore s = 0.33 \text{ m}$$

6.35. A 5-kg mass drops 2 m upon a spring whose modulus is 10 000 N/m. What will be the speed of the block when the spring is deformed 100 mm?

SOLUTION

The mass drops $(2 + 0.1)$ m $= 2.1$ m. The work done by gravity is $9.8 \times 5 \times 2.1 = 102.9$ N·m. The work done by the spring on the mass is negative and equals $\frac{1}{2} kx^2 = \frac{1}{2}(10\,000 \text{ N/m})(0.1 \text{ m})^2 = 50$ N·m.

The kinetic energy of the block increases from zero to $\frac{1}{2} mv^2 = \frac{1}{2}(5)v^2 = 2.5v^2$:

$$U = T_2 - 0 \qquad 102.9 - 50 = 2.5v^2 \qquad \therefore v = 4.6 \text{ m/s}$$

6.36. A mass dropped from rest through 2 m on a spring whose modulus is 3500 N/m causes a maximum shortening in the spring of 200 mm. What is the value of the mass?

SOLUTION

Work done by gravity $= mg(2 + 0.2) = 21.56$ m. Work done by the spring $= -\frac{1}{2} kx^2 = -\frac{1}{2} \times 3500 \times 0.2^2 = -70$ N·m. Since the block starts from rest and ends at rest, the change in kinetic energy is zero. The $U = 0$, or

$$21.56m - 70 = 0 \qquad \therefore m = 3.25 \text{ kg}$$

6.37. Determine the speed of escape, i.e., the initial speed, that must be given to a particle on the earth's surface to project it to an infinite height.

SOLUTION

The particle of weight W is shown in Fig. 6-29 at a distance x from the center of the earth of radius R. The earth's attraction F is known to be inversely proportional to the square of the distance x; that is, $F = -C/x^2$. To determine C, note that the attraction on the earth's surface is the weight W. Thus, $-W = -C/R^2$, $C = WR^2$, and hence $F = -WR^2/x^2$.

The work done in going from $x = R$ to $x = \infty$ is

$$\int_R^\infty F\,dx = \int_R^\infty -\left(\frac{WR^2}{x^2}\right)dx = WR^2\left[\frac{1}{x}\right]_R^\infty = -WR$$

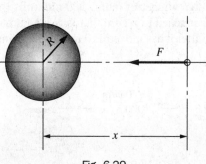

Fig. 6-29

The work done equals the change in kinetic energy. $T_1 = \frac{1}{2}(W/g)v_0^2$ and $T_2 = 0$ (since $v = 0$ when x becomes infinite). Hence,

$$-WR = -\frac{1}{2}\left(\frac{W}{g}\right)v_0^2 \qquad \text{or} \qquad v_0 = \sqrt{2gR}$$

Assuming the diameter of the earth to be 12 700 km, the required speed of escape is calculated to be

$$v_0 = \sqrt{2 \times 9.8 \times 12.7 \times 10^6} = 15\,780 \text{ m/s} \qquad \text{or} \qquad 56\,800 \text{ km/h}$$

6.38. Determine the spring constant k in Fig. 6-30 such that the slender rod AB just reaches the vertical down position when it is released from rest in the horizontal position shown. The spring is stretched 25 mm in the position shown. The mass of the bar is 3.6 kg.

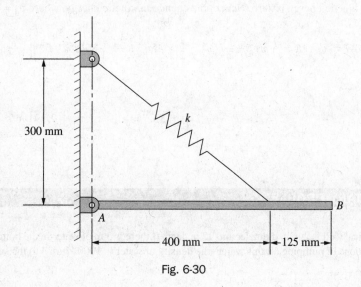

Fig. 6-30

SOLUTION

The unstretched length of the spring is 475 mm. In the vertical position, the spring is stretched $700 - 475 = 225$ mm. The work done by the spring is

$$-\frac{1}{2}k(s_2^2 - s_1^2) = -\frac{1}{2}k(0.225^2 - 0.025^2) = -0.025k \text{ N·m}$$

The work done by the gravity force is

$$mgh = 3.6 \times 9.8 \times \frac{0.525}{2} = 9.26 \text{ N·m}$$

Since the bar starts from rest and ends at rest, the change in the kinetic energy is zero. Hence,

$$U = T_2 - T_1 \qquad -0.025k + 9.26 = 0 \qquad \therefore k = 370 \text{ N/m}$$

6.39. Refer to Fig. 6-31. A 45-kg cylinder of radius 300 mm rolls without slipping under the action of an 360-N force. A spring is attached to a cord that is wound around the cylinder. What is the speed of the center of the cylinder after it has moved 150 mm? The spring is unstretched when the 360-N force is applied.

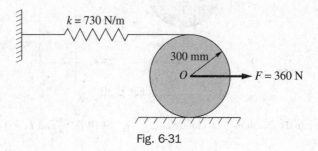

Fig. 6-31

SOLUTION

Since the cylinder rolls without slipping, the spring becomes stretched 300 mm when the center of the cylinder moves 150 mm to the right. Noting that the friction force and normal force under the cylinder do no work (see Problem 6.6), the work is

$$U = -\frac{1}{2}730(s_2^2 - s_1^2) + Fs = -\frac{1}{2}(730)(0.3^2 - 0^2) + 360 \times 0.15 = 21.1 \text{ N·m}$$

The initial kinetic energy is zero. Hence, the change in kinetic energy, where $v_0 = 0.3\omega$ for no slip, is

$$\Delta T = T_2 - T_1 = \frac{1}{2}mv_0^2 + \frac{1}{2}I_o\omega^2 = \frac{1}{2} \times 45v_0^2 + \frac{1}{2}\left(\frac{1}{2} \times 45 \times 0.3^2\right)\left(\frac{v_0}{0.3}\right)^2 = 33.15\,v_0^2$$

Hence,

$$U = T_2 - T_1 \qquad 21.1 = 33.75v_0^2 \qquad \therefore v_0 = 0.791 \text{ m/s}$$

SUPPLEMENTARY PROBLEMS

6.40. A cylindrical well is 2 m in diameter and 12 m deep. If there is 3 m of water in the bottom of the well, determine the work done in pumping all this water (the density of water is 1000 kg/m^3) to the surface.

Ans. 970 kN·m

6.41. Referring to Problem 6.40, what work must be done by a pump that is 60 percent efficient?

Ans. 1620 kN·m

6.42. A person who can push 300 N wishes to roll a 90-kg barrel into a truck that is 900 mm above the ground. How long a board must be used and how much work will the person do in getting the barrel into the truck?

Ans. 2.65 m, 794 N·m

6.43. A 4.5-kg block slides 1.2 m on a horizontal surface. (*a*) If the coefficient of friction is 0.3, what work is done by the block on the surface? (*b*) What work is done by the surface on the block?

Ans. (*a*) $U = 0$, (*b*) $U = 15.9$ N·m

6.44. The 8-kg block is acted upon by a 100-N force as shown in Fig. 6-32. If the coefficient of sliding friction is 0.30, determine the work done by all forces as the block moves 4 m to the right.

Ans. $U = 282$ N·m

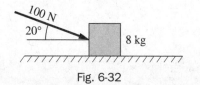

Fig. 6-32

6.45. A 4.5-kg block slides 1.8 m down a plane inclined 40° with the horizontal. Determine the work done by all forces acting on the block. The coefficient of sliding friction is 0.40.

Ans. $U = 26.7$ N·m

6.46. A particle moves along the path $x = 2t$, $y = t^3$, where t is in seconds and distances are in meters. What work is done in the interval from $t = 0$ to $t = 3$ s by a force whose components are $F_x = 2 + t$ and $F_y = 2t^2$? Forces are in newtons.

Ans. $U = 313$ N·m

6.47. A bead that weighs 0.6 N is raised slowly along a frictionless wire, composed of two circular arcs, from A to B as shown in Fig. 6-33. What work is done?

Ans. $U = 0.72$ N·m

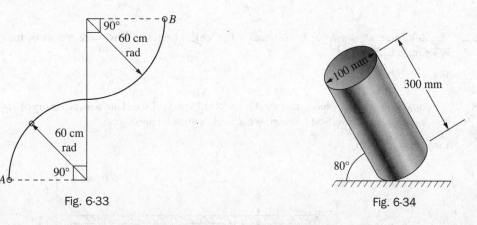

Fig. 6-33   Fig. 6-34

6.48. A freestanding crane has a horizontal boom 75 m long that is 120 m above the ground. If the crane is slowly lifting 35.6 kN of concrete up a distance of 100 m, what work is done? During this lifting, what is the moment that tends to overturn the crane?

Ans. $U = 3560$ kN·m, $M = 2670$ N·m

6.49. The 20-kg solid cylinder shown in Fig. 6-34 is released from rest. Determine the work done by the earth's pull when the bottom hits the floor.

Ans. $U = 1.25$ N·m

6.50. The empty cylindrical tank A in Fig. 6-35 is filled with water by lifting it with a pump from the cubical tank B. Water has a density of 1000 kg/m³. Assume that B was filled at the beginning. How much work is done?

Ans. $U = 176$ kN·m

6.51. The 16-kg mass in Fig. 6-36 drops 2.5 m. It is attached by a light rope to a drum that rotates in frictionless bearings. A constant torque $M = 80$ N·m is supplied to the drum. What work is done on the system?

Ans. $U = 59$ N·m

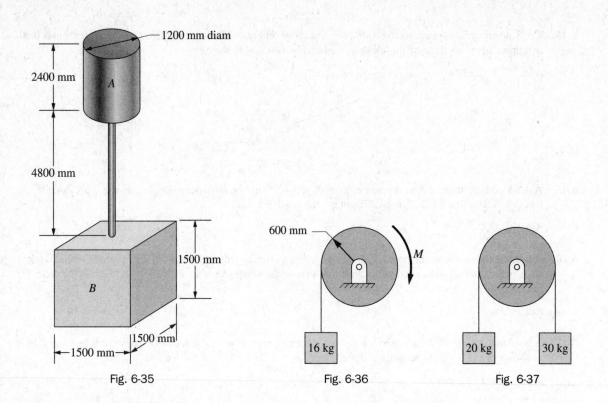

Fig. 6-35   Fig. 6-36   Fig. 6-37

6.52. The 40-kg drum shown in Fig. 6-37 rotates in frictionless bearings. What work is done on the system when the 30-kg mass falls 1m?

Ans. $U = 98$ N·m

6.53. The homogeneous plane object shown in Fig. 6-38 is 25 mm thick and has a mass density of 7840 kg/m³. It falls from a horizontal to a vertical position. What work is done on the object?

Ans. $U = 138$ N·m

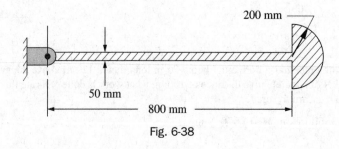

Fig. 6-38

6.54. A couple $M = 2\theta^3 - \theta$ is applied to a shaft that rotates from $\theta = 0°$ to $\theta = 90°$. Determine the work done if M is in N·m.

Ans. $U = 1.81$ N·m

6.55. A force of 30 N will stretch an elastic cord 250 mm. If the force required to stretch the cord varies directly as the deformation, what is the work done in stretching the cord 1500 mm?

Ans.   135 N·m

6.56. A force of 100 N is required to compress a spring through a distance of 100 mm. If the force required to compress the spring varies directly with its deformation, how much work is done in compressing it through 225 mm?

Ans.   25.3 N·m

6.57. At a certain instant during acceleration along a level track, the drawbar pull of a locomotive is 90 kN. What power is being developed if the speed of the train is 60 km/h?

Ans.   1.5 MW

6.58. A 1200-kg automobile climbs a 10 percent grade at a uniform rate of 24 km/h. If the resistance is 100 N/metric ton, what horsepower is the car developing?

Ans.   11.6 hp

6.59. A steam engine raises an 1800-kg mass vertically at the rate of 9 m/s. What is the power of the engine, assuming an efficiency of 70 percent?

Ans.   227 kW

6.60. The power measured by a Prony brake attached to an engine flywheel is 6.3 brake horsepower (bhp). The indicated horsepower (ihp), as measured by means of indicator cards, is 7.1. What is the efficiency of the engine?

Ans.   Eff = 89 percent

6.61. What horsepower is required to raise a 500-kg mass to a height of 2.4 m in 4 s?

Ans.   3.94 hp

6.62. A 50-kg homogeneous cylinder 1200 mm in diameter is rotating at 100 rpm. A torque of 30 N·m is needed to keep this speed constant (overcoming friction). What power is required?

Ans.   314 W

6.63. A 150-mm-diameter pulley is rotating at 2000 rpm. The belt driving it has tensions of 5 and 15 N in the slack and tight sides, respectively. What horsepower is being delivered to the pulley?

Ans.   0.211 hp

6.64. A 0.1 kg body falls 1500 mm to the surface of the earth. What is its kinetic energy as it hits the ground?

Ans. $T = 1.47$ N·m

6.65. A 50-kg body starts from rest and is pulled along the ground by a horizontal force of 300 N. If the coefficient of friction is 0.1 and the force acts for a distance of 2 m and then ceases to act, determine the distance required for the body to come to rest.

Ans.   10.2 m

6.66. A bullet enters a 50-mm-thick plank with a speed of 600 m/s and leaves with a speed of 250 m/s. Estimate the greatest thickness of plank that could be penetrated by the same bullet.

Ans.   60.5 mm

6.67. If the block in Problem 6.5 starts from rest, what is its speed after traversing the 6 m?

Ans.   9.6 m/s

6.68. A 2-kg block slides down a plane inclined 50° with the horizontal. The coefficient of friction between the block and the plane is 0.25. Determine the speed of the block after it has moved 4 m along the plane starting with a velocity of 2 m/s.

Ans. $v = 7.17$ m/s

6.69. A 40-kg block is acted upon by a 150-N horizontal force. The coefficient of friction is 0.25. Determine the speed of the block after it has moved 6 m from rest. Refer to Fig. 6-39.

Ans. $v = 3.95$ m/s

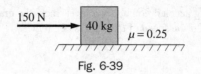

Fig. 6-39

6.70. Determine the kinetic energy possessed by a 100-kg disk that is 500 mm in diameter, 75 mm thick, and rotating at 100 rpm about its center.

Ans.   171 N·m

6.71. A 50-kg, 150-mm-diameter sphere rotates at 120 rpm about an axis 400 mm from its center. What is the kinetic energy of rotation?

Ans.   632 N·m

6.72. A 10-kg cylinder with a radius of 200 mm is rolling on a flat surface with a mass center speed of 2 m/s. What is its kinetic energy?

Ans. $T = 30$ N·m

6.73. A 4-kg sphere has a radius of 1000 mm and a radius of gyration of 600 mm. It is rolling on a horizontal plane with angular speed 3 rad/s. What is the kinetic energy of the sphere?

Ans. $T = 24.5$ N·m

6.74. In Fig. 6-40, the angular velocity of crank CD is 10 rad/s clockwise. The masses of the slender bars are as follows: AB is 2 kg, BC is 4 kg, and CD is 3 kg. Determine the kinetic energy of the system.

Ans.   456 N·m

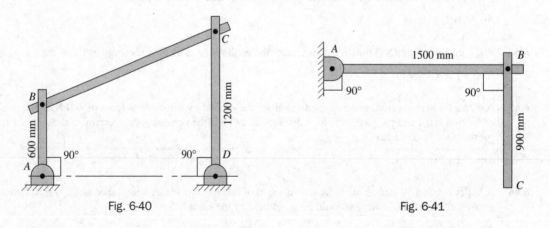

Fig. 6-40   Fig. 6-41

6.75. The slender rods AB and BC in Fig. 6-41 have masses of 5 and 3 kg, respectively. AB rotates with an angular velocity of 8 rad/s clockwise, while BC has an angular velocity of 6 rad/s counterclockwise. Determine the kinetic energy of the system.

Ans.   351 N·m

6.76. A 900-kg solid cylindrical flywheel is 1200 mm in diameter. If the axle is 150 mm in diameter and the coefficient of journal friction is 0.15, find the time required for the flywheel to coast to rest from a speed of 500 rpm.

Ans.   85.5 s

6.77. An electric motor has a 10-kg rotor with radius of gyration $k = 48$ mm. A frictional torque of 0.06 N·m is present. How many revolutions will the rotor make while coming to rest from a speed of 1800 rpm?

Ans. $\theta = 1086$ rev

6.78. A drum rotating 20 rpm is lifting a 1-metric-ton cage connected to it by a cable as shown in Fig. 6-42. If power is cut off, how high will the cage rise before coming to rest? Assume frictionless bearings.

Ans. $h = 102$ mm

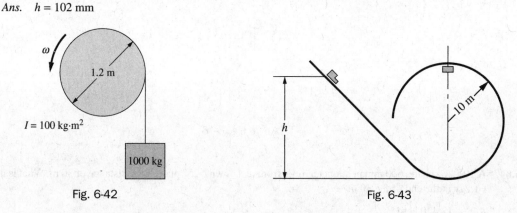

Fig. 6-42   Fig. 6-43

6.79. A car rolls from rest without any friction down the smooth incline and then moves inside the loop shown in Fig. 6-43. Determine the least value of h so that the car will remain in contact with the track.

Ans. $h = 25$ m

6.80. A 4-kg homogeneous slender bar is 1 m long. It pivots about one end. When released from rest in the horizontal position, it falls under the action of gravity and a constant retarding torque of 5 N·m. What will be its angular speed as it passes through its lowest position?

Ans. $\omega = 4.19$ rad/s

6.81. The 30-kg drum in Fig. 6-44 has a radius of gyration $k = 800$ mm. Assume no friction and determine the speed of the drum after it has made one revolution starting from rest.

Ans. $\omega = 3.35$ rad/s

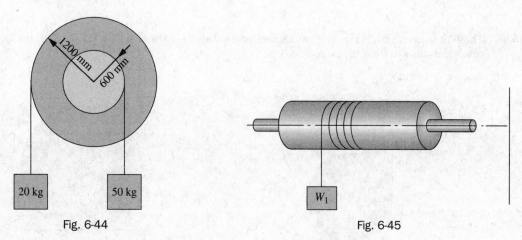

Fig. 6-44   Fig. 6-45

6.82. A well bucket of weight W_1 is attached to a windlass of weight W_2 and radius of gyration k. See Fig. 6-45. If r is the radius of the windlass, how long will it take the bucket to drop a distance s from rest to the water level? Neglect bearing friction and the weight of the rope.

Ans. $t = \sqrt{2s(1 + W_2k^2/W_1r^2)/g}$

6.83. A homogeneous bar of length L is pivoted about a point a distance a from one end, as shown in Fig. 6-46. If the bar is released from rest in the 30° position, what will be the angular speed when the bar is vertical?

Ans. $\omega^2 = \dfrac{0.402g(L - 2a)}{L^2 - 3La + 3a^2}$

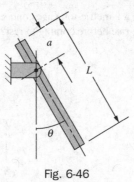

Fig. 6-46

6.84. A 100-kg sphere, 600 mm in diameter, rolls from rest down a 25° plane for a distance of 30 m. What is its kinetic energy at the end of the 30 m?

Ans. 12 420 N·m

6.85. In Problem 6.84, what is the speed of the center of the sphere after it has traveled the 30 m?

Ans. 13.3 m/s

6.86. A sphere rolls a distance s down a plane inclined at angle θ with the horizontal. What is the speed of the sphere if it starts from rest?

Ans. $v = 6.78\sqrt{s \sin \theta}$

6.87. A car has a 900-kg body, four 20-kg wheels, and a 70-kg driver. The wheels are 700 mm in diameter and have a radius of gyration $k = 300$ mm. What will be the speed of the car if it moves from rest 300 m down a 5 percent grade? Neglect any drag.

Ans. 60 km/h

6.88. The 4-kg sphere shown in Fig. 6-47 has a string wrapped around a slot as shown. What will be the speed of the center if it falls 1 m from the rest position?

Ans. $v = 3.74$ m/s

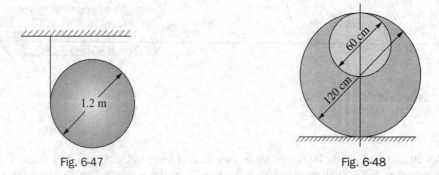

Fig. 6-47   Fig. 6-48

6.89. A 2-kg disk 60 cm in diameter is attached rigidly to an 8-kg disk 120 cm in diameter. If the assembly is released from rest, as shown in Fig. 6-48, what will be the angular speed when the small disk is at the bottom of its travel?

Ans. $\omega = 2.26$ rad/s

6.90. In Fig. 6-49, A has a mass of 7 kg and B has a mass of 4 kg. If B falls 400 mm from rest, determine its speed (*a*) if no friction exists and (*b*) if the coefficient of friction between A and the horizontal plane is 0.20.

Ans. (*a*) $v = 1.69$ m/s, (*b*) $v = 1.36$ m/s

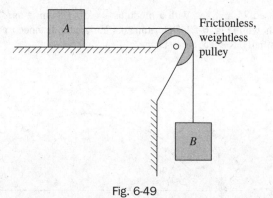

Fig. 6-49

6.91. In the system shown in Fig. 6-50, all ropes are vertical. If the 20-kg weight rises 480 mm from rest, what will be its speed?

Ans. $v = 0.972$ m/s

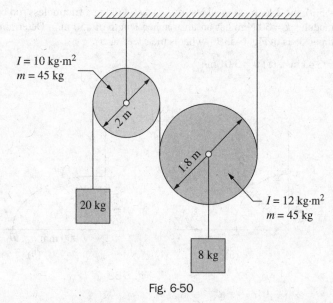

Fig. 6-50

6.92. A 36-Mg freight car moving 8 km/h horizontally hits a bumper with a spring constant of 1750 N/mm. What will be the maximum compression of the spring?

Ans. $d = 320$ mm

6.93. An 8-kg weight slides 150 mm from rest down the 25° plane, shown in Fig. 6-51, where it hits a spring whose modulus is 1800 N/m. The coefficient of friction is 0.20. Determine the maximum compression of the spring.

Ans. $d = 67.6$ mm

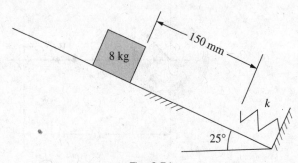

Fig. 6-51

6.94. A spring compressed 75 mm and with a modulus $k = 5000$ N/m is used to propel a 0.05-kg mass from the frictionless tube shown in Fig. 6-52. Determine the horizontal distance r at which the mass will be at the same height as it was initially. Neglect air resistance.

     Ans. $r = 55.4$ m

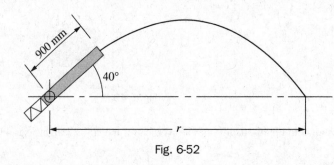

Fig. 6-52

6.95. (*a*) In Fig. 6-53(*a*), the 2-kg mass m slides from rest at A along a frictionless rod bent into a quarter circle. The spring with modulus $k = 30$ N/m has an unstretched length of 450 mm. Determine the speed of m at B. (*b*) If the path is elliptical, as in Fig. 6-53(*b*), what is the speed at B?

     Ans.   (*a*) $v = 3.43$ m/s, (*b*) $v = 3.01$ m/s

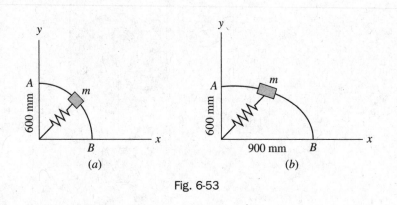

Fig. 6-53

6.96. The pull of the earth on an object varies inversely with the square of the distance from the center of the earth. If W is the weight of the object on the earth's surface (radius R), then the pull at distance ρ is WR^2/ρ^2. (*a*) What work must be done against this gravitational pull to move the particle from the earth's surface to a distance x from the earth's center? (*b*) To infinity? See Fig. 6-54.

     Ans.   (*a*) $U = WR - WR^2/x$, (*b*) $U = WR$

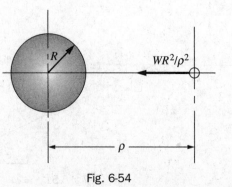

Fig. 6-54

6.97. The 150-kg wheel shown in Fig. 6-55 has a radius of gyration of 360 mm with respect to its center of mass *G*. In the initial position shown, the velocity of *G* is 2 m/s down the plane, and the spring is stretched 150 mm. If the spring modulus is 1000 N/m, what will be the maximum stretch of the spring?

Ans.   1.98 m

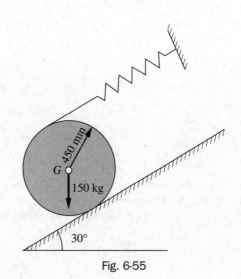

Fig. 6-55

6.98. In Problem 6.38, the bar is started, when it is vertically up and the spring is unstretched, with an angular velocity of 2 rad/s clockwise. What will be the angular velocity when the bar rotates through 180°?

Ans. $\omega = 7.63$ rad/s

6.99. The circular disk in Fig. 6-56 has a mass of 3 kg, and the slender bar *AB* has a mass of 8 kg. The initial angular velocity of the disk is 6 rad/s as shown. What is the value of the moment *M* that will stop the disk when it has rotated 90° counterclockwise? Assume that the roller at *B* is massless.

Ans.   0.54 N·m counterclockwise

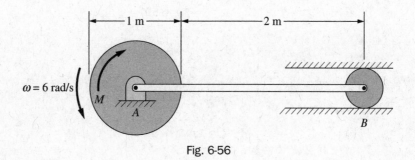

Fig. 6-56

6.100. The plank *AB* in Fig. 6-57 has a mass of 7 kg. The solid cylindrical rollers *D* and *E* are each of mass 5 kg and are 0.5 m in diameter. The plank is released from rest with roller *D* under the end *A* and roller *E* under the mass center *C*. Assuming no slip, determine the velocity of the plank when roller *E* is under end *B*.

Ans.   7.11 m/s

6.101. In Fig. 6-58, the 5-kg slender rod *AB* is pinned at *A* to a 3-kg uniform disk. The rod rests on the horizontal plane at *B*. If the system is released from rest in the position shown, what will be the velocity of the center of the disk when the bar is horizontal? The disk rolls without slipping, and the friction under the bar at *B* can be neglected.

Ans. $v_0 = 1.81$ m/s to the right

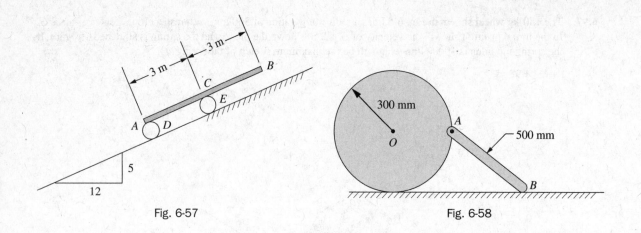

Fig. 6-57   Fig. 6-58

6.102. In Fig. 6-59, gear *A* drives gear *B* under the action of a clockwise moment $M = 28$ N·m. If the two gears are at rest when the moment is applied, what is the angular velocity of gear *B* when gear *A* has turned through 4 rev? Assume the gears *A* and *B* to be disks of mass 3.6 and 14.4 kg, respectively.

Ans. $\omega_B = 27$ rad/s

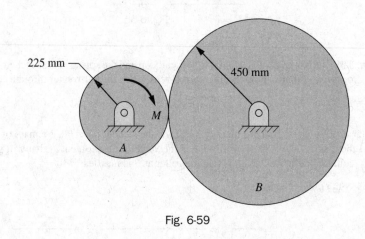

Fig. 6-59

CHAPTER 7

Impulse and Momentum

7.1 Impulse-Momentum Relation for a Particle

The momentum of a particle was defined in Section 3.1 as

$$\mathbf{L} = m\mathbf{v} \tag{1}$$

where m = mass of particle
$\quad \mathbf{v}$ = velocity of particle

The vector sum of the external forces acting on a particle is equal to the time rate of change of the momentum $\mathbf{L}$:

$$\sum \mathbf{F} = \frac{d(m\mathbf{v})}{dt} = \frac{d\mathbf{L}}{dt} = \dot{\mathbf{L}} \tag{2}$$

Integration over the time interval during which the velocity of the particle changes from $\mathbf{v}_1$ to $\mathbf{v}_2$ results in the expression

$$\int_{t_1}^{t_2} \sum \mathbf{F}\, dt = \int_{\mathbf{L}_1}^{\mathbf{L}_2} d\mathbf{L} = \mathbf{L}_2 - \mathbf{L}_1 = m\mathbf{v}_2 - m\mathbf{v}_1 \tag{3}$$

The left-hand side of Eq. (3) is called the *impulse* $\mathbf{I}$ of the resultant force ($\sum \mathbf{F}$) during the time interval from t_1 to t_2. Thus, the impulse is equal to the change in momentum during this time interval, i.e.,

$$\mathbf{I} = \mathbf{L}_2 - \mathbf{L}_1 \tag{4}$$

7.2 Impulse-Momentum Relation for an Assemblage of Particles

The vector sum of the external forces acting on an assemblage of n particles equals the time rate of change of the momentum of a mass m that is equal to the sum of the masses of the n particles and that possesses a velocity equal to that of the mass center of the n particles:

$$\sum \mathbf{F} = \frac{d(m\bar{\mathbf{v}})}{dt} = \frac{d\bar{\mathbf{L}}}{dt} \tag{5}$$

where $\sum \mathbf{F}$ = sum of external forces acting on group of particles
$\quad m$ = mass of all n particles
$\quad \bar{\mathbf{v}}$ = velocity of mass center of group of n particles

As in the case of one particle, the above equation may be integrated as

$$\mathbf{I} = \int_{t_1}^{t_2} \sum \mathbf{F}\,dt = \int_{\bar{\mathbf{v}}_1}^{\bar{\mathbf{v}}_2} d(m\bar{\mathbf{v}}) = m\bar{\mathbf{v}}_2 - m\bar{\mathbf{v}}_1 = \Delta\bar{\mathbf{L}} \tag{6}$$

This states that the impulse $\mathbf{I}$ of all the forces acting in the stated time interval is equal to the change in momentum of a mass m, as stated in the first sentence of this section. Note that the vector $\Delta\bar{\mathbf{L}}$ does not, in general, pass through the mass center of the assemblage of particles.

7.3 Angular Momentum

The *angular momentum* $\mathbf{H}_O$ (also called *moment of momentum*) is the moment about any point O of the linear momentum vector $\mathbf{L}$. In Fig. 7-1, O can be any point, fixed or moving. Thus,

$$\mathbf{H}_O = \boldsymbol{\rho} \times \mathbf{L} = \boldsymbol{\rho} \times (m\mathbf{v}) \tag{7}$$

where $\boldsymbol{\rho}$ = radius vector of particle P relative to O
$\quad\;\; \mathbf{v}$ = absolute velocity of P (tangent to path)

The sum of the moments about a *fixed point* O of the external forces acting on a *particle* is equal to the time rate of change of the angular momentum $\mathbf{H}_O$; that is,

$$\sum \mathbf{M}_O = \frac{d\mathbf{H}_O}{dt} = \dot{\mathbf{H}}_O \tag{8}$$

From Eq. (7), $\mathbf{H}_O = \mathbf{r} \times (m\mathbf{v})$, where $\mathbf{r}$ is the radius vector in a newtonian frame of particle m and $\mathbf{v}$ = absolute velocity of the particle. Taking the time derivative,

$$\frac{d\mathbf{H}_O}{dt} = \dot{\mathbf{r}} \times (m\mathbf{v}) + \mathbf{r} \times (m\dot{\mathbf{v}})$$

Since $\dot{\mathbf{r}} = \mathbf{v}$, $\mathbf{v} \times \mathbf{v} = 0$, $m\dot{\mathbf{v}} = m\mathbf{a} = \sum \mathbf{F}$, and $\mathbf{r} \times (m\mathbf{a}) = \mathbf{r} \times (\sum \mathbf{F}) = \sum \mathbf{M}_O$, we obtain $d\mathbf{H}_O/dt = \sum \mathbf{M}_O$.
Equation (8) may be integrated as follows:

$$\int_{t_1}^{t_2} \sum \mathbf{M}_O\,dt = \int_{\mathbf{H}_1}^{\mathbf{H}_2} d\mathbf{H}_O = \mathbf{H}_2 - \mathbf{H}_1 = \mathbf{r} \times (m\mathbf{v}_2 - m\mathbf{v}_1) \tag{9}$$

The integral on the left is the *angular impulse* acting throughout the time interval t_1 to t_2, and the right-hand side of Eq. (9) is the change that occurs in angular momentum during this time interval.

Equation (8) can be applied as follows to an assemblage of particles. The sum of the moments about point O of the external forces acting on an *assemblage* of n particles is equal to the time rate of change of the angular momentum $\mathbf{H}_O$ about this point O, only if (a) point O is at rest, or (b) the mass center of the n particles is at rest, or (c) the velocities of O and the mass center are parallel (certainly true if O is the mass center). See Problem 7.1, which proves Eq. (8) holds for an assemblage of particles.

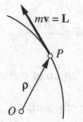

Fig. 7-1 Vectors used to define angular momentum $\mathbf{H}_O$.

7.4 Relative Angular Momentum

The relative angular momentum $\mathbf{H}'_O$ is the moment about any point O of the product of the mass of the particle and the time rate of change of the radius vector $\boldsymbol{\rho}$ of the particle relative to O (see Fig. 7-2):

$$\mathbf{H}'_O = \boldsymbol{\rho} \times (m\dot{\boldsymbol{\rho}}) \tag{10}$$

where $\boldsymbol{\rho}$ = radius vector of particle P relative to O
$\qquad \dot{\boldsymbol{\rho}}$ = time rate of change of $\boldsymbol{\rho}$

The sum of the moments about O of the external forces acting on an assemblage of n particles is equal to the time rate of change of the relative angular momentum $\mathbf{H}'_O$ about this point O, i.e.,

$$\sum \mathbf{M}_O = \frac{d\mathbf{H}'_O}{dt} \tag{11}$$

only if (*a*) O is the mass center of the n particles, or (*b*) O is a point of constant velocity (or at rest), or (*c*) O is a point with an acceleration vector that passes through the mass center (see Problems 7.2 and 7.3).

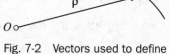

Fig. 7-2 Vectors used to define relative angular momentum $\mathbf{H}'_O$.

7.5 Corresponding Scalar Equations

For a body in translation (all particles have the same velocity), Eq. (5) can be replaced by the scalar equations

$$\sum (\text{Imp})_x = \Delta L_x = m(v'_x - v_x) \tag{12}$$

$$\sum (\text{Imp})_y = \Delta L_y = m(v'_y - v_y) \tag{13}$$

where $\sum(\text{Imp})_x$, $\sum(\text{Imp})_y$ = impulses of external forces in x and y directions
$\qquad m$ = mass of body
$\qquad v'_x, v'_y$ = final velocities of body in x and y directions
$\qquad v_x, v_y$ = initial velocities of body in x and y directions

For a body in rotation about a fixed axis, the above equations become

$$\sum (\text{Ang Imp})_O = \Delta H_O = I_O(\omega' - \omega) \tag{14}$$

where $\sum(\text{Ang Imp})_O$ = angular impulse of external forces about axis of rotation through O
$\qquad I_O$ = moment of inertia of body about axis of rotation
$\qquad \omega'$ = final angular velocity of body
$\qquad \omega$ = initial angular velocity of body

For proof see Problem 7.4.

For a body in plane motion, the above equations become

$$\sum (\text{Imp})_x = \Delta L_x = m(\bar{v}'_x - \bar{v}_x) \tag{15}$$

$$\sum (\text{Imp})_y = \Delta L_y = m(\bar{v}'_y - \bar{v}_y) \tag{16}$$

$$\sum (\text{Ang Imp})_G = \Delta \bar{H} = \bar{I}(\omega' - \omega) \tag{17}$$

where $\sum(\text{Imp})_x, \sum(\text{Imp})_y$ = impulses of external forces in x and y directions

$\dot{m}$ = mass of body

$\bar{v}'_x, \bar{v}'_y$ = final velocities of mass center in x and y directions

$\bar{v}_x, \bar{v}_y$ = initial velocities of mass center in x and y directions

$\sum(\text{Ang Imp})_G$ = angular impulse of external forces about axis through mass center G

$\bar{I}$ = moment of inertia of body about mass center G

ω' = final angular velocity of body

ω = initial angular velocity of body

For proof, see Problem 7.5.

Alternatively, for a body in general plane motion, if the axis of angular momentum is not the center of mass, the scalar angular momentum becomes

$$H_O = I_O \omega + m \bar{x} v_{Oy} - m \bar{y} v_{Ox}$$

where H_O = angular momentum about an axis through O

I_O = moment of inertia about an axis through O

ω = angular velocity of body

m = mass of body

$\bar{x}, \bar{y}$ = coordinates of mass center

v_{Ox}, v_{Oy} = components of velocity of axis O

This formulation of the angular momentum is of particular value in problems involving eccentric collisions.

7.6   Units

Unit	SI	U.S. Customary
Mass	kg	slug = lb-s^2/ft
Impulse	N·s	lb-s
Momentum	kg·m/s = N·s	slug-ft/s = lb-s
Angular impulse	N·m·s	lb-s-ft
Angular momentum	(kg·m^2)(rad/s) = N·m·s	(slug-ft^2)(rad/s) = lb-s-ft

7.7   Conservation of Momentum

Conservation of momentum in a given direction occurs if the sum of the external forces in that direction is zero. This follows because there is then no impulse in that direction, and hence no change in momentum can occur.

7.8   Conservation of Angular Momentum

Conservation of angular momentum about an axis occurs if the sum of the moments of the external forces about that axis is zero. This follows because there is then no angular impulse about that axis, and hence no change in angular momentum can occur.

7.9   Impact

Impact covers the cases where the time intervals during which the forces act are quite small and usually indeterminate. The surfaces of two colliding bodies have a common normal, which is the line of impact.

(*a*) Direct impact occurs if the initial velocities of the two colliding bodies are along the line of impact.

(*b*) Direct central impact occurs if the mass centers in (*a*) are also along the line of impact.

(*c*) Direct eccentric impact occurs if the initial velocities are parallel to the normal to the striking surfaces but are not collinear.

(*d*) Oblique impact occurs if the initial velocities are not along the line of impact.

In direct central impact of the two bodies, the *coefficient of restitution* is the ratio of the relative velocity of separation of the two bodies to their relative velocity of approach. Thus,

$$e = \frac{v_2 - v_1}{u_1 - u_2} = -\frac{v_2 - v_1}{u_2 - u_1} \tag{18}$$

where e = coefficient of restitution

u_1, u_2 = velocities of bodies 1 and 2, respectively, before impact ($u_1 > u_2$ for collision to occur if both are moving in same direction)

v_1, v_2 = velocities of bodies 1 and 2, respectively, after impact

Note: When the impact is oblique, the *normal* components of the velocities are used in the above formula.

Since during impact the same force acts on each body (equal and opposite reaction), the sum of the momenta before impact must equal the sum of the momenta after impact, i.e., momentum is conserved. This relation is expressed as

$$m_1 u_1 + m_2 u_2 = m_1 v_1 + m_2 v_2 \tag{19}$$

7.10 Variable Mass

Suppose at time t a mass m is moving along a straight line with an absolute speed v (see Fig. 7-3). Further suppose a mass dm immediately in front of mass m is moving along the same straight line with absolute speed u. If the mass m absorbs the mass dm in a time interval dt, then the combined mass $m + dm$ will move with speed $v + dv$.

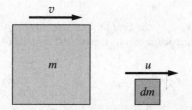

Fig. 7-3   Combination of two masses.

The momentum of the system at time t is $mv + dm\, u$. The momentum of the system at time $t + dt$ is $(m + dm)(v + dv)$. The change in momentum is then

$$dL = (m + dm)(v + dv) - (mv + dm\, u)$$

$$= mv + m\, dv + dm\, v + dm\, dv - mv - dm\, u$$

Since the magnitude of $dm\, dv$ is of second order, we shall drop the term. Dividing by dt,

$$\frac{dL}{dt} = m\frac{dv}{dt} + \frac{dm}{dt}(v - u) \tag{20}$$

If the mass is released (decrease in mass), then dm/dt will be negative.

The above formula was shown for straight-line motion, but is of a more general nature.

Since the sum of the forces acting equals the time rate of change of momentum,

$$\sum F = \frac{dL}{dt} = m\frac{dv}{dt} + \frac{dm}{dt}(v - u) \tag{21}$$

7.1. Given an assemblage of n particles of masses m_1, m_2, m_3, ..., m_n, show that the sum of the moments about point O of the external forces equals the time rate of change of the angular momentum of the group of particles about this point O only if (*a*) point O is at rest, or (*b*) the mass center of the n particles is at rest, or (*c*) the velocities of O and the mass center are parallel (which is certainly true if O is the mass center).

SOLUTION

Figure 7-4 shows the ith particle of the group with mass m_i. The given point O has a position vector $\mathbf{r}_O$ relative to a newtonian frame of reference: O' is fixed. The position vector of P, relative to the fixed frame, is $\mathbf{r}_P$. Let $\mathbf{H}_O$ be the angular momentum relative to O of all the particles of which m_i is representative. Thus,

$$\mathbf{H}_O = \sum_{i=1}^{n} \boldsymbol{\rho}_i \times (m_i \mathbf{v}_i) \tag{1}$$

Taking the time derivative of Eq. (1),

$$\dot{\mathbf{H}}_O = \sum_{i=1}^{n} \dot{\boldsymbol{\rho}}_i \times (m_i \mathbf{v}_i) + \sum_{i=1}^{n} \boldsymbol{\rho}_i \times \frac{d}{dt}(m_i \mathbf{v}_i) \tag{2}$$

From the figure, $\mathbf{r}_p = \mathbf{r}_O + \boldsymbol{\rho}_i$ and hence $\dot{\mathbf{r}}_\rho = \dot{\mathbf{r}}_O + \dot{\boldsymbol{\rho}}_i$. Substituting for $\dot{\boldsymbol{\rho}}_i$ into (2),

$$\dot{\mathbf{H}}_O = \sum_{i=1}^{n} (\dot{\mathbf{r}}_P - \dot{\mathbf{r}}_O) \times (m_i \mathbf{v}_i) + \sum_{i=1}^{n} \boldsymbol{\rho}_i \times (m_i \dot{\mathbf{v}}_i) \tag{3}$$

Expand the first term on the right of (3) into

$$\sum_{i=1}^{n} \dot{\mathbf{r}}_P \times (m_i \mathbf{v}_i) - \sum_{i=1}^{n} \dot{\mathbf{r}}_O \times (m_i \mathbf{v}_i) \tag{4}$$

Since $\dot{\mathbf{r}}_P$ is the absolute velocity $\mathbf{v}_i$ of P, the first term of (4) is zero. Also, since $\dot{\mathbf{r}}_O$ does not change during the summation and is therefore independent of i, the second term of (4) can be written

$$\dot{\mathbf{r}}_O \times \sum_{i=1}^{n} (m_i \mathbf{v}_i) \qquad \text{or} \qquad \dot{\mathbf{r}}_O \times (m\bar{\mathbf{v}})$$

where $\bar{\mathbf{v}}$ is the velocity of the mass center.

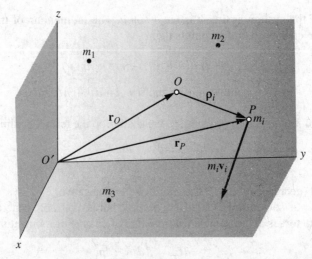

Fig. 7-4

The last term of Eq. (3) is equal to $\sum \mathbf{M}_O$, the sum of the moments of the external forces on all the particles. Thus, (3) may now be written as

$$\dot{\mathbf{H}}_O = -\dot{\mathbf{r}}_O \times (m\overline{\mathbf{v}}) + \sum \mathbf{M}_O \tag{5}$$

Equation (5) indicates that the time rate of change of the angular momentum about O equals $\sum \mathbf{M}_O$ only if $-\mathbf{r}_O \times (m\overline{\mathbf{v}})$ is zero. This occurs when (a) O is fixed, that is, $\dot{\mathbf{r}}_O = 0$; (b) $\overline{\mathbf{v}} = 0$; or (c) $\dot{\mathbf{r}}_O$ and $\overline{\mathbf{v}}$ are parallel (cross product of parallel vectors is zero). If O is the mass center, then $\dot{\mathbf{r}}_O = \overline{\mathbf{v}}$ and $\dot{\mathbf{r}}_O \times m\overline{\mathbf{v}} = 0$.

7.2. Given an assemblage of n particles of masses $m_1, m_2, m_3, \ldots, m_n$, show that the sum of the moments about O of the external forces equals the time rate of change of the moment of *relative* momentum of the group of particles about this point O only if (a) point O is the mass center of the n particles, or (b) point O has constant velocity (or is at rest), or (c) point O has an acceleration vector that passes through the mass center.

SOLUTION

Figure 7-5 shows the ith particle of the group with mass m_i. The given point O has a position vector $\mathbf{r}_O$ relative to a newtonian (inertial) frame of reference, that is, O' is fixed. The position vector of P relative to the fixed frame is $\mathbf{r}_P$. Let $\mathbf{H}'_O$ be the relative angular momentum with respect to O of all the particles of which m_i is representative. Then, from Eq. (10),

$$\mathbf{H}'_O = \sum_{i=1}^{n} \boldsymbol{\rho}_i \times (m_i \dot{\boldsymbol{\rho}}_i) \tag{1}$$

Taking the time derivative of Eq. (1),

$$\dot{\mathbf{H}}'_O = \sum_{i=1}^{n} \dot{\boldsymbol{\rho}}_i \times (m_i \dot{\boldsymbol{\rho}}_i) + \sum_{i=1}^{n} \boldsymbol{\rho}_i \times (m_i \ddot{\boldsymbol{\rho}}_i) \tag{2}$$

From the figure, $\mathbf{r}_P = \mathbf{r}_O + \boldsymbol{\rho}_i$ and hence $\ddot{\mathbf{r}}_P = \ddot{\mathbf{r}}_O + \ddot{\boldsymbol{\rho}}_i$. Substituting for $\ddot{\boldsymbol{\rho}}_i$ in (2) and noting that the first term on the right of (2) is zero ($\dot{\boldsymbol{\rho}}_i \times \dot{\boldsymbol{\rho}}_i = 0$), we have

$$\dot{\mathbf{H}}'_O = \sum_{i=1}^{n} \boldsymbol{\rho}_i \times (m_i \ddot{\mathbf{r}}_P) - \sum_{i=1}^{n} \boldsymbol{\rho}_i \times (m_i \ddot{\mathbf{r}}_O) \tag{3}$$

The last term in (3) may be written

$$\left(\sum_{i=1}^{n} m_i \boldsymbol{\rho}_i \right) \times \ddot{\mathbf{r}}_O$$

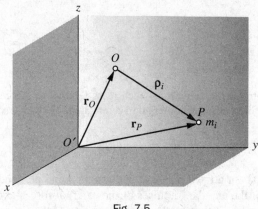

Fig. 7-5

since $\ddot{\mathbf{r}}_O$ does not change as the sum is taken over the n particles. Also,

$$\sum_{i=1}^{n} \boldsymbol{\rho}_i \times (m_i \ddot{\mathbf{r}}_P) = \sum \mathbf{M}_O \quad \text{and} \quad \sum_{i=1}^{n} m_i \boldsymbol{\rho}_i = m\bar{\boldsymbol{\rho}}$$

where $\bar{\boldsymbol{\rho}}$ is the position vector relative to O of the mass center. Then

$$\dot{\mathbf{H}}'_O = \sum \mathbf{M}_O - m\bar{\boldsymbol{\rho}} \times \ddot{\mathbf{r}}_O \tag{4}$$

The last term in (4) is zero if (a) O is the mass center ($\bar{\boldsymbol{\rho}} = 0$), (b) O has constant velocity ($\ddot{\mathbf{r}}_O = 0$), or (c) O has an acceleration $\ddot{\mathbf{r}}_O$ passing through the mass center, i.e., along $\bar{\boldsymbol{\rho}}$ (the cross product of parallel vectors is zero).

7.3. Four equal masses m are spaced at the quarter points of a thin massless rim of radius R. Show that the angular momentum relative to the mass center O using absolute velocities is the same as that obtained by using the relative velocities of the masses to the mass center O.

SOLUTION

Let v = speed of the mass center as the rim rolls to the right. Figure 7-6(a) shows the rim with the masses m_2 and m_4 in a vertical line. In Fig. 7-6(b), the momentum **L** of each mass is shown using the absolute velocities of each mass. Thus, $v_1 = \sqrt{2}\,v$ at 315°, while $v_3 = \sqrt{2}\,v$ at 45°. Of course, $v_2 = 2v$, and v_4 is that of the instant center with no absolute velocity.

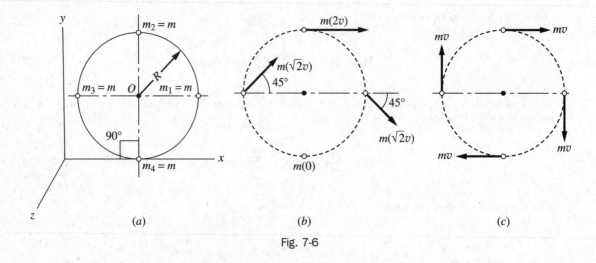

(a)   (b)   (c)

Fig. 7-6

The moment of each momentum is in the negative z direction. Thus, summing in 1, 2, 3, and 4 order, we can write

$$\mathbf{H}_O = -\left[m_1\left(\sqrt{2}\,v\right)\left(\tfrac{1}{2}\sqrt{2}R\right) + m_2(2v)R + m_3\left(\sqrt{2}\,v\right)\left(\tfrac{1}{2}\sqrt{2}R\right) + 0 \right]\mathbf{k} = -(4mvR)\mathbf{k}$$

Figure 7-6(c) illustrates the velocity of each mass relative to the center multiplied by the mass. The moments about O of these relative momentum vectors are

$$\mathbf{H}_O = -(4mvR)\mathbf{k} \qquad \text{(as before)}$$

7.4. For a body rotating about a fixed axis that is through O and perpendicular to the plane of the paper, show that the sum of the angular impulses of the external forces about the fixed axis is equal to the change in $I_O\omega$, where $I_O\omega$ is the angular momentum $\mathbf{H}_O$ of the entire body.

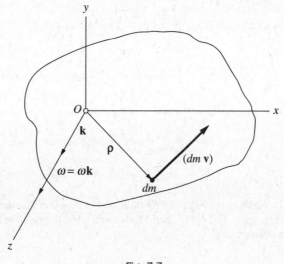

Fig. 7-7

SOLUTION

In Fig. 7-7, dm represents any differential mass with position vector $\boldsymbol{\rho}$ in the plane of the paper. The angular momentum of dm is $\boldsymbol{\rho} \times (dm\ \mathbf{v})$. The angular momentum $\mathbf{H}_O$ for the entire body is

$$\mathbf{H}_O = \int \boldsymbol{\rho} \times (dm\ \mathbf{v})$$

But $\mathbf{v} = \boldsymbol{\omega} \times \boldsymbol{\rho}$ is in the plane of the paper and has magnitude $\rho\omega$ because the vectors $\boldsymbol{\omega}$ and $\boldsymbol{\rho}$ are at right angles. Also, $\boldsymbol{\rho} \times (\boldsymbol{\omega} \times \boldsymbol{\rho})$ is directed out of the paper and has magnitude $\rho^2\omega$ because $\boldsymbol{\rho}$ and $\mathbf{v}$ are at right angles. Thus,

$$\mathbf{H}_O = \int \rho^2 \omega\ dm\ \mathbf{k}$$

where $\mathbf{k}$ is the unit vector perpendicular to the paper and directed toward the reader.

Since ω and $\mathbf{k}$ do not change with dm, they may be taken outside the integral sign. Also, $\int \rho^2 dm = I_O$. Hence,

$$\mathbf{H}_O = I_O \omega \mathbf{k}$$

Then, using Eq. (8), we have

$$\sum \mathbf{M}_O = \frac{d\mathbf{H}_O}{dt} = \frac{d(I_O\omega)}{dt}\mathbf{k}$$

Since the moments of forces and $\boldsymbol{\omega}$ are in the $\mathbf{k}$ direction, the equation can be written in scalar form as

$$\int \left(\sum M_O\, dt \right) = \Delta H_O = I_O(\omega' - \omega)$$

where ω and ω' are the initial and final angular speeds and $\int (\sum M_O\, dt) = \sum (\text{Ang Imp})_O$.

7.5. For a rigid body of mass m in plane motion (assume the plane of the paper is the plane of motion), show that Eqs. (15), (16), and (17) are true. Refer to Fig. 7-8.

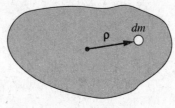

Fig. 7-8

SOLUTION

Since a rigid body is an assemblage of particles that remain at constant distances from one another, Eq. (6) applies. Thus,

$$\int_{t_1}^{t_2} \sum \mathbf{F}\,dt = \Delta \mathbf{L} = m\bar{\mathbf{v}}_2 - m\bar{\mathbf{v}}_1$$

This vector equation is equivalent to the two scalar equations

$$\sum (\text{Imp})_x = m(\bar{v}'_x - \bar{v}_x) \tag{1}$$

$$\sum (\text{Imp})_y = m(\bar{v}'_y - \bar{v}_y) \tag{2}$$

To derive Eq. (17), make use of Eq. (11). As indicated in Problem 7.2, the mass center is one of the points that can be selected to apply Eq. (11). Then $\mathbf{H}'_O$ becomes $\bar{\mathbf{H}}'$, the relative angular momentum of the rigid body about the mass center, and

$$\bar{\mathbf{H}}' = \int \boldsymbol{\rho} \times (dm\,\dot{\boldsymbol{\rho}})$$

Because of the rigid-body constraint, the vector $\boldsymbol{\rho}$ (from the mass center to the element of mass dm) can change only in direction but not in magnitude. Hence, $\dot{\boldsymbol{\rho}}$ is perpendicular to $\boldsymbol{\rho}$ (it is in the plane of the paper) and of magnitude $\rho\omega$. Then $\boldsymbol{\rho} \times (dm\,\dot{\boldsymbol{\rho}})$ is perpendicular to the plane of the paper and of magnitude $\rho^2\omega$. Thus, the vector equation may be replaced by the scalar equation

$$\bar{H}' = \omega \int \rho^2 dm = \bar{I}\omega$$

and Eq. (11) becomes

$$\sum M_G = \bar{H}' = \frac{d(\bar{I}\omega)}{dt}$$

which can now be expressed as

$$(\text{Ang Imp})_G = \int \sum M_G\,dt = \Delta\bar{H} = \bar{I}(\omega' - \omega) \tag{3}$$

7.6. A thin rim of mass m and radius R is rolling without slipping on a horizontal plane, as shown in Fig. 7-9(a). A horizontal force of magnitude P is applied at the top. Show that the sum of the external forces about the mass center G, when equated to the time rate of change of the relative momentum about G, yields the same result as obtained by equating the sum of the moments of the external forces about the instant center A to the time rate of change of the relative angular momentum about A.

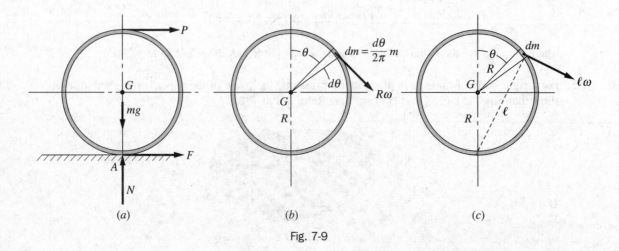

(a)   (b)   (c)

Fig. 7-9

SOLUTION

The free-body diagram in Fig. 7-9(a) shows the normal force N, the friction force F, the applied force P, and the weight mg concentrated at the mass center G. Scalar equations will be used in this analysis.

(a) Figure 7-9(b) shows a differential mass dm of the rim at an angle θ with the vertical. The mass dm is that part of the rim subtended by the angle $d\theta$; hence, $dm = m\,d\theta/2\pi$. The speed of dm relative to the mass center G is $R\omega$ as shown. The relative angular momentum about G of dm is thus $dm\,R^2\omega$. The relative angular momentum of the entire rim is

$$H'_G = \int_0^{2\pi} mR^2\omega \frac{d\theta}{2\pi} = mR^2\omega$$

The sum of the moments of the external forces about G (considering clockwise positive) is

$$\sum M_G = PR - FR$$

But for any group of particles (in this case the rim with mass center speed $v = r\omega$),

$$\sum F_x = G_x \qquad \text{or} \qquad P + F = \frac{d}{dt}(mR\omega)$$

From this,

$$F = -P + \frac{d}{dt}(mR\omega)$$

The equation $\sum M_G = \dot{H}'_G$ yields

$$PR - \left[-P + \frac{d}{dt}(mR\omega)\right]R = \frac{d}{dt}(mR^2\omega) \qquad \text{or} \qquad 2PR = \frac{d}{dt}(2mR^2\omega)$$

(b) To use the instant center A as the center, let ℓ be the distance from A to the mass dm as shown in Fig. 7-9(c). The speed of dm relative to A (which is at rest) is $\ell\omega$. It is perpendicular to the line ℓ as shown. The line has length

$$\ell = \sqrt{R^2 + R^2 + 2RR\cos\theta}$$

The moment of relative momentum of dm about A is $dm(2R^2 + 2R^2\cos\theta)\omega$. Using $dm = m\,d\theta/2\pi$,

$$H'_A = \int_0^{2\pi} mR^2\omega(2 + 2\cos\theta)\frac{d\theta}{2\pi} = 2mR^2\omega$$

The sum of the moments of the external forces about A is

$$\sum M_A = 2PR = \dot{H}'_A = \frac{d}{dt}(2mR^2\omega)$$

This is the same result as determined in part (a).

It is interesting to note that since $2mR^2$ is the moment of inertia I_A of the thin rim about the instant center, the above equation can be written

$$\sum M_A = \frac{d}{dt}(I_A\omega) = I_A\alpha$$

7.7. If the mass center of an assemblage of particles is at rest, Eq. (8) is true and any point O may be used. For two particles of equal mass m mounted on a weightless rim of radius R rotating about the center of the rim, show that $\sum M_O = (d/dt)(2mR^2\omega)$. Refer to Fig. 7-10.

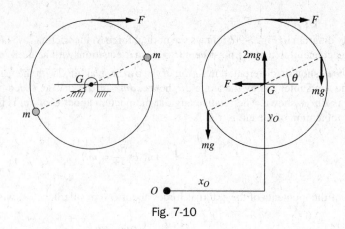

Fig. 7-10

SOLUTION

The bearing reactions are F to the left and $2mg$ up. Summing moments about any point O, $\sum M_O = FR$ is independent of the moment center O:

$$H_O = m(R\omega \cos\theta)(x_O + R\cos\theta) + m(R\omega \sin\theta)(y_O + R\sin\theta)$$

$$- m(R\omega \cos\theta)(x_O - R\cos\theta) - m(R\omega \sin\theta)(y_O - R\sin\theta) = 2mR^2\omega$$

Hence, $FR = (d/dt)\,(2mR^2\omega)$, the same equation as when moments are taken relative to mass center G.

7.8. A 45-kg object is pushed for 5 s by a horizontal force F over a horizontal plane where the coefficient of friction is 0.2. During this time interval, the speed changes from 2 to 4 m/s. Determine the value of the force F.

SOLUTION

The normal force equals the weight of 441 N. The frictional force opposing motion is $0.2 \times 441 = 88.2$ N. The impulse in the horizontal direction is equal to the change in momentum. If the forces are constant, the impulse is $(\sum F)\Delta t$. So, using Eq. (6),

$$(F - 88.2)(5) = 45(4 - 2) \qquad \therefore F = 106.2 \text{ N}$$

7.9. A 10-kg block slides from rest down a plane inclined 30° with the horizontal. Assuming a coefficient of kinetic friction between the block and the plane of 0.3, what will be the speed of the block at the end of 5 s?

SOLUTION

This problem may, of course, be solved by previous methods. However, with time as one of the quantities given, the impulse-momentum method is the simplest. As in Problem 7.8, this is a problem of translation. Draw a free-body diagram indicating all external forces acting on the block (see Fig. 7-11). Since only motion along the plane is considered, one impulse-momentum equation will suffice. Using Eq. (6), we have

$$\left(\sum F\right)\Delta t = m(v_2 - v_1) \qquad [98\sin 30° - 0.3(98\cos 30°)](5) = 10(v_2 - 0) \qquad \therefore v_2 = 11.8 \text{ m/s}$$

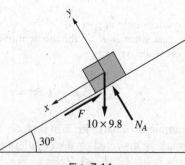

Fig. 7-11

7.10. A 40-kg block rests on a horizontal plane. It is acted upon by a horizontal force that varies from zero according to the law $F = 98t$. If the force acts for 5 s, what is the speed of the block? The coefficient of static friction is 0.25, and the coefficient of kinetic friction is 0.20.

SOLUTION

The force F increases until it reaches the limiting value of static friction; that is, $F = 98t = 0.25 \times 40 \times 9.8$. Thus, at $t = 1$ s, the block begins to move with friction now equal to $0.2 \times 40 \times 9.8 = 78.4$ N, which remains constant.

The impulse horizontally equals the change in momentum horizontally. Thus, Eq. (6) or (12) provides

$$\sum (Imp)_x = \Delta L_x$$

$$\int_1^5 (98t - 78.4)\, dt = 40(v_2 - 0) \qquad \therefore v_2 = 21.6 \text{ m/s}$$

7.11. A 1.5-kg block starts from rest on a smooth horizontal plane under the action of a horizontal force F that varies according to the equation $F = 30t - 50t^2$. Determine the maximum speed.

SOLUTION

The impulse-momentum equation results in

$$\sum (Imp)_x = \Delta L_x \qquad \int_0^t F\, dt = \int_0^t (30t - 50t^2)\, dt = 1.5(v_2 - 0)$$

from which $v_2 = 10t^2 - 11.11t^3$. To find the maximum value of v_2, determine the value of t at which $dv_2/dt = 20t - 33.33t^2 = 0$. This gives $t = 0.6$ s, and hence maximum $v_2 = 1.2$ m/s.

7.12. The 40-kg block shown in Fig. 7-12 is moving up initially with a speed of 2.5 m/s. What constant value of P will result in an upward speed of 5 m/s in 12 s? Assume that the weightless pulleys are frictionless and that the coefficient of friction between the blocks and the plane is 0.10.

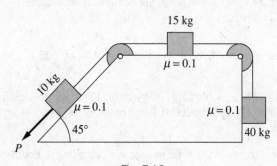

Fig. 7-12

SOLUTION

Sum impulses and momenta along the line of travel of the system. For example, the force P and a component of the gravitational force on the 10-kg block have positive impulses in the line of travel of the 10-kg block, while the friction acting on the 10-kg block has a negative impulse. The cord tension acts on both the 10- and 15-kg masses in opposite directions; therefore its linear impulse along the line of travel is zero. Proceeding in this way, the impulse-momentum equation becomes

$$\sum (Imp) = \Delta L$$

$$[P + 9.8 \times 10 \sin 45° - 0.10(9.8 \times 10 \cos 45°) - 0.10(9.8 \times 15) - 9.8 \times 40](12) = (10 + 15 + 40)(5 - 2.5)$$

from which $P = 358$ N.

7.13. A 4-kg mass is at rest in a smooth horizontal slot. It is struck a 10-N blow that lasts for 0.04 s. Two seconds after the start of the first blow, a second blow of −10 N is delivered and lasts for 0.02 s. What will be the speed of the body after 3 s?

SOLUTION

The forces in this problem are represented in a plot against time, as shown in Fig. 7-13. The impulse for any length of time larger than 2.02 s is the algebraic sum of the two areas, that is, $10(0.04) - 10(0.02) = 0.2$ N·s. Then

$$\sum \text{Imp} = \Delta L \qquad \text{or} \qquad 0.2 = 4(v_2 - 0) \qquad \therefore v_2 = 0.05 \text{ m/s}$$

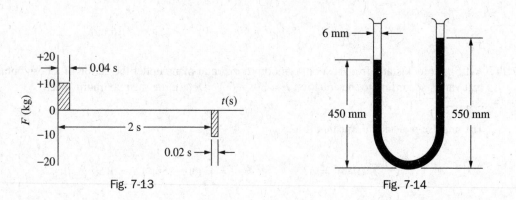

Fig. 7-13   Fig. 7-14

7.14. In Fig. 7-14, the mercury in the left column of the manometer is falling at the rate of 25 mm/s. The left column is 450 mm long, and the right column is 550 mm long. What is the vertical momentum of the mercury? The manometer has a 6-mm inside diameter. Mercury weighs 133 kN/m³.

SOLUTION

Neglecting the vertical momentum in the bend in the tube, it is evident that the momentum of the left column is down while that of the right column is up. The net upward momentum is that of a 100-mm, 6-mm-diameter column moving 0.025 m/s:

$$\text{Momentum} = \left[\frac{1}{4}\pi(0.006)^2(0.1) \right](133 \times 10^3/9.8) \times 0.025 = 9.59 \times 10^{-4} \text{ N·s(or kg·m/s)}$$

7.15. A flywheel of mass 2000 kg and radius of gyration 1200 mm rotates about a fixed center O from rest to an angular speed of 120 rpm in 200 s. What moment M is necessary?

SOLUTION

The angular impulse changes the angular momentum according to Eq. (14):

$$\sum (\text{Ang Imp})_O = \Delta H_O = I_O(\omega_2 - \omega_1)$$

$$M(200) = [2000(1.2)^2](240\pi/60 - 0) \qquad M = 181 \text{ N·m}$$

Make sure that the units check (use kg = N·s²/m).

7.16. A pendulum consists of a bob of mass m and a slender rod of negligible mass. See Fig. 7-15. Show that the differential equation of the motion is $\ddot{\theta} + (g/\ell)\sin\theta = 0$.

SOLUTION

The free-body diagram shows the pendulum displaced an angle θ from the vertical position. The angular momentum $\mathbf{H}_O$ of the bob relative to the support is $I\dot{\theta}\mathbf{k}$, where $\mathbf{k}$ is the unit vector perpendicular to the page, with the arrow pointing toward the reader.

The only force with a moment is the gravitational force acting on the bob; the moment is clockwise or negative. Then

$$\sum \mathbf{M}_O = \frac{d\mathbf{H}_O}{dt} \qquad \text{or} \qquad -mg(\ell \sin\theta)\mathbf{k} = \frac{d}{dt}(I\dot\theta\mathbf{k})$$

Since $I = m\ell^2$ for the bob, we obtain the scalar equation $\ddot\theta + (g/\ell)\sin\theta = 0$.

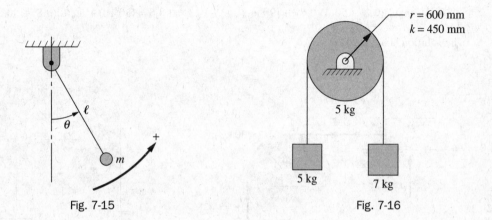

Fig. 7-15   Fig. 7-16

7.17. In Fig. 7-16, a massless rope carries two masses of 5 and 7 kg when hanging on a pulley of mass 5 kg, radius 600 mm, and radius of gyration 450 mm. How long will it take to change the speed of the masses from 3 to 6 m/s?

SOLUTION A

Let T_1 and T_2 be the tensions in the rope supporting the 5- and 7-kg masses, respectively. The useful impulse-momentum equations are as follows, where (1) applies to the 5-kg mass, (2) applies to the 7-kg mass, and (3) applies to the pulley:

$$(T_1 - 5 \times 9.8)t = 5(6 - 3) \tag{1}$$

$$(7 \times 9.8 - T_2)t = 7(6 - 3) \tag{2}$$

$$(T_2 - T_1)(0.6)t = \overline{I}(\omega_{\text{final}} - \omega_{\text{initial}}) = 1.01(10 - 5) \tag{3}$$

where $\overline{I} = mk^2 = 5(0.45)^2 = 1.01 \text{ kg·m}^2$, $\omega_{\text{final}} = 6/0.6 = 10$ rad/s, and $\omega_{\text{initial}} = 3/0.6 = 5$ rad/s. Solve (1), (2), and (3) simultaneously to obtain $t = 2.27$ s.

SOLUTION B

Realizing that the moment about the center of the pulley of the linear momentum of either hanging mass is the angular momentum of the mass, a system approach may be used to solve this problem. Thus,

$$(\text{Initial Ang Mom})_O + (\text{Ang Imp})_O = (\text{Final Ang Mom})_O \tag{4}$$

The tensions do not occur in the system impulse expression because they occur in pairs and thus cancel each other. Also, only the two gravitational forces on the two hanging masses have angular impulses about the center of the pulley. The equation becomes

$$(I_O\omega_1 + 5v_1 \times 0.6 + 7v_1 \times 0.6) + (7 \times 9.8t - 5 \times 9.8t)0.6 = (I_O\omega_2 + 5v_2 \times 0.6 + 7v_2 \times 0.6) \tag{5}$$

or

$$(1.01 \times 5 + 5 \times 3 \times 0.6 + 7 \times 3 \times 0.6) + 11.76t = (1.01 \times 10 + 5 \times 6 \times 0.6 + 7 \times 6 \times 0.6) \tag{6}$$

Hence, $t = 2.27$ s.

7.18. Refer to Fig. 7-17. Determine the mass of B necessary to cause the 50-kg mass A to change its speed from 4 to 8 m/s in 6 s. Assume that the drum rotates in frictionless bearings.

SOLUTION

Apply the system solution, noting that the initial angular momentum about the drum center O plus the angular impulses about O of all the external forces is equal to the final angular momentum about O. Note that $N = 50 \times 9.8$ and hence $F = 0.25 \times 50 \times 9.8 = 123$ N. The angular speeds of the drum are 4/0.8 and 8/0.8, respectively (5 rad/s and 10 rad/s).

$$[30(5) + m_B(4)(0.8) + 50(4)(0.8)] + [9.8m_B(0.8) - 123 \times 0.8]6 = [30(10) + m_B(8)(0.8) + 50(8)(0.8)]$$

The solution is $m_B = 20.5$ kg.

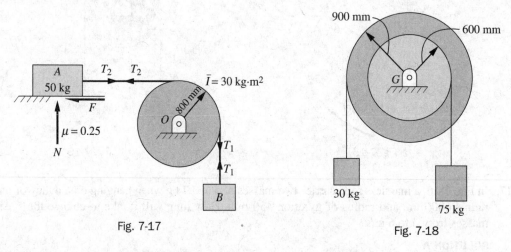

Fig. 7-17   Fig. 7-18

7.19. The drum shown in Fig. 7-18 consists of two homogeneous cylinders integrally connected. The smaller cylinder has a mass of 115 kg and the larger a mass of 230 kg. How much time will elapse before the drum changes its speed from 100 to 300 rpm?

SOLUTION

The moment of inertia of the drum equals

$$\frac{1}{2}(230)(0.9^2) + \frac{1}{2}(115)(0.6^2) = 113.8 \text{ kg·m}^2$$

The linear impulse of the 75-kg mass has a moment about G that is clockwise and has a magnitude of $0.6(75 \times 9.8)t = 441t$ N·s·m. The linear impulse of the 30-kg mass has a counterclockwise moment about G and equals $-0.9(30 \times 9.8)t = -265\, t$ N·s·m. The total angular impulse is the sum of the two values, or 176 N·s·m. This total angular impulse acting on the system is equal to the total change in angular momentum of the system with reference to G.

The change in angular momentum of the drum equals

$$I(\omega_2 - \omega_1) = 113.8\left(300 \times \frac{2\pi}{60} - 100 \times \frac{2\pi}{60}\right) = 2380 \text{ N·s·m}$$

The linear momentum of each weight is multiplied by its arm to determine the moment of its momentum (angular momentum) about G. The 75-kg mass is moving down, and thus its linear momentum has a clockwise (positive) moment of momentum about G. This change in angular momentum equals

$$0.6\left[75\left(0.6 \times \frac{600\pi}{60} - 0.6 \times \frac{200\pi}{60}\right)\right] = 565 \text{ N·s·m}$$

Similarly, the change in linear momentum of the 30-kg mass is up, and thus the mass has a positive (clockwise) moment about G. Its value is

$$0.9\left[30\left(0.9 \times \frac{600\pi}{60} - 0.9 \times \frac{200\pi}{60}\right)\right] = 509 \text{ N·s·m}$$

The total angular impulse is equal to the total change in angular momentum of the system, or

$$176t = 2380 + 565 + 509 \qquad \therefore\ t = 19.6\ \text{s}$$

7.20. Refer to Fig. 7-19. A cylinder of radius r, mass m, and moment of inertia $\bar{I}$ (about the center) rolls from rest down a plane inclined at an angle θ with the horizontal. What is the speed of its center at time t from rest?

SOLUTION

The free-body diagram shows all external forces acting on the cylinder. Do not make the mistake of assuming that the friction force F is equal to the product of the coefficient of friction and the normal force N. Since the cylinder is in plane motion, the impulse-momentum equations for this type of motion are

$$\sum (\text{Imp})_x = \Delta L_x \tag{1}$$

and

$$\sum (\text{Ang Imp})_O = \Delta H_O \tag{2}$$

Note that F is the only external force with a moment about the center O. Equations (1) and (2) become

$$(mg\sin\theta - F)t = m(\bar{v} - 0) \tag{3}$$

$$Frt = \bar{I}(\omega - 0) \tag{4}$$

In these equations, the friction force F and the speed $\bar{v}$ are unknown ($\omega = \bar{v}/r$, since no slipping is assumed). From (4), $F = \bar{I}\omega/(rt)$. Substitute this into (3), with $\omega = \bar{v}/r$, to obtain

$$\bar{v} = \frac{mg\sin\theta}{m + \bar{I}/r^2}\,t$$

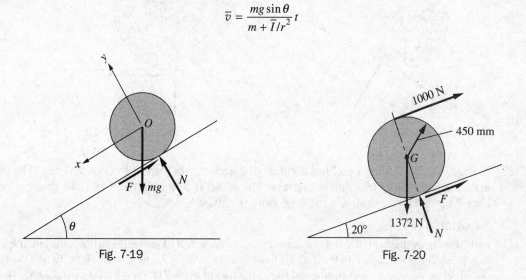

Fig. 7-19   Fig. 7-20

7.21. A solid homogeneous cylinder 900 mm in diameter with a mass of 140 kg is rolled up a 20° incline by a force of 1000 N, applied parallel to the plane. Assuming no slipping, determine the speed of the cylinder after 6 s if the initial speed is zero. Refer to Fig. 7-20.

SOLUTION

Draw the free-body diagram of the cylinder, assuming that the frictional force F acts up the plane. The impulse-momentum equations of plane motion are

$$\sum (\text{Imp})_{\parallel} = \Delta L_{\parallel} \qquad \text{or} \qquad (1000 + F - 1372\sin 20°)6 = 140\bar{v} \tag{1}$$

$$\sum (\text{Ang Imp})_G = \bar{I}\omega \qquad \text{or} \qquad [1000(0.45) - F(0.45)]6 = \frac{1}{2}(140)(0.45^2)\left(\frac{\bar{v}}{0.45}\right) \tag{2}$$

A simultaneous solution yields $\bar{v} = 43.7$ m/s.

7.22. Refer to Fig. 7-21. A solid cylinder starts up the 30° inclined plane with a mass center speed of 8 m/s. If it rolls freely, how long does it take to reach its highest point?

SOLUTION

The initial speed $\bar{v}$ of the mass center G is 8 m/s. The initial angular velocity $\omega = 8/0.5 = 16$ rad/s. The final speed is zero. The free-body diagram shows the forces acting on the cylinder. Apply the two impulse-momentum equations of plane motion:

$$\sum (\text{Imp})_x = \Delta L_x \qquad \text{or} \qquad (F - mg\sin 30°)t = m(0 - 8) \qquad (1)$$

$$\sum (\text{Ang Imp})_G = \Delta H_G \qquad \text{or} \qquad (-0.5F)t = \frac{1}{2}m(0.5)^2(0 - 16) \qquad (2)$$

A simultaneous solution gives $t = 2.25$ s.

The solution may also be found using the instant center as the reference point. The one equation needed is

$$\sum (\text{Ang Imp})_I = \Delta H_I \qquad \text{or} \qquad -(0.5\sin 30)mgt = \frac{3}{2}(m)(0.5)^2(0 - 16)$$

from which $t = 2.25$ s.

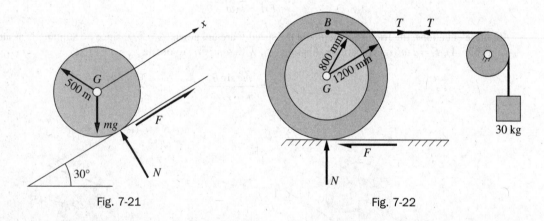

Fig. 7-21   Fig. 7-22

7.23. In Fig. 7-22, the 75-kg wheel has a radius of gyration with respect to G of 900 mm. The pulley is massless and runs in frictionless bearings. The wheel is rolling initially 10 rad/s counterclockwise. How long will it take until it is rolling 6 rad/s clockwise?

SOLUTION

In the free-body diagram, friction is assumed to act to the left. By kinematic relations, the initial speed of G is $v_G = -1.2(10) = -12$ m/s (to the left). Its final speed is $1.2(6) = 7.2$ m/s, to the right. To find the speeds of the 30-kg mass, determine the initial and final velocities of point B on the wheel: $\mathbf{v}_B = \mathbf{v}_{B/G} + \mathbf{v}_G$. Hence, the initial speed of B is $-0.8\omega - 12 = -0.8(10) - 12 = -20$ m/s. The final speed of B is, by similar reasoning, 12 m/s. Thus, the initial speed of the 30-kg mass is 20 m/s up, and its final speed is 12 m/s down.

In the equations, the following sign conventions are used. For the wheel, to the right is positive and clockwise is positive. For the 30-kg mass in (3) below, down is positive:

$$\sum (\text{Imp})_h = \Delta L_h \qquad \text{or} \qquad (T - F)t = 75[7.2 - (-12)] \qquad (1)$$

$$\sum (\text{Ang Imp})_G = \Delta H_G \qquad \text{or} \qquad (0.8T + 1.2F)t = 75(0.9)^2[6 - (-10)] \qquad (2)$$

$$\sum (\text{Imp})_v = \Delta L_v \qquad \text{or} \qquad (9.8 \times 30 - T)t = 30[12 - (-20)] \qquad (3)$$

The simultaneous solution is $t = 7.86$ s.

7.24. Refer to Fig. 7-23. A solid 200-mm-diameter cylinder with a mass of 44 kg is spinning 36 rad/s clockwise about its horizontal axis when it is dropped suddenly onto a horizontal plane. Assuming a coefficient of friction of 0.15, determine the speed $\bar{v}$ of the mass center G when pure rolling begins. Through what distance has the mass center moved before this occurs?

SOLUTION

The free-body diagram shows contact when the plane is first contacted. The same force system will act on the cylinder until pure rolling is attained. The frictional force F acts (1) to decrease the angular speed and (2) to increase the mass center speed from zero to its value $\bar{v} = r\omega$ when pure rolling occurs.

Here $N = W = 431$ N, $F = 0.15 N = 64.7$ N, $m = 44$ kg, and $\bar{I} = \frac{1}{2}mr^2 = \frac{1}{2} \times 44 \times 0.1^2 = 0.44$ kg·m^2. The impulse-momentum equations are

$$\sum (\text{Imp})_h = \Delta L_h \qquad \text{or} \qquad Ft = m(\bar{v} - 0) \qquad (1)$$

$$\sum (\text{Ang Imp})_G = \Delta H_G \qquad \text{or} \qquad -Frt = \bar{I}(\omega - 36) \qquad (2)$$

In (2), assume that the clockwise direction is positive. The equations become

$$64.7t = 44\bar{v} \qquad \text{and} \qquad -64.7 \times 0.1t = 0.44(\omega - 36)$$

where $\omega = \bar{v}/r = 10\bar{v}$. The simultaneous solution is thus $\bar{v} = 1.8$ m/s, $t = 0.83$ s. Then $s = \frac{1}{2}(\bar{v} + 0)t = 0.747$ m.

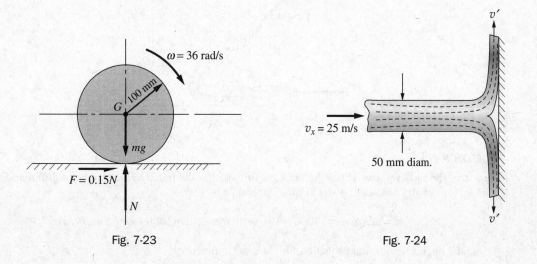

Fig. 7-23   Fig. 7-24

7.25. A 50-mm-diameter stream of water moving 25 m/s horizontally strikes a flat vertical plate, as shown in Fig. 7-24. After striking, the water moves essentially parallel to the plate. What is the force exerted on the plate by the stream of water?

SOLUTION

Consider all the particles of water in time interval Δt. The total mass m of these particles that hits the plate in Δt is

$$m = Av(\Delta t)\rho = \pi \times 0.025^2 \times 25\Delta t \times 1000$$

$$= 49.1\Delta t \text{ kg}$$

where A = area of cross section of stream
       v = speed of stream
       ρ = density of water (1000 kg/m^3)

Let P = force of plate on the mass m of water. Then

$$\sum F_x(\Delta t) = \Delta L_x = m(v'_x - v_x)$$

$$P(\Delta t) = 49.1(\Delta t)(0 - 25)$$

from which $P = -1227$ N. The force of the water on the plate is $+1227$ N, that is, to the right.

7.26. Rework Problem 7.25, but assume that the plate is moving to the right with a speed of 6 m/s.

SOLUTION

Again consider the mass of all particles of water in time Δt. The speed of the water relative to the plate is $v = 25 - 6 = 19$ m/s. Then $m = Av(\Delta t)\rho = 37.3(\Delta t)$, and using the impulse-momentum equation in the x direction, we obtain

$$P(\Delta t) = 37.3(\Delta t)(6 - 25)$$

where the final speed of water is that of the plate. Solving, the force of the plate on the water is $P = -709$ N. Of course, the force of the water on the plate is $+709$ N, that is, to the right.

7.27. A stream of water with cross-sectional area of 2000 mm² and moving 10 m/s horizontally strikes a fixed blade curved as shown in Fig. 7-25. Assuming that the speed of the water relative to the blade is constant (no friction is considered), determine the horizontal and vertical components of the force of the blade on the stream of water.

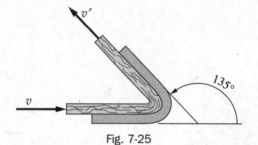

Fig. 7-25

SOLUTION

Note that the final velocity $\mathbf{v}'$ has the same magnitude v as the initial velocity $\mathbf{v}$ but a different direction. The mass m of all particles of water in time interval Δt is

$$m = Av(\Delta t)\rho = (2000 \times 10^{-6}\text{m}^2)(10 \text{ m/s})(\Delta t)(1000 \text{ kg/m}^3) = 20(\Delta t)$$

Using the impulse-momentum equation for the x and y directions,

$$\sum F_x(\Delta t) = \Delta L_x = m(v'_x - v_x) \quad \text{or} \quad P_x(\Delta t) = (20\,\Delta t)(-10\cos 45° - 10)$$

$$\sum F_y(\Delta t) = \Delta L_y = m(v'_y - v_y) \quad \text{or} \quad P_y(\Delta t) = (20\,\Delta t)(+10\sin 45° - 0)$$

where P_x and P_y are, respectively, the x and y components of the blade force. Solving, $P_x = -340$ N (to the left on the water) and $P_y = +140$ N (up on the water).

7.28. A 70-kg tank rests on platform scales. A vertical jet empties water into the tank with a speed of 8 m/s. Its cross-sectional area is 156 mm². What will be the scale reading 1 min later?

SOLUTION

The water from the jet exerts a continuous force F on the bottom of the tank and thus to the scale. From the impulse-momentum equation, considering the water at any time t as a free body, we may write $\sum(\text{Imp})_v = \Delta L_v$ or $Ft = m(0 - 8)$, where $m = Avt\rho$. Then

$$Ft = 156 \times 10^{-6}(8)t(1000)(0 - 8) \quad \text{or} \quad F = -9.98 \text{ N}$$

The negative sign indicates that an upward force (from the scale) is required to stop the water.

At $t = 60$ s, the tank holds (156×10^{-6}) (8) (60) $(9.8 \times 1000) = 734$ N of water. Thus, the scale reading after 1 min will be $9.98 + 734 + 70 \times 9.8 = 1430$ N.

7.29. A 60-kg person sitting in a stationary 70-kg iceboat fires a gun, discharging a 57-g bullet horizontally aft (along the fore-and-aft line). If the bullet speed leaving the gun is 365 m/s, what will be the speed of the boat after the bullet is fired, if we assume no friction?

SOLUTION

Since the impulse of the entire system is zero, the total momentum remains zero. This is true because the action on the bullet is equal to the reaction on the person, gun, and boat. The initial momentum is zero. The momentum after the bullet is fired must also be zero. Assuming the bullet speed to be positive, we can write

$$(m_{\text{boat}} + m_{\text{person}})v + m_{\text{bullet}}(365) = 0$$

$$(70 + 60)v + \frac{57}{1000} \times 365 = 0 \qquad \therefore v = -0.16 \text{ m/s}$$

The minus sign indicates that the boat and person move in the opposite direction from that of the bullet.

7.30. A 60-g bullet was fired horizontally into a stationary 50-kg sandbag suspended on a rope 900 mm long, as shown in Fig. 7-26. It was calculated from the observed angle θ that the bag with the bullet embedded in it swung to a height of 30 mm. What was the speed of the bullet as it entered the bag?

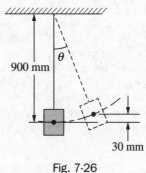

900 mm

θ

30 mm

Fig. 7-26

SOLUTION

Let

v_1 = speed of bullet before impact
v_2 = speed of (bag + bullet) after impact = $\sqrt{2gh} = \sqrt{2(9.8)(0.03)} = 0.767$ m/s

Momentum of system before impact = momentum of system after impact:

$$0.06v_2 + 0 = (0.06 + 50) \times 0.767$$

from which $v_1 = 640$ m/s.

7.31. As another example of conservation of momentum, consider the recoil of guns.

ANALYSIS

Initially the projectile is at rest in the gun. The charge explodes, pushing the projectile from the gun and at the same time pushing back with the same force on the gun. Since no external forces act during this explosion, the sum of the momenta of the projectile forward and the gun backward must be the same as the initial momentum of the system, i.e., zero. Thus,

$$m_p v_p + m_g v_g = 0$$

where m_p, m_g = masses of projectile and gun, respectively
v_p, v_g = speeds of projectile and gun, respectively, immediately after explosion

Solving for speed of recoil, $v_g = -(m_p/m_g)v_p$. The minus sign indicates that the gun moves in the opposite direction to the projectile.

The greater the mass of the gun, the less will be the speed of recoil. Hence, the energy to be absorbed by the recoil spring or other devices will be correspondingly less. Of course, the weight is limited by the need for mobility.

7.32. A 60-g bullet moving with a speed of 500 m/s strikes a 5-kg block moving in the same direction with a speed of 30 m/s. What is the resultant speed v of the bullet and the block, assuming the bullet to be embedded in the block?

SOLUTION

Momentum of system before impact = momentum of system after impact

$$0.06(500) + 5(30) = (0.06 + 5) \times v$$

$$v = 35.6 \text{ m/s}$$

7.33. A spring unstretched is 200 mm long and connected to the two masses shown in Fig. 7-27 and compressed 80 mm. If the system is released on a frictionless horizontal plane, what will be the speed of each block when the spring is again at its unstretched length? The spring constant is 2000 N/m.

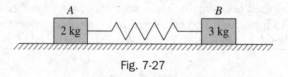

Fig. 7-27

SOLUTION

Since the same spring force acts on the two masses, but in opposite directions, the total linear impulse on the system of both masses is zero. Hence, the momentum of the two masses is constant, i.e., zero, and $2v_A + 3v_B = 0$. From this equation, $v_A = -(3/2)v_B$ at all times.

Another equation involving v_A and v_B is necessary. The work done by the spring in expanding to its original length (by virtue of its potential energy) is equal to the change in kinetic energy of both masses. As the spring expands a distance x from its compressed position, its compressive force = $2000(0.08 - x)$ N.

Hence, the total work done is

$$U = \int_0^{0.08} 2000(0.08 - x)\,dx = 6.4 \text{ N·m}$$

Equate this to the final kinetic energy (initial is zero):

$$6.4 = \frac{1}{2}(2)v_A^2 + \frac{1}{2}(3)v_B^2$$

Solve simultaneously with the previous equation to obtain $v_B = 1.31$ m/s (to the right) and $v_A = 1.96$ m/s (to the left).

7.34. Figure 7-28 shows a remote-controlled 5-kg object at a distance of 1.5 m from the center of a horizontal "weightless" turntable that is turning 2 rad/s about a vertical axis. An opposing moment $M = 3t$ N·m is applied to the shaft. Determine (*a*) how long it will take for the turntable to reach a speed of 1.5 rad/s and (*b*) how far the object must be moved, if the moment is removed, to bring the turntable back to a speed of 2 rad/s.

SOLUTION

(*a*) The angular momentum of the object is the product of its linear momentum mv and its radial distance r. The angular impulse equals the change in angular momentum. Hence, using $v = r\omega$, we have

$$\int_0^t -3t\,dt = 5(1.5 \times 1.5)(1.5) - 5(1.5 \times 2)(1.5) \qquad \therefore t = 1.94 \text{ s}$$

(*b*) To solve the second part of the problem, use conservation of angular momentum to determine the necessary radius *r* at which to place the object:

$$5(1.5)(1.5)(1.5) = 5(1.5)(2)r \qquad \therefore r = 1.125 \text{ m}$$

The object must be moved from 1.5 to 1.125 m, or 0.375 m toward the center.

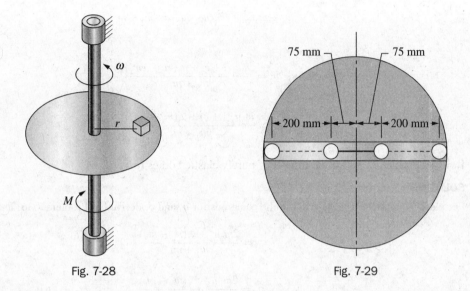

Fig. 7-28   Fig. 7-29

7.35. Two small spheres, each with a mass of 2 kg, are connected by a light string as shown in Fig. 7-29. The horizontal platform is rotating under no external moments at 36 rad/s when the string breaks. Assuming no friction between the spheres and the groove in which they ride, determine the angular speed of the system when the spheres hit the outer stops. The moment of inertia *I* for the disk is 0.5 kg·m^2.

SOLUTION

The initial moment of inertia I_i is that of the disk and the two spheres at distance 0.075 m from the center. Hence,

$$I_i = 0.5 + 2(2)(0.075^2) = 0.5225 \text{ kg·m}^2$$

The initial angular momentum is $I_i\omega_i$, or 0.5225(36). The final moment of inertia I_f is

$$I_f = 0.5 + 2(2)(0.275^2) = 0.8025 \text{ kg·m}^2$$

Since no external moments act on the system, the angular momentum is conserved; then

$$I_f\omega_f = I_i\omega_i \qquad 0.8025\omega_f = 0.5225(36) \qquad \therefore \omega_f = 23.4 \text{ rad/s}$$

7.36. Solve the following impact equations for unknowns v_1 and v_2:

$$e = \frac{v_2 - v_1}{u_1 - u_2} \tag{1}$$

$$m_1u_1 + m_2u_2 = m_1v_1 + m_2v_2 \tag{2}$$

where e = coefficient of restitution
     u_1, u_2 = speed of bodies 1 and 2, respectively, before impact
    v_1, v_2 = speeds of bodies 1 and 2, respectively, after impact
   m_1, m_2 = masses of bodies 1 and 2, respectively

SOLUTION

Multiply (1) by $(u_1 - u_2)m_1$ to obtain

$$em_1(u_1 - u_2) = m_1 v_2 - m_1 v_1 \qquad (3)$$

Add (2) and (3) to get

$$m_1 u_1 + m_2 u_2 + em_1(u_1 - u_2) = (m_2 + m_1)v_2 \qquad (4)$$

Then,

$$v_2 = \frac{m_1 u_1(1 + e) + u_2(m_2 - em_1)}{m_2 + m_1}$$

Similarly,

$$v_1 = \frac{m_2 u_2(1 + e) + u_1(m_1 - em_2)}{m_2 + m_1}$$

7.37. Rework Problem 7.36 for the impact of purely elastic bodies ($e = 1$).

SOLUTION

For elastic impact, substitute $e = 1$ in the equations for v_2 and v_1 derived in Problem 7.36. There results

$$v_1 = \frac{2m_2 u_2 + u_1(m_1 - m_2)}{m_2 + m_1} \qquad (1)$$

$$v_2 = \frac{2m_1 u_1 + u_2(m_2 - m_1)}{m_2 + m_1} \qquad (2)$$

Of special interest is the case when $m_1 = m_2 = m$. Then the equations become

$$v_1 = \frac{2mu_2 + 0}{m + m} = u_2 \qquad (3)$$

$$v_2 = \frac{2mu_1 + 0}{m + m} = u_1 \qquad (4)$$

The above equations explain what takes place when a moving coin strikes an identical stationary coin on a frictionless surface. The final speed of the moving coin will be the initial speed of the other coin (in this case zero), while the final speed of the formerly stationary coin will be the initial speed of the moving coin. In other words, the moving coin stops and the stationary one assumes its speed. The same thing is true of a straight row of coins. The first one (farthest away from the moving coin) moves away while the others remain stationary.

7.38. What happens during inelastic impact?

SOLUTION

This is the case when one body absorbs the other or clings to it. Common sense indicates that they have a common final speed. Note that if $e = 0$ is substituted into the equations for v_1 and v_2 in Problem 7.36, this is actually so:

$$v_1 = \frac{m_2 u_2 + m_1 u_1}{m_2 + m_1} \qquad v_2 = \frac{m_1 u_1 + m_2 u_2}{m_2 + m_1} \qquad \therefore v_1 = v_2$$

7.39. Two billiard balls meet centrally with speeds of 2 and −4 m/s. What will be their final speeds after impact if the coefficient of restitution is assumed to be 0.8?

SOLUTION

Let subscript 1 refer to the 2 m/s ball and subscript 2 to the −4 m/s ball. The masses are assumed equal ($m_1 = m_2 = m$). Hence, $u_1 = 2$ m/s and $u_2 = -4$ m/s. The problem is to determine v_1 and v_2. Work directly from the following fundamental equations rather than apply the results of Problem 7.36:

$$e = \frac{v_2 - v_1}{u_1 - u_2} \qquad 0.8 = \frac{v_2 - v_1}{2 - (-4)} \qquad \therefore v_2 - v_1 = 4.8 \text{ m/s} \qquad (1)$$

$$m_1u_1 + m_2u_1 = m_1v_1 + m_2v_2 \qquad m(2) + m(-4) = mv_1 + mv_2 \qquad \therefore v_1 + v_2 = -2 \text{ m/s} \qquad (2)$$

Solve the two equations simultaneously to get $v_1 = -3.8$ m/s and $v_2 = 1.8$ m/s.

7.40. A ball rebounds vertically from a horizontal floor to a height of 20 m. On the next rebound, it reaches a height of 14 m. What is the coefficient of restitution between the ball and the floor?

SOLUTION

Let subscript 1 refer to the ball. The initial and final speeds of the floor, u_2 and v_2, are zero since it is assumed that the floor remains stationary during impact. The second time the ball hits the floor it has fallen from a height of 20 m. Its speed u_1 is

$$u_1 = \sqrt{2gh} = \sqrt{2(9.8)(20)} = 19.8 \text{ m/s}$$

By similar reasoning, the speed v_1 of rebound may be found. The ball with speed v_1 rises to a height of 14 m. Hence, its starting speed v_1 is $v_1 = \sqrt{2(9.8)(14)} = 16.6$ m/s. Now apply the following equation of impact, where downward speeds are chosen positive:

$$e = \frac{v_2 - v_1}{u_1 - u_2} = \frac{0 - (-16.6)}{19.8 - 0} = 0.84$$

The value of e could be determined by considering the square roots of the successive heights to which the ball bounced. This is true since the value of e in this case is actually the ratio of the speeds which depend on the square roots of the heights $e = \sqrt{14/20} = 0.84$.

7.41. A 2-kg ball moving horizontally with a speed of 4 m/s, as shown in Fig. 7-30(*a*), hits the bottom of the rigid, slender 2.5-kg bar, which pivots about its top. If the coefficient of restitution is 0.7, determine the angular speed of the bar and the linear speed of the ball just after impact.

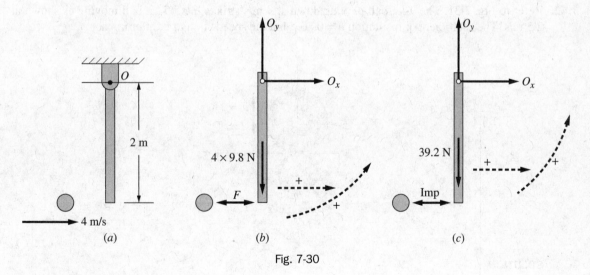

Fig. 7-30

SOLUTION

During the first phase of the impact, from time 0 to t_1, the linear impulse acting between the ball and bar will cause the sphere to slow down and the lower end of the bar to speed up to a common speed u. The free-body diagram for this first phase is shown in Fig. 7-30(*b*). The linear impulse of the interacting force F causes the change in momentum of the ball, as indicated in (1) below, where to the right is considered positive:

$$-\int_0^{t_1} F \, dt = \Delta L = 2(u - 4) \qquad (1)$$

This linear impulse has a moment about the pivot point O of the bar, and causes a change in the angular momentum of the bar as indicated in (2) below, where counterclockwise is considered positive:

$$\int_0^{t_1} F(2)\,dt = \Delta H_O = I_O(\omega - 0) \tag{2}$$

where $I_O = \frac{1}{3}mL^2 = \frac{1}{3}2.5(2^2) = 3.33$ kg·m^2, $\omega = (u/2)$ rad/s. Substituting these values into (2) and dividing by 2, we have

$$\int_0^{t_1} F\,dt = 0.833u \tag{3}$$

Adding (1) and (3), we find $u = 2.82$ m/s. Thus, at the end of this first phase (compression phase) the ball and the lower end of the bar have a common speed of 2.82 m/s to the right. The bar then has an angular speed $\omega = 0.842/2 = 0.421$ rad/s counterclockwise.

The impulse acting during this phase can be calculated from (1) as

$$\int_0^{t_1} F\,dt = -2(2.82 - 4) = 2.35 \text{ N·s}$$

In the second phase (restitution phase), the linear impulse acting is only 0.7 of that acting in the compression phase, as shown in Fig. 7-30(c). Proceeding as before, using v as the speed of the ball immediately after impact and ω' as the angular speed of the bar immediately after impact, the equations for the ball and bar are, respectively,

$$-0.7(2.35) = 2(v - 2.82) \tag{4}$$

$$0.7(2.35)(2) = 3.33\left(\omega' - \frac{2.82}{2}\right) \tag{5}$$

from which $v = 2.72$ ft/s to the right and $\omega' = 2.40$ rad/s counterclockwise.

7.42. Refer to Fig. 7-31. The 9-kg ball moving down at 3 m/s strikes the 5.5-kg ball moving as shown at 2.5 m/s. The coefficient of restitution $e = 0.80$. Find the speeds v_1 and v_2 after impact.

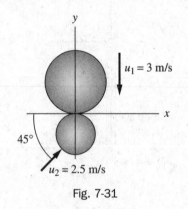

Fig. 7-31

SOLUTION

To determine the y components of the final speeds, apply the two equations

$$e = \frac{(v_2)_y - (v_1)_y}{(u_1)_y - (u_2)_y} \qquad \text{and} \qquad m_1(v_1)_y + m_2(v_2)_y = m_1(u_1)_y + m_2(u_2)_y$$

Then

$$0.80 = \frac{(v_2)_y - (v_1)_y}{-3 - 1.77} \qquad \text{and} \qquad 9(v_1)_y + 5.5(v_2)_y = 9(-3) + 5.5(+1.77)$$

whose solution is $(v_2)_y = -3.56$ m/s, i.e., down; $(v_1)_y = 0.26$ m/s, i.e., up.

In the x direction, the 5.5-kg ball will continue to the right with undiminished speed. Thus, $(v_2)_x = 1.77$ m/s to the right, and $(v_1)x = 0$.

To summarize, the 9-kg ball will rebound up with a speed of 0.26 m/s, and the 5.5-kg ball will move to the right and down with components of 1.77 and 3.56 m/s, respectively.

7.43. Refer to Fig. 7-32. A body at rest is free to move on a frictionless tabletop. It is struck by an impact force F at a distance d from its mass center G. Investigate the motion and locate the instantaneous center of the body in terms of d and its radius of gyration about its mass center.

SOLUTION

Employing the impulse-momentum equations for plane motion,

$$\sum (Imp)_h = \Delta L_h \qquad \text{or} \qquad \int F \, dt = m(\bar{v} - 0) \qquad (1)$$

$$\sum (Ang \, Imp)_G = \Delta H_G \qquad \text{or} \qquad \int Fd \, dt = \bar{I}(\omega - 0) \qquad (2)$$

Since the distance d is constant, it may be removed from under the integral sign. Then $\bar{v} = \int F \, dt / m$ and $\omega = (d/\bar{I}) \int F dt$. But $\bar{I} = m\bar{k}^2$; hence, $\omega = (d/m\bar{k}^2) \int F \, dt$, where $\bar{k}$ is the radius of gyration of the body with respect to the mass center.

To locate the instantaneous center—a point about which the body tends to pivot only—select any point O at a distance b from the mass center on a line perpendicular to the action line of the impact force F, as shown in the figure. The velocity of O is $\mathbf{v}_O = \mathbf{v}_{O/G} + \bar{\mathbf{v}}$. But $\bar{v}$ has been found in terms of the linear impulse. Also, $v_{O/G} = b\omega$, where ω has just been defined in terms of the linear impulse. Note that for the object shown, $\bar{v}$ is to the right (+) and ω is counterclockwise. Hence, $v_{O/G}$ is to the left (−).

Substituting in the formula,

$$v_O = -b\omega + \bar{v} = \frac{-bd}{m\bar{k}^2} \int F \, dt + \frac{1}{m} \int F \, dt$$

To make O an instantaneous center, v_O must be zero. Set the above expression equal to zero and solve to obtain $b = \bar{k}^2/d$.

This may be viewed slightly differently. Suppose the distance d was chosen originally as $\bar{k}^2/b$. Then point O will remain at rest. When this occurs, the force F is said to be applied through the *center of percussion P.*

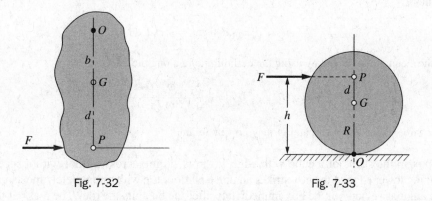

Fig. 7-32   Fig. 7-33

7.44. Apply the results of the preceding problem to find the height h above the plane at which a solid cylinder of radius R should be struck by a horizontal impact force F in order to have no sliding at the point of contract O. See Fig. 7-33.

SOLUTION

According to the theory, F should be struck through the center of percussion P in order that the point of contact O be the instantaneous center. Hence, $h = d + R = \bar{k}^2/R + R$. But for a cylinder,

$$\bar{k}^2 = \frac{\bar{I}}{m} = \frac{1}{2} \frac{mR^2}{m} = \frac{1}{2} R^2 \qquad \therefore h = \frac{1}{2} R + R = \frac{3}{2} R$$

7.45. The sphere of mass m_1 in Fig. 7-34(a) is moving to the right on a frictionless horizontal plane with a velocity u_1 when it strikes (at right angles) a slender bar of mass m_2 that has a mass-center velocity u_2 to the right and an initial angular velocity ω_i clockwise. If the coefficient of restitution is e, set up the equations needed to solve for the final velocity v_1 of the sphere, the final mass-center velocity v_2 of the bar, as well as the final angular velocity ω_f of the bar. The moment of inertia of the bar relative to the mass center G is $\bar{I}$.

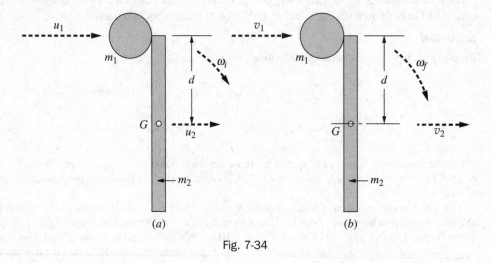

Fig. 7-34

SOLUTION

Figure 7-34(b) depicts the conditions immediately after impact. Note that the initial velocity of the point of impact on the bar is given by the kinematic relation $u_2 + \omega_i d$. Similar reasoning shows its final velocity to be $v_2 + \omega_f d$.

Linear momentum is conserved; hence, calling to the right positive, there results

$$m_1 u_1 + m_2 u_2 = m_1 v_1 + m_1 v_2 \tag{1}$$

Angular momentum relative to the mass center G is conserved; hence, calling clockwise positive, there results

$$m_1 u_1 d + \bar{I} \omega_1 = m_1 v_1 d + \bar{I} \omega_f \tag{2}$$

A third equation is found by using the definition of e along the impact line:

$$e = -\frac{(v_2 + \omega_f d) - v_1}{(u_2 + \omega_i d) - u_1} \tag{3}$$

The three unknowns v_1, v_2, and ω_f may now be found.

7.46. A box in the shape of a cube with edge 1.5 m is sliding across a floor with a speed of 4 m/s. The leading lower edge of the box strikes an upraised floor tile with a completely inelastic impact. What is the angular velocity of the box immediately after the box hits the tile? The mass of the box is 9 kg.

SOLUTION

Since the impact force on the box acts through the leading lower edge, the angular momentum about that edge is conserved. Because the impact is completely inelastic, the leading edge has zero velocity during impact, and so the motion subsequent to the impact is rotation about O, the leading lower edge of the box. Hence, as shown in Fig. 7-35,

$$H_O = H'_O$$

$$mv\left(\frac{1}{2}s\right) = I_O \omega$$

But

$$I_O = \bar{I} + md^2 \qquad \text{or} \qquad I_O = \frac{1}{6}ms^2 + m\left[\left(\frac{1}{2}s\right)^2 + \left(\frac{1}{2}s\right)^2\right]$$

$$I_O = \frac{2}{3}ms^2 = \frac{2}{3}(9)(1.5)^2 = 13.5 \text{ kg·m}^2$$

Substituting in the conservation of momentum equation yields

$$(9)(4)(0.75) = 13.5\omega \qquad \therefore \omega = 2 \text{ rad/s clockwise}$$

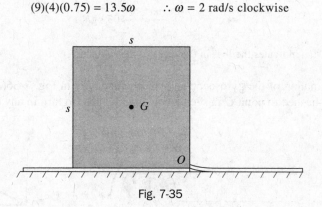

Fig. 7-35

7.47. A rocket and its fuel have an initial mass m_0. Fuel is burned at a constant rate $dm/dt = C$. The gases exhaust at a constant speed relative to the rocket. Neglecting air resistance, find the speed of the rocket at time t.

SOLUTION

Let

m = mass of rocket and remaining fuel at time t; that is, $m = m_0 - Ct$
v = speed of rocket
u = speed of gases

We shall assume vertical motion with up being positive. The only external force to the system at time t is the weight mg which is negative. Our equation is

$$\sum F = m\frac{dv}{dt} - \frac{dm}{dt}(v - u)$$

where $v - u$ is the relative speed of the rocket to the gases and is a constant K. Then

$$-mg = m\frac{dv}{dt} - CK \qquad -(m_0 - Ct)g = (m_0 - Ct)\frac{dv}{dt} - CK \qquad \therefore \frac{dv}{dt} = -g + \frac{CK}{m_0 - Ct}$$

Integrating the last equation,

$$\int_0^v dv = \int_0^t -g\,dt + \int_0^t \frac{CK}{m_0 - Ct}\,dt$$

or

$$v = -gt + \left[-K\ln(m_0 - Ct)\right]_0^t = -gt - K\ln(m_0 - Ct) + K\ln m_0 = K\ln\frac{m_0}{m_0 - Ct} - gt$$

7.48. An empty rocket has a mass of 2000 kg. It is fired vertically up with a fuel load of 8500 kg. The exit speed of the exhaust gases relative to the nozzle is a constant 2000 m/s. At what rate, in kg/s, must the gases be discharged at the beginning if the desired acceleration is 9.8 m/s² up? Assume that the nozzle exhausts at atmospheric pressure.

SOLUTION

Selecting up as positive, we insert values into the equation

$$\sum F = m\frac{dv}{dt} - \frac{dm}{dt}(v - u)$$

to obtain

$$-10\,500 \times 9.8 = 10\,500(9.8) - \frac{dm}{dt}(-2000)$$

Solving,

$$\frac{dm}{dt} = -103 \text{ kg/s}$$

The negative sign indicates the loss in the mass of the system.

7.49. Analyze the motion of the gyroscope shown schematically in Fig. 7-36(*a*) as a spinning wheel and rotor that is attached to point *O* in such a way that it is free to turn in any direction about *O*.

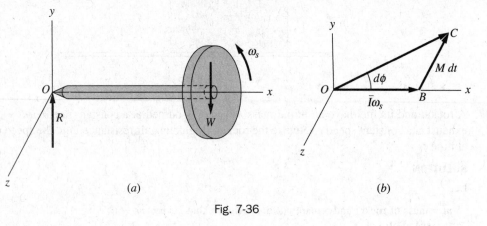

(*a*)   (*b*)

Fig. 7-36

ANALYSIS

Assume that the wheel is spinning with a large angular velocity ω_s about the *x* axis, called the spin axis. Also assume that when the system is released, the external forces acting on it are the weight *W* of the wheel (neglect weight of rotor) and the upward reaction *R* at *O*. The weight *W* exerts a torque on the system about the *z* axis, called the moment axis. It is known by actual experiment that the wheel does not fall but turns about the *y* axis, the axis of precession.

The explanation for this follows. Before the system is released, the angular momentum of the system is equal to the product of its moment of inertia and its angular velocity ω_s, that is, $\overline{I}\omega_s$. This can be represented by a vector **OB,** as shown in Fig. 7-36 (*b*), directed along the *x* axis to the right. (According to the right-hand rule, the thumb points to the right when the fingers are curled in the direction of ω_s.)

Immediately upon release, there is an external torque **M** acting on the system. It acts clockwise about the *z* axis when viewed from the positive end of that axis, and its magnitude is equal to the product of the weight and the perpendicular distance from the *z* axis to the center of mass. For a short time interval *dt*, the impulse of this torque is **M** *dt*. Its units are the same as the units of angular momentum, and it may be represented according to the right-hand rule by a vector along or parallel to the *z* axis and acting in the negative direction. Figure 7-36 (*b*) illustrates this vector **BC**. But this angular impulse **M** *dt* is the change in the angular momentum of the system. The new angular momentum is the vector sum of **OB** and **BC**. This line **OC** is in the *xz* plane but at an angle $d\phi$ with the original *x* axis. Note that the new axis of spin is along *OC*. Using scalar quantities and substituting $d\phi$ for $\tan d\phi$, there results

$$d\phi = \frac{Mdt}{\overline{I}\omega_s} \qquad \text{or} \qquad M = \overline{I}\omega_s\frac{d\phi}{dt}$$

But $d\phi/dt$ is the angular speed of the spin axis about the precession axis, say, ω_p.

The equation of the gyroscope is $M = \overline{I}\omega_s\omega_p$. Note that the spin axis remains in the *xz* plane, provided ω_s remains large.

7.50. An 1800-kg rotor with a radius of gyration 900 mm is mounted with its geometric axis along the fore-and-aft line of a ship that is turning 1 rpm counterclockwise as viewed from above. The rotor turns 300 rpm counterclockwise when observed from the rear of the ship. Assuming the center-to-center distance between bearings to be 1050 mm, determine the front and rear bearing reactions on the rotor.

SOLUTION

A free-body diagram of the rotor is shown in Fig. 7-37 with the x axis as the fore-and-aft line of the ship. The front and rear reactions are R_F and R_R, respectively. According to the right-hand rule, the angular momentum $\bar{I}\omega_s$ about the x axis is drawn to the left along the x axis. To precess in the given direction about the y axis, the rotor must be acted upon by an angular impulse ($\mathbf{M}\,dt$) as shown.

According to the right-hand rule, this moment (caused by the reactions) must be counterclockwise when viewed from the positive end of the z axis. The magnitude of this moment is $M = (1.05/2)\,(R_F - R_R)$.

The values given are $\omega_s = 2\pi(300/60) = 31.4$ rad/s, $\omega_p = 2\pi(1/60) = 0.105$ rad/s, and $\bar{I} = mk^2 = 1800(0.9^2) = 1458$ kg·m². Then, from Problem 7.49, $M = \bar{I}\omega_s\omega_p = (1458)(31.4)(0.105) = 4810$ N·m and $R_F - R_R = M\,(2/1.05) = 9160$ N.

A vertical sum yields $R_F + R_R = 1800 \times 9.8$ N. The solution is

$$R_F = 13.4 \text{ kN} \qquad \text{and} \qquad R_R = 4.24 \text{ kN}$$

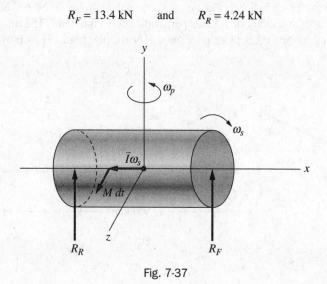

Fig. 7-37

7.51. A satellite in circular orbit about a celestial body with the same mass and radius of the earth ejects a pod tangent to its path. If the satellite is orbiting 800 km above the body, what must be the speed of the pod such that it will strike the body at an angle of 60° with respect to the surface? Refer to Fig. 7-38. The celestial body possesses no atmosphere.

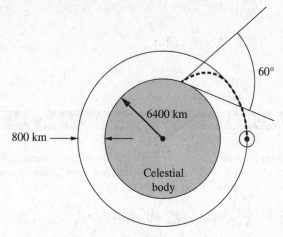

Fig. 7-38

SOLUTION

Since the gravitational force on the pod acts through the center of the body, the angular momentum of the pod about the body center is conserved. So $\Delta H_O = 0$ and $mv_1(R + 800) = mv_2 \cos 60° \, (R)$, where R is the radius of 6400 km. Hence, $v_2 = 2.25v_1$.

Also, from the work-energy equation, $U = \Delta T$:

$$\int_{R+800}^{R} \left(-\frac{GMm}{r^2} \right) dr = \frac{1}{2} m(v_2^2 - v_1^2) \qquad \text{where} \qquad GM = 4 \times 10^{14} \, \text{m}^3/\text{s}^2$$

$$GM \left(\frac{1}{R} - \frac{1}{R + 800} \right) \frac{1}{1000} = \frac{1}{2}(v_2^2 - v_1^2) = 2.531 v_1^2$$

$$v_1 = 1660 \text{ m/s}$$

7.52. One end of an elastic band, shown schematically in Fig. 7-39, is attached to a fixed pin. The other end is attached to a 0.23-kg ball. The elastic constant of the band is 60 N/m. The band is stretched 300 mm and released with a velocity of 1.5 m/s perpendicular to the band. The unstretched length of the band is 600 mm, and motion takes place on a smooth horizontal plane. How close d to the fixed pin will the ball pass?

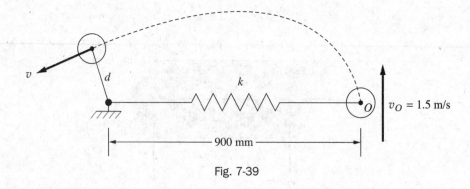

Fig. 7-39

SOLUTION

Since the force of the band on the ball acts through point O, the angular momentum about O is conserved. Hence, $(mv_0) (0.9) = (mv)d$. Or, $vd = 1.35$. Furthermore, $U = \Delta T$ or

$$-\frac{1}{2} k(s^2 - s_0^2) = \frac{1}{2} m(v^2 - v_0^2)$$

$$-\frac{1}{2}(60)(0^2 - 0.3^2) = \frac{1}{2}(0.23)(v^2 - 1.5^2) \qquad \therefore v = 5.1 \text{ m/s}$$

Substituting in $vd = 1.35$ yields $d = 0.27$ m.

SUPPLEMENTARY PROBLEMS

7.53. In Problem 7.6, show that the selection of the top point as the moment center yields $2PR = (d/dt)(4mR^2\omega)$, which does not agree with the results obtained in that problem. [It should not agree, because the top point does not fit the conditions imposed on point O in Eq. (11).]

7.54. A 50-kg mass falls freely from rest for 4 s. Find its momentum at that time.

    Ans.   1960 kg·m/s = 1960 N·s

7.55. A 450-kg pile-driver hammer is dropped from a height of 8 m. If the time required to stop the pile driver is 1/20 s, determine the average force acting during that time.

Ans. $F_{av} = 11.3$ kN

7.56. A body is thrown upward with an initial speed of 15 m/s. Find the time required for the body to attain a speed of 6 m/s downward.

Ans. 2.14 s

7.57. A 25-kg body is projected up a 30° plane with initial speed of 6 m/s. If the coefficient of friction is 0.25 determine the time required for the body to have an upward speed of 3 m/s.

Ans. 0.53 s

7.58. A 1000-kg sphere with a diameter of 3 m is rotating about a diameter at 600 rpm. What angular torque will bring the sphere to rest in 3 min?

Ans. $T = 314$ N·m

7.59. A 50-kg block rests on a horizontal floor. It is acted upon by a horizontal force that varies from zero according to the law $F = 75\,t$. If the force acts for 10 s, what is the speed of the block? Assume that the coefficient of static and kinetic friction is 0.25.

Ans. 50.6 m/s

7.60. The magnitude of a force applied tangentially to the rim of a 1200-mm-diameter pulley varies according to the law $F = 0.03t$. Determine the angular impulse of the force about the axis of rotation for the interval 0 to 35 s.

Ans. 11.0 kg·m^2/s

7.61. A 140-kg disk rotating at 1000 rpm has a diameter of 900 mm. Find its angular momentum.

Ans. 1480 kg·m^2/s

7.62. A bob of mass m travels in a circular path of radius R on a smooth horizontal plane under the restraint of a string which passes through a hole O in the plane, as shown in Fig. 7-40. If the angular speed of the bob is ω_1 when the radius is R, what will be the angular speed if the string is pulled from underneath until the radius of the path is $\frac{1}{2}R$? Find the ratio of the final tension in the string to its initial value.

Ans. 8

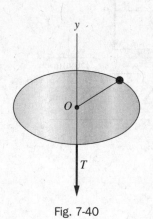

Fig. 7-40

7.63. A 12-kg disk with a radius of gyration equal to 600 mm is subjected to a torque $M = 2t$ N·m. Determine the angular speed of the disk 2 s after it started from rest. Assume frictionless bearings.

Ans. $\omega = 0.93$ rad/s

7.64. A 10-kg block is acted upon by a horizontal force P that varies according to the graph shown in Fig. 7-41. The block moves from rest on a smooth horizontal surface. What will be the position and speed after 7 s?

Ans. $v = 8.64$ m/s, $s = 23.9$ m

7.65. A 6-kg rotor with radius of gyration 360 mm comes to rest in 85 s from a speed of 180 rpm. What was the frictional torque that stopped the rotor?

Ans. $T = 0.172$ N·m

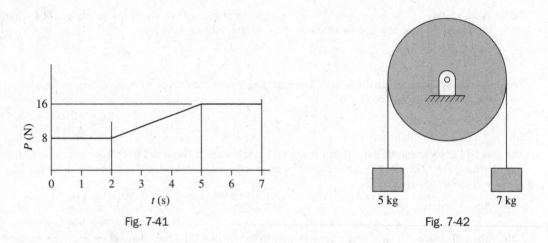

Fig. 7-41   Fig. 7-42

7.66. The two masses shown in Fig. 7-42 are connected by a light inextensible cord that passes over a homogeneous 4-kg cylinder having a diameter of 1200 mm. How long will it take the 7-kg mass to move from rest to a speed of 2 m/s^2.

Ans. $t = 1.43$ s

7.67. A 100-kg rotor has a radius of gyration of 700 mm. The rotor is rotating at 6 rad/s counterclockwise when the force P is applied to the brake, as shown in Fig. 7-43. What is the magnitude of P if the rotor stops in 10 s? The coefficient of friction between the brake and rotor is 0.4.

Ans. $P = 13$ N

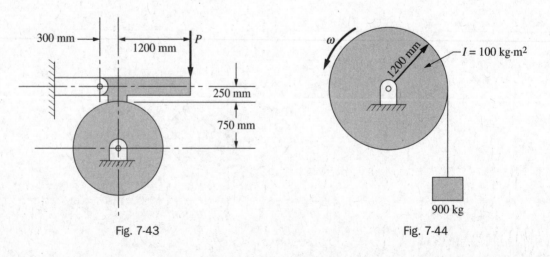

Fig. 7-43   Fig. 7-44

7.68. A drum rotating at 20 rpm is lifting a 900-kg cage connected to it by a light inextensible cable as shown in Fig. 7-44. If power is cut off, how long will it take the cage to come to rest? Assume frictionless bearings.

Ans. $t = 0.152$ s

7.69. Two homogeneous disks of weights W_1 and W_2 and radii r_1 and r_2 are free to turn in frictionless bearings in the fixed vertical frame shown in Fig. 7-45. The upper disk W_1 is turning with angular velocity ω clockwise when it is dropped onto the lower disk W_2. Assuming that the coefficient of friction μ is constant, determine the time until slipping stops. What are the angular velocities at that time?

Ans. $t = \dfrac{r_1\omega}{2\mu g(1 + W_1/W_2)}$, $\omega_1 = \dfrac{\omega}{1 + W_2/W_1}$ clockwise, $\omega_2 = \dfrac{r_1}{r_2}\left(\dfrac{\omega}{1 + W_2/W_1}\right)$ counterclockwise

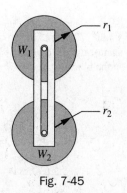

Fig. 7-45

7.70. A rotor with a moment of inertia of 435 kg·m^2 is supported by a shaft that turns in two bearings. The friction in the bearings produces a resisting moment of 28 N·m. After the power is cut off, the rotor slows to 180 rpm in 150 s. What was the original speed of the rotor?

Ans. 272 rpm

7.71. A table fan with a mass m is at rest on a horizontal tabletop. The coefficient of friction between the fan and the top is μ. Let A be the area of air drawn through the fan and ρ the mass density of the air. If v is the velocity of the downstream air, derive an expression for the theoretical maximum velocity before there is slip of the fan on the tabletop.

Ans. $v = \left(\sqrt{\mu mg}\right)/A\rho$

7.72. Water issues at the rate of 20 m/s from a 100-mm-diameter nozzle and strikes a vertical wall, as shown in Fig. 7.46. Determine the force of the water on the wall.

Ans. 3140 N

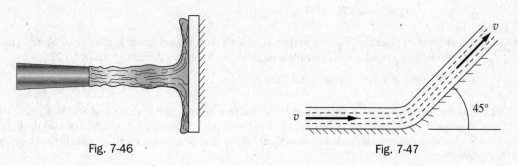

Fig. 7-46   Fig. 7-47

7.73. A jet of water 50 mm in diameter exerts a force of 1200 N on a flat stationary plate perpendicular to the stream. What is the nozzle speed of the jet?

Ans. 24.7 m/s

7.74. A jet of water issues from a 25-mm-diameter nozzle at 30 m/s. Determine the total force against a curved stationary vane where the direction of the jet changes 45° as shown in Fig. 7-47. Assume no friction.

Ans. 129 N

7.75. A jet of water leaves a nozzle at a speed of 10 m/s. It impinges at right angles against a vertical plate, which is stopped from moving by a horizontal force of 400 N. What is the diameter of the nozzle?

Ans. 71.4 mm

7.76. The force on a curved stationary vane that changes the direction of a jet of water by 45° is 2000 N. If the nozzle speed of the jet is 30 m/s, calculate the diameter of the nozzle.

Ans. 98.3 mm

7.77. Refer to Fig. 7-48. A stream of water 2700 mm^2 in cross section and flowing horizontally with a speed of 30 m/s splits into two equal parts against the fixed blade. Assuming no friction between the water and the blade, determine the force of the water on the blade.

Ans. $F_x = 4150$ N

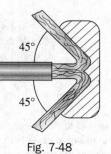

45°

45°

Fig. 7-48

7.78. In the preceding problem, assume that the stream splits so that two-thirds of the water moves to the top portion of the blade and one-third to the lower portion of the blade. Determine the reaction of the water on the blade.

Ans. $F_x = 4150$ N to the right, $F_y = 573$ N up

7.79. A nozzle with a 1700-mm^2 cross-sectional area discharges a horizontal stream of water with a speed of 25 m/s against a fixed vertical plate. What is the force against the plate?

Ans. 1060 N to the right

7.80. A stream of water with a 2000-mm^2 cross-sectional area moving at 10 m/s strikes a stationary curved blade, as shown in Fig. 7-47. Assuming that friction is negligible, find the horizontal and vertical components of the force of the blade on the stream.

Ans. $F_x = 58.6$ to the left, $F_y = 141$ up

7.81. A horizontal jet of water issues left to right from a 100-mm-diameter nozzle at a rate of 30 m/s. Determine the force exerted on a fixed curved vane if the direction of the stream is changed through 120°.

Ans. $F_x = 10.6$ kN to the left, $F_y = 6.12$ kN up

7.82. Two homogeneous sliding disks A and B are mounted on the same shaft. Disk A is a 50-kg disk 1000 mm in diameter, 50 mm thick, and at rest. Disk B is a 100-kg disk 1000 mm in diameter, 100 mm thick, and rotating clockwise at 600 rpm. If the disks are suddenly engaged so as to rotate together, what is their common angular speed?

Ans. 400 rpm

7.83. In Problem 7.82, what is the percent loss in kinetic energy of the system because the disks are coupled together?

Ans. 33.3 percent

7.84. A 32-kg child is standing in a 45-kg boat that is initially at rest. If the child jumps horizontally front to aft with a speed of 2 m/s relative to the boat, determine the speed of the boat.

Ans. 0.831 m/s

7.85. After a flood, an 18-kg goat finds itself adrift on one end of a 25-kg log, 2 m long. As the other end touches shore, the goat maneuvers to that end. When it gets there, how far is it from shore? Assume that the log is at right angles to the shore and the water is almost calm after the storm.

Ans. 0.837 m

7.86. A 0.09-kg bullet is fired from a 6.4-kg rifle with a muzzle speed of 300 m/s. What is the speed of the rifle recoil?

Ans. 4.22 m/s

7.87. A 50-Mg gun fires a 500-kg projectile. If the recoil apparatus exerts a constant force of 400 kN and the gun moves back 200 mm, calculate the muzzle speed of the projectile.

Ans. 179 m/s

7.88. A 270-kg projectile is shot with an initial velocity of 680 m/s from a 90-Mg gun. What is the velocity of recoil of the gun?

Ans. $v_R = 2.04$ m/s

7.89. In Fig. 7-49, a series of n identical balls is shown on a smooth horizontal surface. If number 1 moves horizontally with speed u into number 2, which in turn collides with number 3, etc., and if the coefficient of restitution for each impact is e, determine the speed of the nth ball.

Ans. $v_n = (1 + e)^{n-1} u / 2^{n-1}$

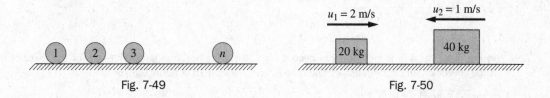

| Fig. 7-49 | Fig. 7-50 |

7.90. In Fig. 7-50, the 20-kg block is moving to the right with a speed of 2 m/s. The 40-kg block is moving to the left with a speed of 1 m/s. If the coefficient of restitution is assumed to be 0.4, determine the speeds immediately after impact.

Ans. $v_{20} = 0.8$ m/s to the left, $v_{40} = 0.4$ m/s to the right

7.91. A 3-kg ball and a 5-kg ball of the same diameter move on a smooth horizontal plane along a straight line with speeds of 5 and −3 m/s, respectively. Determine their speeds after impact if the impact is (*a*) inelastic, (*b*) elastic, and (*c*) such that the coefficient of restitution is 0.4.

Ans. (*a*) 0, 0; (*b*) −5, 3 m/s; (*c*) −2, −1.2 m/s

7.92. A 20-kg block *A* moving 12 m/s horizontally to the right meets a 16-kg block *B* moving 8 m/s horizontally to the left. If the coefficient of restitution is $e = 0.7$, determine the speeds of *A* and *B* immediately after impact.

Ans. $v_A = 3.11$ m/s to the left, $v_B = 10.9$ m/s to the right

7.93. A 900-kg car traveling at 50 km/h overtakes a 700-kg car traveling at 25 km/h in the same direction. What is their common speed after coupling? What is the loss in kinetic energy?

Ans. 10.85 m/s, 9500 N·m

7.94. A 30-kg ball *A* moving to the right with a speed of 30 m/s strikes squarely a 10-kg ball *B* that has a speed of 7 m/s in the opposite direction. If the coefficient of restitution is 0.6, determine the speeds of the balls after impact.

Ans. $v_A = 15.2$ m/s to the right, $v_B = 37.4$ m/s to the right

7.95. A ball falls freely from rest 5 m above a smooth plane angled at 30° to the horizontal. If $e = 0.5$, to what height will the ball rebound?

Ans. $h = 0.08$ m

7.96. A ball falls 6 m from rest. It hits a horizontal plane and rebounds to a height of 5 m. Determine the coefficient of restitution.

Ans. $e = 0.91$

7.97. A ball is dropped onto a smooth horizontal floor, from which it bounces to a height of 9 m. On the second bounce, it attains a height of 6 m. What is the coefficient of restitution between the ball and the floor?

Ans. $e = 0.82$

7.98. A 4-kg mass falls 150 mm onto a 2-kg platform mounted on springs whose combined $k = 800$ N/m. If the impact is fully plastic ($e = 0$), determine the maximum distance the platform moves down from its initial position. See Fig. 7-51.

Ans.   102 mm

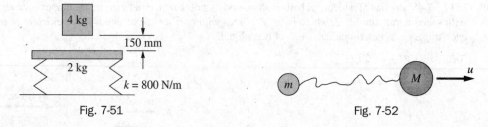

Fig. 7-51   Fig. 7-52

7.99. In Fig. 7-52, mass M is moving with speed u when the string becomes taut. What will be the speed of mass m if the coefficient of restitution is e? The masses are on a smooth horizontal plane.

Ans. $v = Mu(1 + e)/(M + m)$

7.100. A mass m_1 moving with speed u strikes a stationary mass m_2 hanging on a string of length L, as shown in Fig. 7-53. If the coefficient of restitution is e, determine (a) the speed of each mass immediately after the impact and (b) the height h to which mass m_2 will rise.

Ans.   (a) $v_1 = u\dfrac{m_1 - em_2}{m_1 - m_2}$, $v_2 = u\dfrac{m_1(1 + e)}{m_1 + m_2}$; ($b$) $h = \dfrac{u^2 m_1^2 (1 + e)^2}{(m_1 + m_2)^2 2g}$

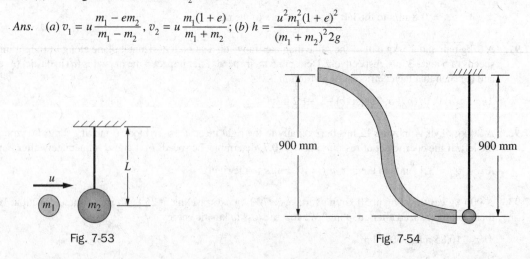

Fig. 7-53   Fig. 7-54

7.101. A 1-kg ball traverses the frictionless tube shown in Fig. 7-54, falling a vertical distance of 900 mm. It then strikes a 1-kg ball hanging on a 900-mm cord. Determine the height to which the hanging ball will rise (a) if the collision is perfectly elastic, (b) if the coefficient of restitution is 0.7.

Ans.   (a) 900 mm, (b) 650 mm

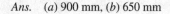

7.102. Refer to Fig. 7-55. A ball thrown from position *A* against a smooth vertical circular wall rebounds and hits position *B* at the other end of the diameter through *A*. Show that the coefficient of restitution is equal to the square of the tangent of angle θ. The triangle is horizontal.

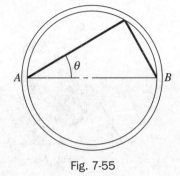

Fig. 7-55

7.103. A billiard ball moving at 4 m/s strikes a smooth horizontal plane at an angle of 35° as shown in Fig. 7-56. If the coefficient of restitution is 0.6, what is the velocity with which the ball rebounds?

Ans. $v = 3.55$ m/s, $\theta = 22.8°$

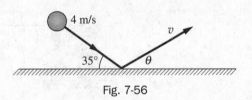

Fig. 7-56

7.104. The 500-g sphere shown in Fig. 7-57 is moving on a smooth horizontal plane with a velocity of 5 m/s. It strikes the end of a homogeneous slender 1.8-kg bar that is 4 m long. If the horizontal bar is initially at rest and the coefficient of restitution is 0.6, determine the speed of the sphere immediately after impact.

Ans. 4.06 m/s

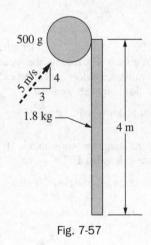

Fig. 7-57

7.105. Solve Problem 7.46 if the collision between the box and the tile is perfectly elastic.

Ans. $\omega = 4$ rad/s clockwise

7.106. A slender, horizontal bar is falling downward at 3 m/s when its right-hand end strikes the edge of a table. If the coefficient of restitution between the table and the bar is 0.45, determine the angular velocity of the bar immediately after impact. The 1.4-kg bar is 900 mm long.

Ans. $\omega = 7.25$ rad/s counterclockwise

7.107. In Fig. 7-58, a large drum is filled with liquid weighing 7850 N/m^3. The drum and liquid weigh 890 N and are at rest on a horizontal sheet of ice for which the coefficient of friction is 0.05. If a 100-mm plug 900 mm below the surface of the liquid is suddenly removed, will the drum move?

Ans. Yes

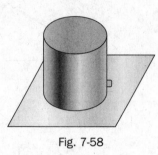

Fig. 7-58

7.108. A tank of liquid with a mass density of ρ rests on a cart. If h is the height of the fluid above the orifice, what horizontal force F is necessary to hold the open tank at rest when the fluid starts issuing from the nozzle? The cross-sectional area of the orifice is A (see Fig. 7-59).

Ans. $F = 2Ah\rho g$

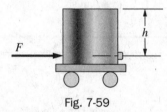

Fig. 7-59

7.109. A rocket has a mass of 3000 kg empty. If it is projected vertically up from the earth with a fuel load of 7000 kg, calculate the initial acceleration. Assume that the gases exhaust at 2000 m/s relative to the rocket and that the initial rate of fuel burning is 150 kg/s.

Ans. $a = 20.2$ m/s^2

7.110. The rotor of a gyroscope is a 100-mm-diameter cylinder with a mass of 0.18 kg. It is mounted horizontally midway between bearings 150 mm apart. The rotor is turning at 9000 rpm in a clockwise direction when viewed from the rear. The assembly is turning at 2 rad/s about a vertical axis in a clockwise direction when viewed from above. What are the bearing reactions on the rotor shaft?

Ans. $R_F = 3.71$ N up, $R_R = 1.94$ N down

7.111. Refer to Fig. 7-60. A solid 150-mm-diameter wheel that is 50 mm thick rotates at 628 rad/s. Neglect the mass of the shaft attached to it. Assume a density of 7700 kg/m^3. Determine the speed of precession.

Ans. $\omega_p = 1.11$ rad/s clockwise about the y axis when viewed from above

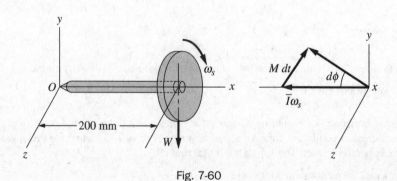

Fig. 7-60

7.112. A space shuttle is to link up with a space laboratory orbiting at a constant altitude of 400 km. At an altitude of 80 km and speed of 4000 m/s, the shuttle engine is shut off. What must be the angle that the velocity makes with the vertical, at shutoff, if the shuttle is to arrive tangent to the orbit of the laboratory? Neglect any drag effects.

Ans. 56.8°

7.113. A satellite is launched parallel to the surface of the earth at an altitude of 640 km. The speed of launch is 33 000 km/h. Determine the maximum altitude reached by the satellite.

Ans. 13 600 km

7.114. Repeat Problem 7.113 if the satellite is launched outward at 60° from the vertical.

Ans. 2400 km

CHAPTER 8

Mechanical Vibrations

8.1 Definitions

Mechanical vibration of a system possessing masses and elasticity is motion about an equilibrium position that repeats itself in a definite time interval.

The *period* is the time interval for the vibration to repeat itself.

A *cycle* is each repetition of the entire motion completed during the period.

The *frequency* is the number of cycles in a unit of time.

Free vibrations occur in a system not acted upon by external disturbing forces.

Natural frequency is the frequency of a system undergoing free vibrations.

Forced vibrations occur in a system acted upon by periodic external disturbing forces.

Resonance occurs when the frequency of the periodic external disturbing forces vibrations coincides with or at least approaches the natural frequency of the system.

Transient vibrations disappear with time. Free vibrations are transient in character.

Steady-state vibrations continue to repeat themselves with time. Forced vibrations are examples of steady-state vibrations.

8.2 Degrees of Freedom

The degrees of freedom of a system depend on the number of variables (coordinates) needed to describe its motion. For example, in Fig. 8-1(*a*), the motion of the mass on a spring that is assumed to vibrate only in a vertical line can be described with one coordinate, and thus it possesses a single degree of freedom. A bar supported by the two springs in Fig. 8-1(*b*) needs two variables as shown, and therefore it possesses 2 degrees of freedom.

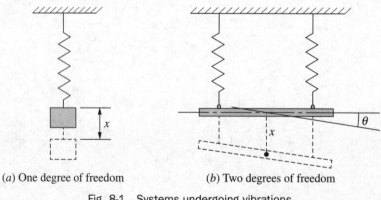

(*a*) One degree of freedom   (*b*) Two degrees of freedom

Fig. 8-1 Systems undergoing vibrations.

8.3 Simple Harmonic Motion

Simple harmonic motion, as shown by Eq. (11) in Chapter 2, can be represented by a sine or cosine function of time. Thus,

$$x = X \sin \omega t \tag{1}$$

is an equation of simple harmonic motion. It could represent the projection on a diameter of a vector of length X as the tip of the vector X rotates on a circular path with a constant angular velocity ω (radians per second). For this motion,

x = length of projection
X = length of rotating vector
ω = circular frequency, rad/s
$\tau = 2\pi/\omega$ = period, s
$f = \omega/2\pi$ = frequency, cycles per second (Hz)

Bodies vibrate with simple harmonic motion or some combination of simple harmonic motions with different frequencies and amplitudes.

8.4 Multicomponent Systems

A multicomponent system is analyzed by replacing it with an equivalent system of masses, springs, and damping devices. The differential equations of this idealized system, when solved, will approximate the desired results. Engineering judgment will suggest proper modifications to fit the actual system.

The problems that follow illustrate free vibrations with and without damping and forced vibrations with and without damping. Only viscous damping (where the damping force is proportional to the velocity of the body) is considered. However, it is good to point out two other types of damping: (1) coulomb damping, which is independent of the velocity and arises because of sliding of the body on dry surfaces (its force is thus proportional to the normal force between the body and the surface on which it slides), and (2) solid damping, which occurs as internal friction within the material of the body itself (its force is independent of the frequency and proportional to the maximum stress induced in the body itself).

8.5 Units

The following is a table of the SI units and U.S. Customary units used in mechanical vibrations problems.

Quantity	SI	U.S. Customary
Length	m	in
Time	s	s
Velocity	m/s	ft/s or in/s
Acceleration	m/s^2	ft/s^2 or in/s^2
Mass	kg	slug
Mass moment of inertia	$kg \cdot m^2$	$slug\text{-}ft^2$ or $lb\text{-}s^2\text{-}in$
Force	N	lb
Spring	N/m	lb/in
Damping constant	kg/s	lb-s/in

SOLVED PROBLEMS

Free Vibrations—Linear

8.1. A mass m hangs on a vertical spring whose spring constant or modulus is k. Assuming that the mass of the spring may be neglected, study the motion of the mass if it is released at a distance x_0 below the equilibrium position with an initial velocity v_0 downward.

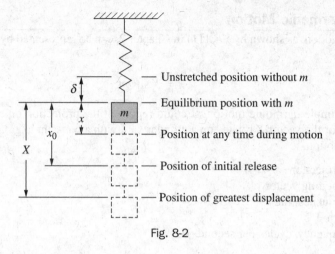

Fig. 8-2

SOLUTION

In Fig. 8-2, various positions of the mass m are shown. The distances are exaggerated for clarity. The value X is the amplitude of the motion. Of course, the mass will also rise to a height X above the equilibrium position.

The tension in the spring is equal to the product of the spring modulus k and the distance the spring is stretched or compressed from its unstretched position (without m). Draw free-body diagrams of the mass in its equilibrium position and at the position x below equilibrium. Note that the position x of the mass at any time is expressed from its position of static equilibrium.

In Fig. 8-3, no acceleration is shown, since the system is in equilibrium. Hence, $T = k\delta = mg$. Consequently, $\delta = mg/k$.

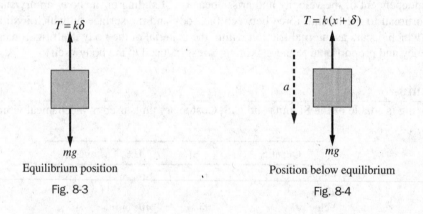

Fig. 8-3   Equilibrium position

Fig. 8-4   Position below equilibrium

In Fig. 8-4, assume that displacements below the equilibrium position are positive. Since the direction of the acceleration is unknown, assume it to be positive. A negative sign would then mean it is directed up. Apply Newton's laws to this free-body diagram and obtain

$$\sum F_v = ma \qquad \text{or} \qquad mg - T = ma \qquad (1)$$

Substitute $T = k(\delta + x)$ and $mg = k\delta$ to get $k\delta - k\delta - kx = ma$. Since $a = d^2x/dt^2$, the equation of motion becomes

$$\frac{d^2x}{dt^2} + \frac{k}{m}x = 0 \qquad (2)$$

It should be apparent that whenever x is positive (below the equilibrium position), the acceleration (which is then negative) is directed up. This means that as the mass moves down to its lowest position, the acceleration is up or the mass is decelerating. Just after reaching the bottom and starting up, the displacement is still positive and the acceleration is still directed oppositely, i.e., up. Therefore, the mass will accelerate up to the equilibrium position. Above this position, the displacement is negative and the acceleration is directed down to the equilibrium position. Hence, up to its top point of travel, the mass decelerates, and between the top point and the position of equilibrium it accelerates. The acceleration is always directed toward the equilibrium position.

Assume that the solution of the above second-order differential equation has the form

$$x = A\sin\omega t - B\cos\omega t \tag{3}$$

where ω is the circular frequency in rad/s. To determine whether this is a solution, take the second derivative with respect to time (d^2x/dt^2) and substitute it into the differential equation. Note that

$$\frac{dx}{dt} = A\omega\cos\omega t - B\omega\sin\omega t \qquad \text{and} \qquad \frac{d^2x}{dt^2} = -A\omega^2\sin\omega t - B\omega^2\cos\omega t = -\omega^2 x$$

Substitute the value of d^2x/dt^2 into the equation of motion to obtain $-\omega^2 x + (k/m)x = 0$. Hence, ω must equal $\sqrt{k/m}$ if the assumed value of x is to be a solution. Then the solution is thus far

$$x = A\sin\sqrt{\frac{k}{m}}t + B\cos\sqrt{\frac{k}{m}}t \tag{4}$$

Note that a cycle of motion will be completed at intervals of 2π rad, i.e., when $(\sqrt{k/m})\tau = 2\pi$, where τ is the period or time for one cycle. Hence, $\tau = 2\pi\sqrt{m/k}$. The frequency is the inverse of the period, or

$$f = \frac{1}{\tau} = \frac{1}{2\pi}\sqrt{\frac{k}{m}} \tag{5}$$

As pointed out previously, $\delta = mg/k$; hence, the above formulas may also be written $\tau = 2\pi\sqrt{\delta/g}$ and $f = (1/2\pi)\sqrt{g/\delta}$.

The constants of A and B in (4) should be evaluated on the basis of the initial conditions given in the problem. Here it is assumed that at $t = 0$, $x = x_0$, and $v = v_0$. Substitute these values of x and v into the equations for these variables, but be sure also to substitute the time $t = 0$:

$$x_0 = A\sin(\omega\cdot 0) + B\cos(\omega\cdot 0)$$

$$v_0 = A\omega\cos(\omega\cdot 0) - B\omega\sin(\omega\cdot 0)$$

The first equation yields $B = x_0$, and the second equation gives $A = v_0/\omega$.

Hence, the solution is

$$x = \frac{v_0}{\omega}\sin\omega t + x_0\cos\omega t \tag{6}$$

The solution may be written in another form:

$$x = X\cos(\omega t - \phi) \tag{7}$$

where the amplitude $X = \sqrt{(v_0/\omega)^2 + (x_0)^2}$ and the phase angle $\phi = \tan^{-1}(v_0/x_0\omega)$. Note that $\omega = \sqrt{k/m}$.

This problem is the most commonly used example of free vibrations, which theoretically would continue indefinitely after the mass is set in motion. Since only one variable has been used to describe the motion, the system possesses 1 degree of freedom. Many additional problems can now be solved by reducing them to this type. In other words, replace the actual elastic suspension by an equivalent spring attached to a vibrating body.

8.2. A 30-kg machine is mounted on 40-kg platform, which in turn is supported by four springs. Assume that each spring carries one-fourth of the load, and determine the period and frequency of vibration. Each spring has a spring constant of 3000 N/m.

SOLUTION

$$f = \frac{1}{2\pi}\sqrt{\frac{k}{m}} = \frac{1}{2\pi}\sqrt{\frac{3000}{70/4}} = 2.08 \text{ cps (Hz)}$$

$$\tau = \frac{1}{f} = \frac{1}{2.08} = 0.480 \text{ s}$$

8.3. Solve Problem 8.1 using the conservation of energy theorem.

SOLUTION

This theorem states that the sum of the potential energy V and kinetic energy T of the system is a constant provided the system is conservative (no friction or damping is assumed at this point in the discussion).

At any distance x below the equilibrium position, the spring tension is $mg + kx$. Hence, the potential energy V_s of the spring is equal numerically to the work done by this force in stretching the spring:

$$V_s = \int_0^x (mg + kx)\,dx = mgx + \frac{1}{2}kx^2$$

During this same displacement x, the mass has lost potential energy of the amount mgx. Then the total potential energy of the system is $\frac{1}{2}kx^2$.

At a distance x below equilibrium, the kinetic energy T of the mass, which is moving with a velocity dx/dt, is $T = \frac{1}{2}m(dx/dt)^2$. Since the spring is assumed to be massless, its kinetic energy is zero. Therefore the kinetic energy of the system is that of the mass only.

The conservation of energy states that $T + V = $ constant, or that $d/dt(T + V) = 0$. Then

$$\frac{d}{dt}\left[\frac{1}{2}m\left(\frac{dx}{dt}\right)^2 + \frac{1}{2}kx^2\right] = 0 \qquad \text{or} \qquad \frac{1}{2}m(2)\left(\frac{dx}{dt}\right)\frac{d^2x}{dt^2} + \frac{1}{2}k2x\frac{dx}{dt} = 0$$

This reduces to the same differential equation obtained in Problem 8.1:

$$\frac{d^2x}{dt^2} + \frac{k}{m}x = 0$$

Hence, the frequency is

$$\sqrt{\frac{k}{m}} \qquad \text{rad/s} \qquad \text{or} \qquad \frac{1}{2\pi}\sqrt{\frac{k}{m}} \qquad \text{Hz}$$

Note that in a linear differential equation of the type shown above with the second-order term having a coefficient 1, the circular frequency ω in rad/s equals the square root of the coefficient of the x term ($\omega = 2\pi f$, where ω is in rad/s and f is in hertz).

8.4. A small mass m is fastened to a vertical wire that is under tension T, as shown in Fig. 8-5. What will be the natural frequency of vibration of the mass if it is displaced laterally a slight distance and then released? Assume the weight of the mass to be small compared to the tension and that the distance x is small so the wire tensions remain essentially unchanged.

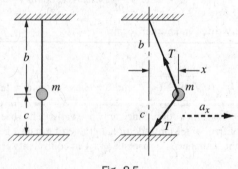

Fig. 8-5

SOLUTION

Assume the mass m at some time during the motion is at a distance x to the right of equilibrium. In the horizontal direction, this mass is acted upon by the components of the two tensions T shown. The tension T will change by just a small increment, only the direction changes significantly. For *small* displacements, these x components of the tensions will be Tx/b and Tx/c, both acting to the left, or negative, if we assume x to the right is positive. Hence, the differential equation of motion is

$$-\frac{Tx}{c} - \frac{Tx}{b} = ma_x = m\frac{d^2x}{dt^2} \qquad \text{or} \qquad \frac{d^2x}{dt^2} + \frac{1}{m}\left(\frac{T}{c} + \frac{T}{b}\right)x = 0$$

Note that this differential equation is entirely similar, except for the coefficient of the x term, to the spring equation. Hence, as indicated in Problem 8.3, the frequency is

$$f = \frac{1}{2\pi}\sqrt{\frac{T}{m}\left(\frac{b+c}{bc}\right)} \qquad \text{Hz}$$

8.5. A cylinder floats as shown in Fig. 8-6. The cross-sectional area of the cylinder is A, and the mass is m. What will be the frequency of oscillation if the cylinder is depressed somewhat and released? The mass density of the liquid is ρ. Neglect the damping effects of the liquid as well as the inertia effects of the moving liquid.

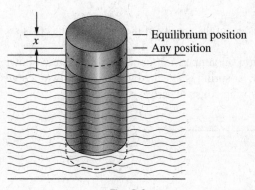

Fig. 8-6

SOLUTION

When the cylinder is a distance x below its equilibrium position, it is acted upon by a buoyant force equal in magnitude to the gravitational force on the displaced liquid. This is exactly analogous to the spring in Problem 8.1. Using Newton's laws, the unbalanced force, which is up for a downward displacement, is equated to the product of the mass and its acceleration. Call a downward displacement positive. Then the differential equation of motion is

$$-\rho g A x = m\frac{d^2x}{dt^2} \qquad \text{or} \qquad \frac{d^2x}{dt^2} + \frac{\rho g A}{m}x = 0$$

Thus, the frequency is

$$f = \frac{1}{2\pi}\sqrt{\frac{\rho g A}{m}} \qquad \text{Hz}$$

8.6. Liquid of density ρ and total length ℓ is used in a manometer, as shown in Fig. 8-7. A sudden increase in pressure on one side forces the liquid down. Upon release of the pressure, the liquid oscillates. Neglecting any frictional damping, what will be the frequency of vibration?

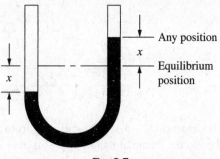

Fig. 8-7

SOLUTION

Assume the liquid to be a distance x below the equilibrium position in the left column and, of course, a distance x above the equilibrium position in the right column.

The unbalanced force tending to restore equilibrium is the gravitational force on a column of the liquid $2x$ high. This force is $2xA\rho g$, where A is the area of the cross section of the liquid. The total mass of liquid in motion is $\rho A \ell$. Using Newton's law,

$$-2xA\rho g = \rho A \ell \frac{d^2 x}{dt^2} \qquad \text{or} \qquad \frac{d^2 x}{dt^2} + \frac{2g}{\ell} x = 0$$

and

$$f = \frac{1}{2\pi} \sqrt{\frac{2g}{\ell}} \qquad \text{Hz}$$

Note that f is independent of the density of the liquid and the cross-sectional area.

8.7. Determine the natural undamped frequency of the system shown in Fig. 8-8. The bar is assumed massless, and the lower spring is attached to the bar midway between the points of attachment of the upper springs. The spring constants are as shown.

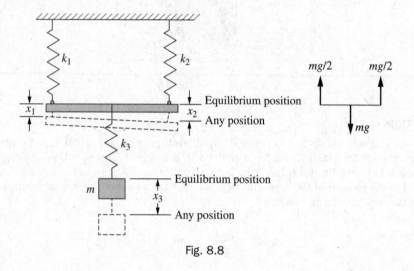

Fig. 8.8

SOLUTION

The total static displacement δ of the mass m is equal to the stretch x_3 of spring k_3 plus the average of x_1 and x_2, which are the extensions of springs k_1 and k_2, respectively. This problem may be solved by substituting this value of δ into the equation $f = (1/2\pi)\sqrt{g/\delta}$ (refer to Problem 8.1). Note that the entire gravitational force mg is transferred through spring k_3 whereas $\frac{1}{2}mg$ is transferred through each spring k_1 and k_2 to the support. The latter fact can be ascertained by referring to the free-body diagram of the bar shown to the right of the figure. Hence,

$$x_3 = \frac{mg}{k_3} \qquad x_2 = \frac{\frac{1}{2}mg}{k_2} \qquad x_1 = \frac{\frac{1}{2}mg}{k_1}$$

and

$$\delta = x_3 + \frac{1}{2}(x_2 + x_1) = mg\left(\frac{4k_1 k_2 + k_1 k_3 + k_2 k_3}{4k_1 k_2 k_3}\right)$$

Substituting,

$$f = \frac{1}{2\pi} \sqrt{\frac{4k_1 k_2 k_3}{m(4k_1 k_2 + k_1 k_3 + k_2 k_3)}}$$

8.8. In Fig. 8-9, the mass m is suspended by means of spring k_2 from the end of a rigid massless beam that is of length ℓ and attached to the frame at its left end. It is also supported in a horizontal position by a spring k_1 attached to the frame as shown. What is the natural frequency f of the system?

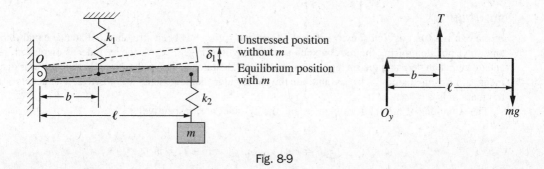

Fig. 8-9

SOLUTION

The horizontal position is the state of equilibrium. Determine the static displacement δ_1 of the end of the beam from its original position when no load is impressed. In the free-body diagram of the beam in the equilibrium position, note that the tension in spring k_1 must be $T = (\ell/b)mg$ to hold the beam in equilibrium. (This can be seen by taking moments about O.) Hence, the spring k_1 elongates under this tension by an amount $(mg\ell/b)k_1$. If a point on the beam a distance b from the left end is displaced this amount $mg\ell/bk_1$, the right end is displaced (using similar triangles) by an amount

$$\delta_1 = \frac{\ell}{b}\frac{mg\ell}{bk_1} = \frac{mg}{k_1}\left(\frac{\ell}{b}\right)^2$$

The total static displacement δ of m is then equal to the sum of δ_1 and the elongation of spring k_2 in transmitting the force mg to the beam. Then

$$\delta = \frac{mg}{k_2} + \left(\frac{\ell}{b}\right)^2\frac{mg}{k_1} = mg\left[\frac{k_1 + (\ell/b)^2 k_2}{k_1 k_2}\right]$$

and

$$f = \frac{1}{2\pi}\sqrt{\frac{g}{\delta}} = \frac{1}{2\pi}\sqrt{\frac{k_1 k_2}{m[k_1 + (\ell/b)^2 k_2]}} \qquad \text{Hz}$$

Free Vibrations—Angular

8.9. What will be the natural frequency of the system shown in Fig. 8-10(a) for small displacements?

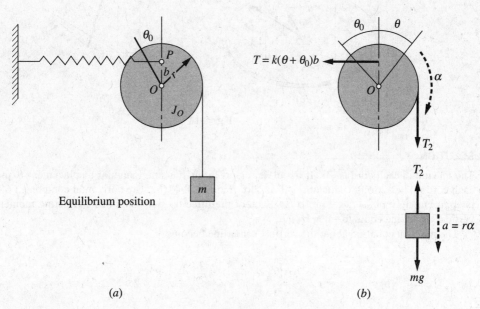

Equilibrium position

(a) $T = k(\theta + \theta_0)b$ $a = r\alpha$ (b)

Fig. 8-10

SOLUTION

In the equilibrium position shown in Fig. 8-10(a), the spring has been stretched a distance equal to $b\theta_0$. In any other position during the motion such as that shown in Fig. 8-10(b), the cylinder is displaced through an additional angle θ and the spring is stretched a total distance of $b(\theta + \theta_0)$. Note that the angular displacements are assumed small. Also assume clockwise angular displacements are positive. The acceleration of the mass m is $a = r\alpha$.

The equations of motion for the mass and the cylinder are, respectively,

$$\sum F = ma \qquad \text{or} \qquad mg - T_2 = mr\alpha \tag{1}$$

$$\sum M_O = J_O\alpha \qquad \text{or} \qquad T_2 r - bk(\theta + \theta_0)b = J_O\alpha \tag{2}$$

where J_O is the centroidal polar moment of inertia.[*] Substitute the value of T_2 from (1) into (2) to obtain

$$(mg - mr\alpha)r - b^2k\theta - b^2k\theta_0 = J_O\alpha$$

But in the equilibrium position with P directly above O, $\sum M_O = 0$; this means that $(k\theta_0 b)b = mgr$. (Note that $T_2 = mg$ when the system is in equilibrium.) The equation of motion for the cylinder becomes

$$(J_O + mr^2)\frac{d^2\theta}{dt^2} + b^2k\theta = 0 \qquad \text{from which} \qquad f = \frac{1}{2\pi}\sqrt{\frac{b^2k}{J_O + mr^2}} \qquad \text{Hz}$$

8.10. A circular homogeneous plate with radius R and weight W is supported by three cords of length ℓ, equally spaced in a circle, as shown in Fig. 8-11(a). Determine the natural frequency of oscillations for small displacements of the plate about its vertical centerline.

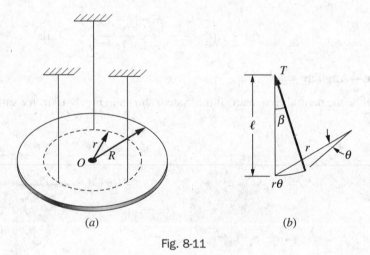

(a)   (b)

Fig. 8-11

SOLUTION

The tension in each cord is $\frac{1}{3}W$. If the plate rotates through a small angular displacement θ, the bottom of each cord rotates through a distance $r\theta$, as shown in Fig. 8-11(b). The horizontal component of the tension is approximately $\frac{1}{3}W\sin\beta = \frac{1}{3}W(r\theta/\ell)$. Each of the three tensions supplies a resisting moment about the vertical centerline equal to $\frac{1}{3}W(r\theta/\ell)r$.

The moment equation about the vertical centerline becomes

$$-3\left(\frac{W}{3}\right)\left(\frac{r\theta}{\ell}\right)r = \frac{1}{2}\left(\frac{W}{g}\right)(R^2)\frac{d^2\theta}{dt^2}$$

---

[*]In Eq. (5) of Chap. 5 this was called the *moment of inertia about the axis of rotation.* The quantities are the same.

This reduces to

$$\frac{d^2\theta}{dt^2} + \frac{2g}{\ell}\left(\frac{r^2}{R^2}\right)\theta = 0$$

and

$$f = \frac{1}{2\pi}\left(\frac{r}{R}\right)\sqrt{\frac{2g}{\ell}} \quad \text{cps}$$

8.11. A homogeneous steel disk 200 mm in diameter and 50 mm thick is rigidly attached to a vertical steel wire 2 mm in diameter and 900 mm long. What is the natural frequency of the system if the vertical wire is simply twisted (not a pendulum)?

SOLUTION

When a wire is twisted through an angle θ, there is a restoring torque $K\theta$. The equation of motion for the disk becomes $-K\theta = J_O\alpha$. Hence,

$$\frac{d^2\theta}{dt^2} + \frac{K}{J_O}\theta = 0$$

and

$$f = \frac{1}{2\pi}\sqrt{\frac{K}{J_O}} \quad \text{Hz}$$

According to the theory developed in strength of materials, the angle of twist for the wire is

$$\theta = \frac{T\ell}{\frac{1}{32}\pi d^4 G}$$

where T = torque, N·m
  ℓ = length of wire, m
  G = shear modulus of elasticity (80 GPa for steel)
  d = diameter, m

Hence, the torsional constant is

$$K = \frac{T}{\theta} = \frac{\pi d^4 G}{32\ell} = \frac{\pi(0.002)^4(80 \times 10^9)}{32(0.9)} = 0.14 \text{ N·m/rad}$$

The moment of inertia of the disk is

$$J_O = \frac{1}{2}mr^2 = \frac{1}{2}\left[\left(\frac{1}{4}\pi\right)(0.2)^2(0.05)(7850)\right](0.1)^2 = 0.062 \text{ kg·m}^2$$

Note that the density of steel used was 7850 kg/m^3. Finally,

$$f = \frac{1}{2\pi}\sqrt{\frac{K}{J_O}} = \frac{1}{2\pi}\sqrt{\frac{0.14}{0.062}} = 0.24 \text{ Hz}$$

8.12. Two heavy masses with moments of inertia J_1 and J_2 are connected by a shaft of small diameter d, as shown in Fig. 8-12. Find an expression for the frequency of vibration due to a twist of one mass relative to the other.

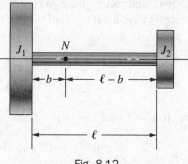

Fig. 8-12

ANALYSIS

If one mass is held and the other rotated and then both are released, the system will oscillate. Since no external torques are assumed to act on the system, the *angular momentum* of the system must be conserved. Thus,

$$J_1\omega_1 + J_2\omega_2 = 0 \quad \text{or} \quad \omega_2 = -\left(\frac{J_1}{J_2}\right)/\omega_1$$

Note that ω_1 and ω_2 represent the varying angular velocities of J_1 and J_2. The angular velocities are time-dependent.

Since the previous equation indicates that the masses always are rotating in opposite directions, there is a section of the shaft that is always at rest. This nodal section N can be used to study the motion of the masses, since each can be treated as a torsional pendulum (see Problem 8.11) relative to that section.

The time for one mass to complete a cycle must equal the time for the other to complete a cycle. If they differed, one mass would eventually rotate in the same direction as the other. But the above equation indicates that the masses always rotate in opposite directions. Since the periods are equal, the number of cycles per second for each must be equal. Thus,

$$f = \frac{1}{2\pi}\sqrt{\frac{K_1}{J_1}} = \frac{1}{2\pi}\sqrt{\frac{K_2}{J_2}}$$

where K_1 and K_2 are the torsional spring constants of the parts of the shaft from the nodal section N to each end. Hence, those constants are related by $K_1/K_2 = J_1/J_2$.

For a cylindrical shaft, $K = \pi d^4 G/32\ell$, where d is its diameter, G is the shearing modulus of elasticity, and ℓ is its length (see Problem 8.11). Assume the nodal section is a distance b from the J_1 mass; then

$$K_1 = \frac{\pi d^4 G}{32b} \quad \text{and} \quad K_2 = \frac{\pi d^4 G}{32(\ell - b)}$$

Hence, $K_1/K_2 = (\ell - b)/b = J_1/J_2$, from which $b = J_2\ell/(J_1 + J_2)$. This locates the nodal section.

For the left portion,

$$f = \frac{1}{2\pi}\sqrt{\frac{K_1}{J_1}}$$

where

$$K_1 = \frac{\pi d^4 G}{32b} = \frac{\pi d^4 G(J_1 + J_2)}{32\ell J_2}$$

Then

$$f = \frac{1}{2\pi}\sqrt{\frac{\pi d^4 G(J_1 + J_2)}{32\ell J_1 J_2}}$$

Since the polar moment of inertia of the area of a circle is $J = \frac{1}{32}\pi d^4$, the above expression may be written

$$f = \frac{1}{2\pi}\sqrt{\frac{JG(J_1 + J_2)}{J_1 J_2 \ell}}$$

Note that J_1 and J_2 refer to the *mass* polar moments of inertia of the cylinders and J refers to the *area* polar moment of inertia of a cross section of the shaft.

8.13. A device has a 120-kg flywheel at each end of a steel shaft. Assume that each flywheel has a radius of gyration of 300 mm. The shaft connecting the two is 600 mm long and has a diameter of 50 mm. Determine the natural frequency of torsional oscillation. Use $G_{\text{sted}} = 80$ GPa.

SOLUTION

The moment of inertia for each flywheel is

$$J_f = mr^2 = 120(0.3)^2 = 10.8 \text{ kg·m}^2$$

The moment of inertia for the cross-sectional area of the shaft is

$$J = \frac{1}{32}\pi(0.05)^4 = 6.14 \times 10^{-7} \text{ m}^4$$

Using the formula from the preceding problem, we have

$$f = \frac{1}{2\pi}\sqrt{\frac{JG(J_1 + J_2)}{\ell J_1 J_2}} = \frac{1}{2\pi}\sqrt{\frac{6.14 \times 10^{-7}(80 \times 10^9)(10.8 + 10.8)}{0.6(10.8)(10.8)}}$$

$$= 19.6 \text{ Hz}$$

You should check the units under the radical to make sure they are s^{-2}.

8.14. A 50-mm-diameter steel shaft 375 mm long is attached at one end to a 138-kg flywheel with a radius of gyration of 150 mm and at the other end to a rotor that weighs 46 kg and has a radius of gyration of 100 mm. Where is the nodal section and what is the natural frequency of torsional oscillation?

SOLUTION

Moment of inertia of the flywheel:

$$J_f = m_f r_f^2 = 138 \times 0.15^2 = 3.11 \text{ kg·m}^2$$

Moment of inertia of the rotor:

$$J_r = m_r r_r^2 = 46 \times 0.1^2 = 0.46 \text{ kg·m}^2$$

Moment of inertia of the shaft cross section:

$$J = \frac{1}{32}\pi d^4 = \frac{1}{32}\pi \times 0.05^4 = 6.14 \times 10^{-7} \text{ m}^4$$

From Problem 8.12, the distance of the nodal section from the flywheel is

$$b = \frac{J_r \ell}{J_f + J_r} = \frac{3.11 \times 0.375}{3.11 + 0.46} = 0.327 \text{ m}$$

$$\text{Frequency } f = \frac{1}{2\pi}\sqrt{\frac{JG(J_f + J_r)}{LJ_f J_r}} = \frac{1}{2\pi}\sqrt{\frac{6.14 \times 10^{-7} \times 80 \times 10^9 (3.11 + 0.46)}{0.375 \times 3.11 \times 0.46}} = 93.3 \text{ Hz}$$

8.15. Figure 8-13(*a*) shows a 5-kg homogeneous bar pivoted at its left end. The right end is supported by a weightless cylinder, which is floating as shown in water. The cylinder has a cross-sectional area of 1250 mm². Neglecting the damping effect of the water and the inertia effect of the moving water, determine the frequency of oscillation if the bar is depressed slightly from its horizontal (equilibrium) position.

SOLUTION

In Fig. 8-13(*b*), the equilibrium position of the horizontal bar is shown. The force at the right end is equal to the weight of the displaced water since the cylinder and rod have negligible weight. The moment equation for any angle θ is

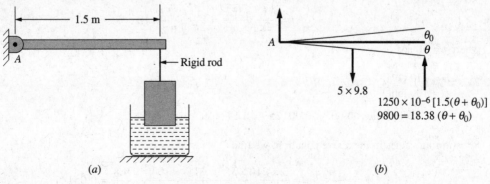

(*a*)   (*b*)

Fig. 8-13

$$\sum M_A = I_A \frac{d^2\theta}{dt^2}$$

or $(5 \times 9.8)0.75 - 18.38(\theta + \theta_0) \times 1.5 = \frac{1}{3} \times 5 \times 1.5^2 \frac{d^2\theta}{dt^2}$

However, in the equilibrium phase, the moments about A equal zero, and hence

$$(5 \times 9.8) \times 0.75 - 18.38\theta_0 \times 1.5 = 0$$

The equation of motion is

$$-\left(\frac{2}{144}\right)50\theta \times 62.4 \times 5 = 2.59\frac{d^2\theta}{dt^2} \qquad -18.38\theta \times 1.5 = 3.75\frac{d^2\theta}{dt^2}$$

which reduces to

$$\frac{d^2\theta}{dt^2} + 7.35\theta = 0 \qquad \therefore \omega_n = \sqrt{7.35} = 2.71 \text{ rad/s}$$

8.16. A reed-type tachometer is composed of small cantilever beams with weights attached to their free ends. If the vibration frequency of a disturbing force corresponds to the natural vibration frequency of one of the reeds, it will vibrate. Since each reed is calibrated, it is possible to determine the frequency of the disturbance immediately. What weight W should be placed on the free end of a spring steel reed 1.25 mm thick, 5 mm wide, and 100 mm long so that its natural frequency is 50 Hz? See Fig. 8-14.

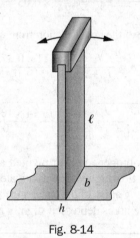

Fig. 8-14

SOLUTION

By the use of theory developed in strength of materials, the static deflection of a cantilever beam due to a concentrated mass m on the free end is $\Delta = mg\ell^3/3EI$, where E is the modulus of elasticity and I is the moment of inertia of the cross-sectional area about the neutral axis. Hence,

$$f = \frac{1}{2\pi}\sqrt{\frac{3EI}{m\ell^3}} = \frac{1}{2\pi}\sqrt{\frac{3EIg}{W\ell^3}} \text{ Hz}$$

where $f = 50$ cps
      $g = 9.8$ m/s^2
      $E = 210$ GPa
      $\ell = 100$ mm
      $I = \frac{1}{12}bh^3 = \frac{1}{12} \times 0.005 \times 0.00125^3 = 8.14 \times 10^{-13}$

Substituting values in the above equation, we find

$$W = \frac{3EIg}{4\pi^2 f^2 \ell^3} = \frac{3 \times 210 \times 10^9 \times 8.14 \times 10^{-13} \times 9.8}{4\pi^2 \times 50^2 \times 0.1^3} = 0.051 \text{ N}$$

Free Vibrations—Plane Motion

8.17. A disk of mass m and radius r rolls without slipping on a horizontal plane. It is attached to a wall by a spring of modulus k. If the disk is displaced to the right and released, derive the differential equation of motion and determine the frequency of oscillation. See Fig. 8-15.

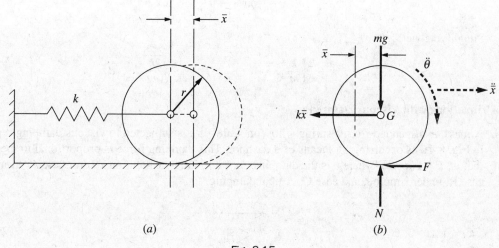

(a)   (b)

Fig. 8-15

SOLUTION

Figure 8-15(b) shows the free-body diagram of the disk. The equations of motion are

$$\sum F_h = -F - k\bar{x} = m\ddot{\bar{x}}$$

$$\sum M_G = rF = \frac{1}{2}mr^2\ddot{\theta}$$

For a nonslip rolling wheel, $\ddot{\bar{x}} = r\ddot{\theta}$. Adding the equations of motion yields

$$-k\bar{x} = m\ddot{\bar{x}} + \frac{1}{2}m\ddot{\bar{x}}$$

Or, rewritten,

$$\ddot{\bar{x}} + \frac{2}{3}\left(\frac{k}{m}\right)\bar{x} = 0$$

so that

$$f = \frac{1}{2\pi}\sqrt{\frac{2k}{3m}} \qquad \text{Hz}$$

Alternatively, the conservation of energy theorem can be used. Consider the center of mass G to be displaced $\bar{x}$ from the equilibrium position. The potential energy V of the spring is given by

$$V = \int k\bar{x}\,dx = \frac{1}{2}k\bar{x}^2$$

The kinetic energy of the rolling disk is

$$T = \frac{1}{2}m\dot{\bar{x}}^2 + \frac{1}{2}I_G\dot{\theta}^2$$

Because energy is conserved, $T + V = $ constant. Or

$$\frac{d(T+V)}{dt} = 0$$

$$\frac{d}{dt}\left[\frac{1}{2}m\dot{\bar{x}} + \frac{1}{2}I_G\dot{\theta}^2 + \frac{1}{2}k(\bar{x})^2\right] = 0 \qquad \text{or} \qquad m\dot{\bar{x}}\ddot{\bar{x}} + I_G\dot{\theta}\ddot{\theta} + k\bar{x}\dot{\bar{x}} = 0$$

But

$$\bar{x} = r\theta \qquad \dot{\bar{x}} = r\dot{\theta}, \ddot{\bar{x}} = r\ddot{\theta} \qquad \text{and} \qquad I_G = \frac{1}{2}mr^2$$

from which

$$m\dot{\bar{x}}\ddot{\bar{x}} + \frac{1}{2}mr^2\left(\frac{\dot{\bar{x}}}{r}\right)\left(\frac{\ddot{\bar{x}}}{r}\right) + k\bar{x}\dot{\bar{x}} = 0$$

Simplifying yields

$$\ddot{\bar{x}} + \frac{2}{3}\left(\frac{k}{m}\right)\bar{x} = 0 \qquad \therefore f = \frac{1}{2\pi}\sqrt{\frac{2k}{3m}} \qquad \text{Hz}$$

Free Vibrations with Viscous Damping

8.18. A mass m suspended from a spring whose modulus is k is subjected to viscous damping represented in Fig. 8-16 as occurring by means of a dashpot. This damping force is proportional to the velocity; that is, $F = c(dx/dt)$, where c is the damping constant. Find the solution $x(t)$ for case A: overdamping, case B: underdamping, and case C: critical damping.

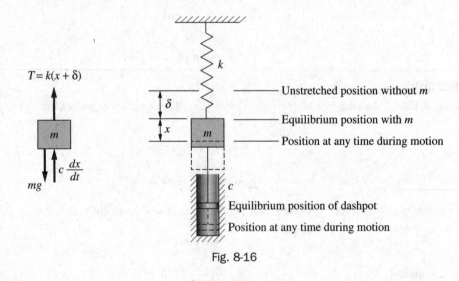

Fig. 8-16

ANALYSIS

Figure 8-16 indicates the essential data. The free-body diagram to the left illustrates all the forces acting on the mass when displaced a distance x below equilibrium and traveling down.

Note, as before, that in the equilibrium position $k\delta = mg$. Note also that the damping force $c(dx/dt)$ opposes motion. The equation of motion ($\sum F = ma$ with an assumed position down) becomes

$$mg - k(x + \delta) - c\frac{dx}{dt} = m\frac{d^2x}{dt^2} \qquad \text{or} \qquad \frac{d^2x}{dt^2} + \frac{c}{m}\frac{dx}{dt} + \frac{k}{m}x = 0 \qquad (1)$$

Assume a solution of this differential equation in the form $x = Ae^{\lambda t}$, where A and λ are nonzero constants. Substitute this value into the equation (noting that $dx/dt = A\lambda e^{\lambda t}$ and $d^2x/dt^2 = A\lambda^2 e^{\lambda t}$) to obtain

$$A\lambda^2 e^{\lambda t} + A\frac{c}{m}\lambda e^{\lambda t} + A\frac{k}{m}e^{\lambda t} = 0 \qquad \text{or} \qquad \left(\lambda^2 + \frac{c}{m}\lambda + \frac{k}{m}\right)e^{\lambda t} = 0$$

The desired solution must be such that the above equations are zero. Since $e^{\lambda t}$ cannot be zero, its coefficient must be zero; that is, $\lambda^2 + (c/m)\lambda + k/m = 0$.

Using the quadratic formula, the two solutions for λ are

$$\lambda_1, \lambda_2 = \frac{-c}{2m} \pm \frac{1}{2m}\sqrt{c^2 - 4km} \qquad (2)$$

The general solution is of the form

$$x = Ae^{\lambda_1 t} + Be^{\lambda_2 t} \tag{3}$$

The radical in (2) may be real, imaginary, or zero depending on the magnitude of the damping coefficient c. The value of c that makes the radical zero is called the *critical damping coefficient* c_c. Its value is

$$c_c = 2\sqrt{km} \tag{4}$$

$$= 2m\,\omega_n \tag{5}$$

where ω_n is the natural frequency of the undamped system. Then (2) becomes

$$\lambda_1, \lambda_2 = -\frac{c}{2m} \pm \frac{1}{2m}\sqrt{c^2 - c_c^2} \tag{6}$$

Three cases arise, depending on the value of c_c.

Case A. Overdamping: $c > c_c$. The radical in (6) is real and both exponents are negative. The general solution of the differential equation (1) is

$$x = Ae^{\lambda_1 t} + Be^{\lambda_2 t} \tag{7}$$

When $t = 0$, $x = Ax^0 + Be^0 = A + B$. The plot of either part of the solution of such motion indicates that the frictional resistance is so large that the mass, after its initial displacement, creeps back to equilibrium without vibrating (see Fig. 8-17). Since there is no period to the motion, it is called *aperiodic*.

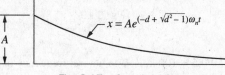

Fig. 8-17 Overdamped motion

Case B. Underdamping: $c < c_c$. The radical is imaginary. Using $i = \sqrt{-1}$, the general solution of (1) can be written as

$$x = e^{-(c/2m)t}(A\sin\omega_d t + B\cos\omega_d t) \tag{8}$$

$$= Xe^{-(c/2m)t}\sin(\omega_d t + \phi) \tag{9}$$

where $\omega_d = \sqrt{4km - c^2}/2m$ and ϕ is the phase angle. A plot of this solution shows the sine curve with its height continuously decreasing because it is multiplied by the exponential function, which decays with time. See Fig. 8-18.

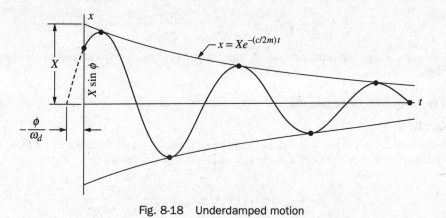

Fig. 8-18 Underdamped motion

Case C. Critical damping: $c = c_c$. The differential equation (1) has two equal roots so the solution is written as

$$x = (A + Bt)e^{-(c/2m)t}$$

The t term is inserted with B because otherwise only one of the two solutions would be found. The graph here is similar to that of case A. The motion is aperiodic.

8.19. In Fig. 8-19, mass m is suspended from a spring whose constant is 3750 N/m and is connected to a dashpot providing viscous damping. The damping force is 50 N when the velocity of the dashpot plunger is 0.5 m/s. The mass of m and the plunger is 6 kg. What will be the frequency of the damped vibrations?

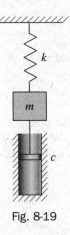

Fig. 8-19

SOLUTION

The damping coefficient $c = 50/0.5 = 100$ kg/s. The natural frequency of the undamped system is

$$\omega_n = \sqrt{\frac{k}{m}} = \sqrt{\frac{3750}{6}} = 25 \text{ rad/s}$$

The critical damping coefficient according to Problem 8.18 is

$$c_c = 2m\omega_n = 2 \times 6 \times 25 = 300 \text{ kg/s}$$

Obviously, $c < c_c$, which, according to Problem 8.18, case B, is underdamping, and oscillations will be present. The solution is

$$x = e^{-(c/2m)t}(A \sin\omega_d t + B \cos\omega_d t)$$

The frequency of the damped vibration is ω_d.

$$\omega_d = \sqrt{\frac{4km - c^2}{2m}} = \sqrt{\frac{4(3750)(6) - 100^2}{2(6)}} = 23.6 \text{ rad/s}$$

Note that the period of the damped vibration is $2\pi/23.6 = 0.27$ s, and the period of the undamped system is $2\pi/25 = 0.25$ s.

8.20. In Problem 8.19, determine the rate of decay of the oscillations.

SOLUTION

This is conveniently expressed by introducing a new term, the logarithmic decrement D, which is the natural logarithm of the ratio of any two successive amplitudes one cycle apart:

$$D = \ln\frac{x_1}{x_2} = \ln\frac{Xe^{-(c/2m)t}\sin(\omega_d t + \phi)}{Xe^{-(c/2m)(t+\tau)}\sin[\omega_d(t + \tau) + \phi]}$$

The numerator indicates a value of x at time t, whereas the denominator gives the value at time $t + \tau$, where τ is the period of motion. Hence, the two amplitudes occur one cycle apart (we neglect the fact that the sine curve is tangent to its envelope $Xe^{-d\omega_n t}$ at a point slightly different from the point of maximum amplitude).

Now $\sin(\omega_d t + \phi) = \sin[\omega_d(t + \tau) + \phi]$, since they are evaluated one cycle or 2π rad apart. Hence the above expression for D becomes

$$D = \ln \frac{e^{-(c/2m)t}}{e^{-(c/2m)(t+\tau)}} = \frac{c}{2m}\tau$$

In Problem 8.19, the period $\tau = 2\pi/\omega_d = 2\pi/23.6 = 0.266$ s.

The logarithmic decrement $D = c\tau/2m = 100(0.266)/(2)(6) = 2.22$.

The ratio of any two successive amplitudes is $e^D = e^{2.22} = 9.21$.

8.21. Set up the differential equation of motion for the system shown in Fig. 8-20. Determine the natural frequency of the damped oscillations.

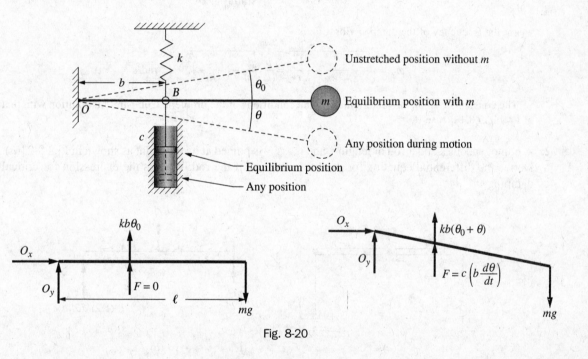

Fig. 8-20

SOLUTION

In linear vibration, we expressed the distance a body moves from equilibrium as a function of time. In rotation such as this, we are interested in expressing the angular displacement of a body in terms of time.

The free-body diagram for any phase during the motion is shown separately. Note that if the arm is moving down, the damping force opposes motion and acts up. It is equal to the product of the damping coefficient c and the velocity of the plunger in the dashpot. This is the velocity of point B on the rod that is a distance b from O and hence has a linear velocity equal to the product of b and the angular velocity $d\theta/dt$ of the rod.

The spring force is the product of k and the total linear displacement of the spring. This displacement is that of point B and hence equals $b(\theta_0 + \theta)$. The force is $kb(\theta_0 + \theta)$.

The equilibrium free-body diagram is also shown. Take moments about O to show $kb^2\theta_0 = mg\ell$. This information will simplify the differential equation of motion.

The sum of moments about O for the free-body diagram for any position is equated to $I_O\alpha$. But for a concentrated mass m, $I_O = m\ell^2$ and $\alpha = d^2\theta/dt^2$. Therefore the equation of motion $\sum M_O = I_O\alpha$ becomes

$$mg\ell - kb^2(\theta_0 + \theta) - cb^2\frac{d\theta}{dt} = m\ell^2\frac{d^2\theta}{dt^2}$$

But $mg\ell = kb^2\theta_0$. Then, simplifying,

$$\frac{d^2\theta}{dt^2} + \frac{cb^2}{m\ell^2}\frac{d\theta}{dt} + \frac{kb^2}{m\ell^2}\theta = 0$$

To solve, let $\theta = e^{\lambda t}$. Then the equation becomes (if $e^{\lambda t}$ is a solution)

$$\lambda^2 e^{\lambda t} + \frac{cb^2}{m\ell^2}\lambda e^{\lambda t} + \frac{kb^2}{m\ell^2}e^{\lambda t} = 0 \qquad \text{or} \qquad \lambda = \frac{-cb^2}{2m\ell^2} \pm \frac{1}{2}\sqrt{\frac{c^2 b^4}{m^2\ell^4} - \frac{4kb^2}{m\ell^2}}$$

Critical damping occurs when the radicand is zero. Hence,

$$c_c = \frac{2}{b}\sqrt{mk}$$

If vibrations occur, the radicand will be negative and the solution will be of the form

$$\theta = Ce^{-(cb^2/2m\ell^2)t}\sin(\omega_d t + \phi)$$

where the frequency of the damped vibration is

$$\omega_d = \sqrt{-\frac{c^2 b^4}{4m^2\ell^4} + \frac{kb^2}{m\ell^2}} = \frac{b}{\ell}\sqrt{\frac{k}{m} - \left(\frac{cb}{2m\ell}\right)^2} \qquad \text{rad/s}$$

The constants C and ϕ are determined by the conditions of the problem. Compare this solution with that of Problem 8.18, case B.

8.22. A homogeneous slender rod of length ℓ and mass w is pinned at its midpoint as shown in Fig. 8-21(a). Derive the differential equation for small oscillations of the rod. What is the expression for critical damping?

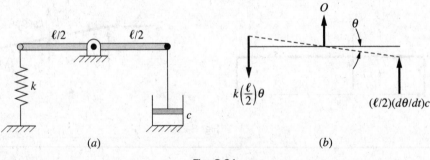

Fig. 8-21

SOLUTION

The free body is shown displaced clockwise through a small angle θ in Fig. 8-21(b). The damping force and the spring tension both resist motion as shown. The moment equation $\sum M_O = I_O\alpha$ about O becomes

$$-\frac{\ell}{2}\left(k\frac{\ell}{2}\theta\right) - \frac{\ell}{2}\left(\frac{\ell}{2}\frac{d\theta}{dt}c\right) = \frac{1}{12}m\ell^2\frac{d^2\theta}{dt^2}$$

Consequently, the differential equation for small oscillations is

$$\frac{d^2\theta}{dt^2} + \frac{3c}{m}\frac{d\theta}{dt} + \frac{3k}{m}\theta = 0$$

To solve, assume that $\theta = e^{\lambda t}$. Hence,

$$\frac{d\theta}{dt} = \lambda e^{\lambda t} \qquad \text{and} \qquad \frac{d^2\theta}{dt^2} = \lambda^2 e^{\lambda t}$$

Substitute to obtain

$$\lambda^2 e^{\lambda t} + \frac{3c}{m}\lambda e^{\lambda t} + \frac{3k}{m}e^{\lambda t} = 0$$

Since $e^{\lambda t}$ will not be zero, this equation is satisfied if

$$\lambda^2 + \frac{3c}{m}\lambda + \frac{3k}{m} = 0$$

The solution is

$$\lambda = -\frac{3c}{2m} \pm \frac{1}{2m}\sqrt{9c^2 - 12km}$$

Critical damping occurs when the radicand in the expression is zero. Hence,

$$9c_c^2 = 12km \qquad \text{or} \qquad c_c = \sqrt{\frac{4km}{3}}$$

Forced Vibrations without Damping

8.23. In Fig. 8-22, a mass m is suspended on a spring whose modulus is k. The mass is subjected to a periodic disturbing force $F\cos\omega t$. Analyze the motion.

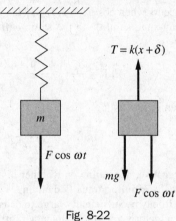

Fig. 8-22

ANALYSIS

The differential equation now has an additional term when compared with the equation of free vibrations:

$$-kx + F\cos\omega t = m\frac{d^2 x}{dt^2}$$

or, see Eq. (2) of Problem 8.1,

$$\frac{d^2 x}{dt^2} + \frac{kx}{m} = \frac{F}{m}\cos\omega t \tag{1}$$

According to the theory of differential equations, the solution of this equation consists of the sum of two parts: the homogeneous solution previously determined in Problem 8.1 for the equation when the right-hand side is set equal to zero (the transient part) and a particular solution that satisfies (1).

Assume that the particular solution, which is also called the *steady-state solution*, is of the form $x_P = X\cos\omega t$. Then $dx/dt = -X\omega\sin\omega t$ and $d^2 x_r/dt^2 = -X\omega^2\cos\omega t$. Substitute x_P into (1):

$$-X\omega^2\cos\omega t + \frac{k}{m}X\cos\omega t = \frac{F}{m}\cos\omega t$$

Hence,

$$X = \frac{F/k}{1 - \omega^2 m/k}$$

Let δ_F be the deflection that the force F would impart to the spring if acting on it statically; that is, $\delta_F = F/k$. Also note that $\omega_n^2 = k/m$, where ω_n is the natural frequency when the disturbing force is absent. Then X may be written

$$X = \frac{\delta_F}{1 - (\omega/\omega_n)^2}$$

For convenience in analysis, let $\omega/\omega_n = r$; then the steady-state solution is

$$x_p = \frac{\delta_F}{1 - r^2} \cos \omega t \tag{2}$$

Note that its frequency is the same as the disturbing frequency.

The entire solution is then

$$x = A \sin \sqrt{\frac{k}{m}}t + B \cos \sqrt{\frac{k}{m}}t + \frac{1}{1 - r^2} \delta_F \cos \omega t \tag{3}$$

The first two terms, representing the free vibrations, are transient in character because some damping is always present to cause those vibrations to decay. Hence, consider only the solution

$$x_F = \delta_F \frac{1}{1 - r^2} \cos \omega t \tag{4}$$

Its maximum value, which occurs when $\cos \omega t = 1$, is $\delta_F/(1 - r^2)$ and is called the *amplitude X*. The ratio of the amplitude of the steady-state solution to the static deflection δ_F that F would cause is called the *magnification factor*. Its value is

$$\text{m.f.} = \frac{\delta_F/(1 - r^2)}{\delta_F} = \frac{1}{1 - r^2} \tag{5}$$

Since $r = \omega/\omega_n = f/f_n$, this can be written

$$\text{m.f.} = \frac{1}{1 - (f/f_n)^2} \tag{6}$$

Its value can be positive or negative, depending on whether f is less than f_n. When $f = f_n$, resonance occurs and the amplitude is theoretically infinite. Actually damping, which is always present, holds the amplitude to a finite amount. However, it may be sufficiently large to cause failure in structures.

A plot of the magnification factor versus the frequency ratio r is shown in Fig. 8-23. The negative value when $r > 1$ indicates that the force F is directed one way while the displacement x is opposite.

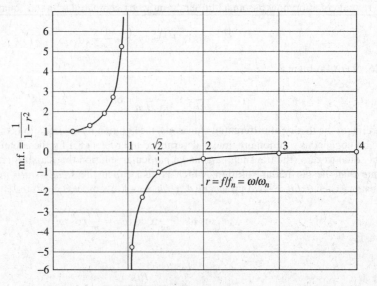

Fig. 8-23

Note that when $r = \sqrt{2}$, the magnification factor is

$$\text{m.f.} = \frac{1}{1 - (\sqrt{2})^2} = -1$$

This means that if the ratio r is made greater than $\sqrt{2}$, the magnitude of the magnification factor will be less than 1. Thus, the disturbing force under these conditions will cause less movement than if it were applied statically.

8.24. A disturbing force of 9 N acts harmonically on a 5-kg mass suspended on a spring whose modulus is 6000 N/m. What will be the amplitude of excursion of the mass if the disturbing frequency is (*a*) 1 Hz, (*b*) 5.40 Hz, (*c*) 50 Hz?

SOLUTION

The natural frequency of the system is

$$f_n = \frac{1}{2\pi}\sqrt{\frac{k}{m}} = \frac{1}{2\pi}\sqrt{\frac{6000}{5}} = 5.51 \text{ Hz}$$

The deflection that the disturbing force would give to the spring if applied statically is $\delta_F = 9/6000 = 1.5$ mm.

(*a*) The frequency ratio $r = f/f_n = 1/5.51 = 0.185$; hence, the amplitude will be

$$X = \frac{\delta_F}{1 - r^2} = \frac{1.5}{1 - 0.0329} = 1.551 \text{ mm}$$

(*b*) The ratio $r = 5.40/5.51 = 0.98$; hence, the amplitude will be

$$X = \frac{1.5}{1 - 0.98^2} = 37.9 \text{ mm}$$

(*c*) The ratio $r = 50/5.51 = 9.01$; hence, the amplitude will be

$$X = \frac{1.5}{1 - 9.01^2} = -0.0184 \text{ mm}$$

Note that in this case the amplitude is opposite to the direction in which the force is exerted but is negligible in magnitude.

8.25. A 30-kg refrigerator unit is supported on three springs, each with a modulus of k N/m. The unit operates at 600 rpm. What should be the value of k if one-twelfth of the disturbing force of the unit is to be transmitted to the supporting box?

SOLUTION

Assume that the disturbing force transmitted is proportional to the amplitude of the motion of the unit. This is logical because the supporting springs transmit forces proportional to their deformation, which equals the amplitude of the motion of the unit.

From Problem 8.23, the ratio of the amplitude of the steady-state motion to the static deflection that the disturbing force would cause (in this case $-\frac{1}{12}$) is equal to $1/(1 - r^2)$. Note that the ratio is negative because the natural frequency of the springs must be less than the disturbing frequency for a reduction to occur, and, thus, according to the graph in Problem 8.23, the magnification factor is below the line.

Hence, in this problem, $-\frac{1}{12} = 1/(1 - r^2)$, from which $r^2 = 13, r = f/f_n = \sqrt{13}$, and $f_n = [600/(60\sqrt{.3})] = 2.77$ cps. Using the mass on one spring, f_n is equated to $(1/2\pi)\,k/m$. Then

$$2.77 = \frac{1}{2\pi}\sqrt{\frac{k}{10}} \qquad \therefore k = 3030 \text{ N/m}$$

Forced Vibrations with Viscous Damping

8.26. In Fig. 8.24(*a*), a mass m is suspended from a spring whose modulus is k. It is also connected to a dashpot that provides viscous damping. Analyze the motion if the mass is subjected to a harmonic disturbing force $F_0 \cos \omega t$.

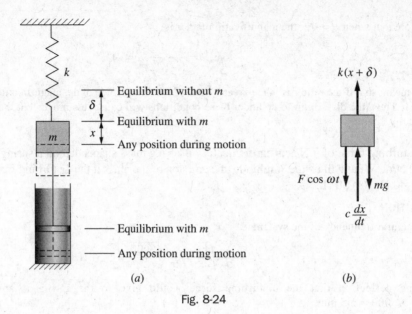

Fig. 8-24

SOLUTION

The free-body diagram in Fig. 8-24(*b*) shows the forces acting on the mass. Assuming that down is positive, the equation of motion is

$$\sum F = ma$$

or

$$mg + F\cos\omega t - k(x + \delta) - c\frac{dx}{dt} = m\frac{d^2x}{dt^2}$$

As in previous problems, the free-body diagram for the equilibrium position shows $mg - k\delta = 0$. Thus, the differential equation becomes

$$\frac{d^2x}{dt^2} + \frac{c}{m}\frac{dx}{dt} + \frac{k}{m}x = \frac{F}{m}\cos\omega t \tag{1}$$

The transient solution will be neglected, as in Problem 8.23. The steady-state solution is quite difficult to find; it is

$$x_P = \frac{F\cos(\omega t - \phi)}{\sqrt{(k - m\omega^2)^2 + (c\omega)^2}} \quad \text{with} \quad \tan\phi = \frac{c\omega/k}{1 - m\omega^2/k} \tag{2}$$

Since usually only the amplitude X of the motion is considered, the equation for the amplitude may be written

$$X = \frac{F}{\sqrt{(k - m\omega^2)^2 + (c\omega)^2}} = \frac{F/k}{\sqrt{(1 - m\omega^2/k)^2 + (c\omega/k)^2}} \tag{3}$$

This is further simplified by noting that F/k is the static deflection δ_F that the disturbing force would give the spring. Also $m/k = 1/\omega_n^2$ where ω_n is the natural undamped frequency of the system (rad/s). The critical damping coefficient (Problem 8.18) is $c_c = 2m\omega_n$.

Defining $d = c/c_c$ and using $r = \omega/\omega_n$, the last term of the radicand is

$$\frac{X}{\delta_F} = \frac{1}{\sqrt{(1 - r^2)^2 + (2rd)^2}} \quad \text{and} \quad \tan\phi = \frac{2rd}{1 - r^2} \tag{4}$$

The ratio X/δ_F is the magnification factor m.f.

In Fig. 8-25, a graph of the magnification factor versus the frequency ratio shows the peaking that occurs near $r = \omega/\omega_n = 1$ and the influence that damping exerts on the heights of these peaks.

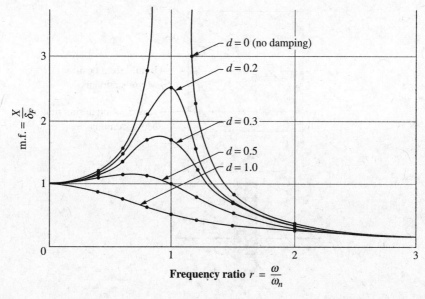

Fig. 8-25

Note that the amplitude peaks in the graph are at a value slightly less than $r = 1$. The exact value may be found by taking the derivative of X/δ_F with respect to r and equating the result to zero:

$$\frac{d}{dr}\left(\frac{X}{\delta_F}\right) = \frac{[2(1 - r^2)(-2r) + 2(2rd)2d]}{\frac{1}{2}\sqrt{[(1 - r^2)^2 + (2rd)^2]}} = 0$$

The radical cannot equal zero; hence,

$$2(1 - r^2)(-2r) + 2(2rd)2d = 0 \qquad \text{or} \qquad r(r^2 + 2d^2 - 1) = 0$$

The solution for the peak amplitude is

$$r = \sqrt{-2d^2 + 1} \tag{5}$$

In Fig. 8-26, a plot is also shown of the phase angle ϕ for various damping coefficients.

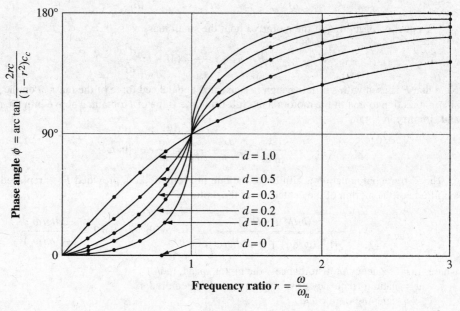

Fig. 8-26

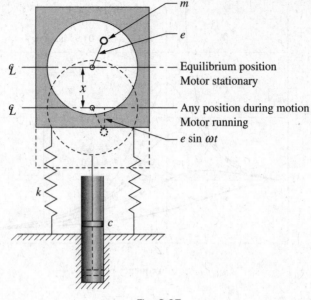

Fig. 8-27

8.27. A small mass m is attached with an eccentricity e to the flywheel of a motor mounted on springs as shown in Fig. 8-27. The effective modulus of the springs is k. The dashpot introduces viscous damping of an amount c. If the total mass of the motor and small mass is M, analyze the motion of the system under the action of the disturbing force caused by the eccentrically mounted mass m. (This problem illustrates the effect of either reciprocating or rotating unbalance.)

ANALYSIS

In the running position shown, the mass m is below the centerline by an amount $e \sin \omega t$. Hence, if the positive direction is assumed as down, the absolute displacement x_1 of the mass m is the sum of its vertical displacement ($e \sin \omega t$) relative to the centerline and the absolute displacement x of the centerline; that is, $x_1 = x + e \sin \omega t$.

Let F = unbalanced force exerted by motor on mass m to impart to it an acceleration d^2x_1/dt^2. Then

$$F = m \frac{d^2 x_1}{dt^2} = m \frac{d^2 (x + e \sin \omega t)}{dt^2} = m \frac{d^2 x}{dt^2} - me\omega^2 \sin \omega t$$

The equation of motion of the motor without the small mass m is

$$-F - kx - c \frac{dx}{dt} = (M - m) \frac{d^2 x}{dt^2}$$

Note that F is used with a negative sign because this unbalanced force of the mass m on the motor is opposite in direction to that of the motor on m. Substitute the value of F obtained above into the motor equation and simplify to obtain

$$\frac{d^2 x}{dt^2} + \frac{c}{M} \frac{dx}{dt} + \frac{k}{M} x = \frac{m}{M} e\omega^2 \sin \omega t$$

This differential equation is similar to the one in Problem 8.26, provided F_0 is replaced by $me\omega^2$. The amplitude of the motion of the centerline of the motor is

$$X = \frac{(m/M)e(\omega/\omega_n)^2}{\sqrt{[1 - (\omega/\omega_n)^2]^2 + (2d\omega/\omega_n)^2}} \qquad \text{with} \qquad \tan\theta = \frac{2d\omega/\omega_n}{1 - (\omega/\omega_n)^2}$$

where ω = frequency of disturbance—the motor speed, rad/s
      ω_n = natural frequency of spring-supported system, rad/s
      d = damping ratio = c/c_c
      θ = phase angle

At resonance, the above equation reduces to $X = (me/M)/2d$. Also, for very large values of ω/ω_n, the same equation reduces within reasonable limits of accuracy to $X = me/M$.

The distance b from the geometric center O of the motor to the center of mass G of the system, i.e., of the motor plus the mass m, can be determined by taking moments about O. Refer to Fig. 8-28. Thus, $bM = em$ or $b = me/M$, which at large values of ω/ω_n is equal to X. Hence, at high motor speeds (ω/ω_n very large), the magnitude of the displacement X of the centerline is equal to the magnitude of b. However, at large values of ω/ω_n the value of θ, as deduced from the equation

$$\tan\theta = \frac{2d\omega/\omega_n}{1-(\omega/\omega_n)^2}$$

approaches $180°$. Hence, the displacement X equals b but is $180°$ out of phase.

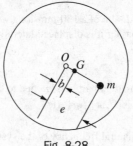

Fig. 8-28

From the geometry of Fig. 8-28, the absolute displacement x_G of the center of mass G equals the sum of the absolute displacement of the centerline and the relative displacement of the center of mass to the centerline, or

$$x_G = X\sin(\omega t - 180°) + b\sin\omega t = -b\sin\omega t + b\sin\omega t = 0$$

The center of mass G stands still at high motor speeds.

8.28. A spring-supported mass is subjected to a disturbing force of varying frequency. A resonant amplitude of 12 mm is observed. Also, at very high disturbing frequencies, an almost fixed amplitude of 1.3 mm is observed. What is the damping ratio d of the system?

SOLUTION

At resonance, from Problem 8.27, $X = (me/M)/2d$; and at high frequencies, $X = me/M$. Thus, $me/M = 1.3$ mm. Substituting into the first equation,

$$X = 12 = \frac{1.3}{2d} \qquad \therefore d = 0.054$$

SUPPLEMENTARY PROBLEMS

8.29. A 5-kg mass vibrates with harmonic motion $x = X\sin\omega t$. If the amplitude X is 100 mm and the mass makes 1750 vibrations per minute, find the maximum acceleration of the mass.

Ans. $a = 3360$ m/s^2

8.30. A cylinder oscillates about a fixed axis with a frequency of 10 cycles per minute. If the motion is harmonic with an amplitude of 0.10 rad, find the maximum acceleration in rad/s^2.

Ans. $\alpha = 0.11$ rad/s^2

8.31. An instrument weighing 19.6 N is fastened to four rubber mounts, each of which is rated at 0.7135 mm deflection per newton loading. What will be the natural frequency of vibration in Hz?

Ans. $f = 8.42$ Hz

8.32. A 125-mm-diameter maple log weighing 7850 N/m^3 is 1500 mm long. If, while it is floating vertically in the water, it is displaced downward from its equilibrium position, what will be the period of oscillation?

Ans. $\tau = 6.89$ s

8.33. Determine the natural frequency of vertical vibration of a horizontal simple beam of length ℓ to which a mass m is fastened at the midpoint. Neglect the mass of the beam. Note that the deflection of m is $mg\ell^3/48EI$.

Ans. $f = (2/\pi)\sqrt{3EI/m\ell^3}$ Hz

8.34. A 100-g mass is fastened to the midpoint of a 150-mm-long vertical wire in which the tension is 15 N. What will be the period of vibration of the mass if it is displaced laterally and then released?

Ans. $\tau = 0.1$ s

8.35. A simple pendulum consists of a small bob of mass m tied to the end of a string of length ℓ. Show that, for small oscillations, the natural frequency is $(1/2\pi)\sqrt{g/\ell}$ Hz.

8.36. Refer to Fig. 8-29. Determine the natural frequency of the system composed of a mass m suspended by a spring, whose constant is k, from the end of a massless cantilever beam of length ℓ. [*Hint:* A unit force applied at m will cause it to deflect a total distance of $1/k + 1/(3EI/\ell^3)$.]

Ans. $f = \dfrac{1}{2\pi}\sqrt{\dfrac{3EIk}{m(3EI + k\ell^3)}}$ Hz

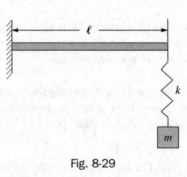

Fig. 8-29

8.37. Determine the natural frequency in Hz of the pendulum system in Fig. 8-30. The springs have constants k_1 and k_2, respectively, and are unstretched in the equilibrium position. Neglect the weight of the stiff rod. (*Hint:* The pendulum in its displaced position is acted on by a gravitational force mg and the sum of the forces in the springs.)

Ans. $f = \dfrac{1}{2\pi}\sqrt{\dfrac{mg\ell + (k_1 + k_2)h^2}{m\ell^2}}$ Hz

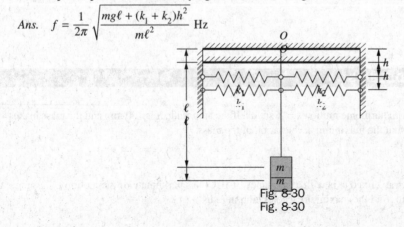

Fig. 8-30

8.38. In Fig. 8-31, the mass $m = 6$ kg. The spring modulus $k = 5000$ N/m. Neglecting the weight of the bell crank, which is a continuous solid object, determine the frequency of the system in cps.

Ans. $f = 1.62$ cps

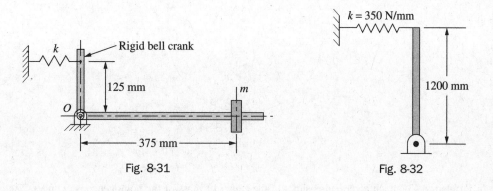

Fig. 8-31   Fig. 8-32

8.39. The slender rod shown in Fig. 8-32 has a mass of 7 kg and is 1200 mm long. The spring constant is 350 N/m. Determine the frequency for small oscillations.

Ans. 1.87 Hz

8.40. A disk with moment of inertia J_O is rigidly attached to a slender shaft (or wire) of torsional stiffness K, where K is the torque necessary to twist the shaft through 1 rad. What will be the frequency of oscillations if the shaft is twisted through a small angle and then released? See Fig. 8-33.

Ans. $f = (1/2\pi)\sqrt{K/J_O}$ Hz

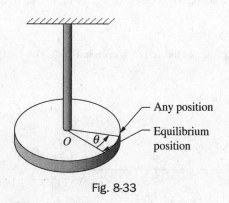

Fig. 8-33

8.41. A steel disk 100 mm in diameter and 3 mm thick is rigidly attached to a steel wire 0.8 mm in diameter and 500 mm long. What is the natural frequency of this torsional pendulum?

Ans. 0.84 Hz

8.42. An engine has a 70-kg flywheel at each end of a 50-mm-diameter steel shaft. Assuming that the equivalent length of shaft between the flywheels is 600 mm, determine the natural frequency of torsional oscillation in cps. The radius of gyration for each flywheel is 220 mm, and $G = 80$ GPa.

Ans. 35 cps

8.43. The homogeneous cylinder shown in Fig. 8-34 has a mass of 60 kg and a diameter of 1200 mm. The system is in equilibrium when the diameter as shown is horizontal. The spring, which is vertical in the phase shown, has a spring constant of $k = 2$ N/mm. Determine the frequency for small oscillations.

Ans. 1.41 cps

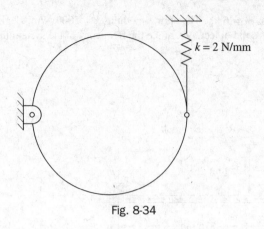

Fig. 8-34

8.44. In Problem 8.43, attach a vertical dashpot to the center of mass of the homogeneous cylinder. The cylinder is initially rotated 5° clockwise and released from rest. The damping coefficient is one-tenth of the critical damping coefficient. Find the damping coefficient and the angular displacement of the cylinder when $t = 2$ s.

Ans. $c = 17$ N·s/m, $\theta = 0.013$ rad clockwise

8.45. In Problem 8.37, replace the spring of modulus k_2 with a horizontal dashpot of damping coefficient c. Derive the expression for the critical damping coefficient.

Ans. $c_c = \dfrac{2m\ell^2}{h^2}\sqrt{\dfrac{kh^2}{m\ell^2} + \dfrac{g}{\ell}}$

8.46. A 3-kg mass is attached to a spring with constant $k = 2.5$ N/mm. Determine the critical damping coefficient.

Ans. $c_c = 173$ N·s/m

8.47. Determine the damped frequency of the system shown in Fig. 8-35.

Ans. $\omega_d = \sqrt{\dfrac{k\ell^2}{mb^2} - \dfrac{c^2}{4m^2}}$ rad/s

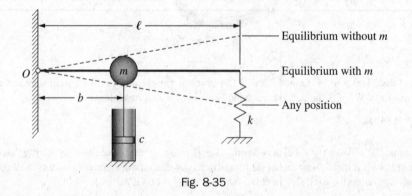

Fig. 8-35

8.48. A vibrating system consists of a 5-kg mass, a spring with constant $k = 3.5$ N/mm, and a dashpot with damping constant $c = 100$ N·s/m. Determine (*a*) the damping factor d; (*b*) the damped natural frequency ω_d; (*c*) the logarithmic decrement D; (*d*) the ratio of any two successive amplitudes.

Ans. (*a*) $d = 0.378$, (*b*) $\omega_d = 24.5$ rad/s, (*c*) $D = 2.56$, (*d*) ratio = 13.0

8.49. Derive the differential equation of motion for the spring-mounted mass m shown in Fig. 8-36. The mass is subjected to a harmonic disturbing force $F \sin \omega t$.

Ans. $\dfrac{d^2x}{dt^2} + \dfrac{(k_A + k_B)}{m}x = \dfrac{F}{m}\sin \omega t$

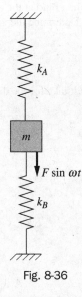

Fig. 8-36

8.50. A 90-kg machine is supported on three springs, each with a constant $k = 10$ kN/m. A harmonic disturbing force of 20 N acts on the machine. Determine (*a*) the resonant frequency and (*b*) the maximum distance the machine moves from equilibrium if the disturbing frequency is 200 cycles per minute. Assume negligible damping.

Ans. (*a*) $f = 2.91$ Hz, (*b*) 2.13 mm

8.51. A spring-supported mass is subjected to a harmonic disturbing force of varying frequency. A resonant amplitude of 20 mm is observed. At very high disturbing frequencies, an almost fixed amplitude of 1.75 mm is observed. What is the damping ratio d of the system?

Ans. $d = 0.044$

8.52. A spring-supported mass is subjected to a harmonic disturbing force of varying frequency. A resonant amplitude of 20 mm is observed. At high disturbing frequencies, an almost fixed amplitude of 2 mm is observed. What is the damping ratio of the system?

Ans. $d = 0.05$

8.53. In Problem 8.49, let $k_A = k_B = 1800$ N/m, $m = 9$ kg, $F = 54$ N, and the forcing frequency $f = 1.2$ Hz. Find the natural frequency, the magnification factor, and the maximum displacement of the mass.

Ans. $f_n = 3.18$ Hz, m.f. $= 1.16$, $x_{max} = 17.5$ mm

8.54. In Problem 8.38, a forcing function $8 \cos \omega t$ is applied vertically to the mass m. Determine the maximum frequency of the forcing function if the magnification factor is to be no more than 2.

Ans. $f = 1.15$ Hz

8.55. A 30-kg motor is supported by four springs each of modulus $k = 5250$ N/m. The system is damped by a dashpot of damping coefficient c. Determine minimum c such that oscillation does not occur.

Ans. $c_{min} = 1590$ kg/s

8.56. A small ship model of mass m is placed in a towing tank and connected to each end of the tank by springs of spring constant k. The model is displaced and allowed to oscillate. It is noted that the period of damped oscillation is $\frac{3}{4}$ s and the ratio of the amplitudes of two successive cycles is $\frac{3}{7}$. Determine the damping coefficient of the fluid in the tank.

Ans. $c = 2.27\,m$

8.57. A spherical mass of 7.5 kg is suspended in a fluid at the end of a spring of constant $k = 700$ N/m. The fluid has a known damping coefficient of 70 kg/s. Determine the displacement of the sphere from equilibrium after 3 s if the initial displacement is 150 mm.

Ans. $x = 11$ mm

8.58. In the preceding problem, how long will it take for the displacement to reach (*a*) $\frac{1}{10}$ of the initial displacement and (*b*) $\frac{1}{100}$ of the initial displacement?

Ans. (*a*) 2.6 s, (*b*) 5.3 s

SI Units

The International System of Units (abbreviated SI) has three classes of units—base, supplementary, and derived. The seven base units and two supplementary units are listed below. Also listed are derived units with and without special names as used in mechanics.

Base Units

Quantity	Unit	Symbol
length	meter	m
mass	kilogram	kg
time	second	s
electric current	ampere	A
temperature	kelvin	K
amount of substance	mole	mol
luminous intensity	candela	cd

Supplementary Units

Quantity	Unit	Symbol
plane angle	radian	rad
solid angle	steradian	sr

Derived Units with Special Names and Symbols

(Used in mechanics)

Quantity	Unit	Symbol	Formula
force	newton	N	$kg \cdot m/s^2$
frequency	hertz	Hz	s^{-1}
energy, work	joule	J	$N \cdot m$
power	watt	W	J/s
stress, pressure	pascal	Pa	N/m^2

Derived Units without Special Names

(Used in mechanics)

Quantity	Unit	Symbol
acceleration	meter per second squared	m/s^2
angular acceleration	radian per second squared	rad/s^2
angular velocity	radian per second	rad/s
area	square meter	m^2
density, mass	kilogram per cubic meter	kg/m^3
moment of force	newton meter	$N \cdot m$
velocity	meter per second	m/s
volume	cubic meter	m^3

SI Prefixes

(Commonly used in mechanics)

Multiplication factor	Prefix	Symbol
$1\ 000\ 000\ 000 = 10^9$	giga	G
$1\ 000\ 000 = 10^6$	mega	M
$1000 = 10^3$	kilo	k
$0.001 = 10^{-3}$	milli	m
$0.000\ 001 = 10^{-6}$	micro	μ
0.01	centi*	c

*discouraged except in cm, cm^2, and cm^3

Conversion Factors

To convert from	to	multiply by
degree (angle)	radian (rad)	1.745 329 E − 02*
foot	meter (m)	3.048 000 E − 01
ft/min	meter per second (m/s)	5.080 000 E − 03
ft/s	meter per second (m/s)	3.048 000 E − 01
ft/s^2	meter per second^2 (m/s^2)	3.048 000 E − 01
ft·lbf	joule (J)	1.355 818 E + 00
ft·lbf/s	watt (W)	1.355 818 E + 00
horsepower	watt (W)	7.456 999 E + 02
inch	meter (m)	2.540 000 E − 02
km/h	meter per second (m/s)	2.777 778 E − 01
kW·h	joule (J)	3.600 000 E + 06
kip (1000 lb)	newton (N)	4.448 222 E + 03
liter	meter^3 (m^3)	1.000 000 E − 03
mile (international)	meter (m)	1.609 344 E + 03
mile (U.S. survey)	meter (m)	1.609 347 E + 03
mi/b (international)	meter per second (m/s)	4.470 400 E − 01
ounce-force	newton (N)	2.780 139 E − 01
ozl·in	newton meter (N·m)	7.061 552 E − 03
pound (lb avoirdupois)	kilogram (kg)	4.535 924 E − 01
slug·ft^2 (moment of inertia)	kilogram meter^2 (kg·m^2)	4.214 011 E − 02
lb/ft^2	kilogram per meter^2 (kg·m^2)	1.601 846 E + 01
pound force (lbf)	newton (N)	4.448 222 E + 00
lbf·ft	newton meter (N·m)	1.335 818 E + 00
lbf·in	newton meter (N·m)	1.139 848 E − 01
lbf/ft	newton per meter (N/m)	1.459 390 E + 01
lbf/ft^2	pascal (Pa)	4.788 026 E + 01
lbf/in	newton per meter (N/m)	1.751 268 E + 02
lbf/in^2 (psi)	pascal (Pa)	6.894 575 E + 03
slug	kilogram (kg)	1.459 390 E + 01
slug/ft^2	kilogram per meter^2 (kg/m^2)	5.153 788 E + 02
ton (2000 lb)	kilogram (kg)	9.071 847 E + 02
W·h	joule (J)	3.600 000 E + 03

*E − 02 means multiply by 10^{-2}

Second Moments of Areas and Mass Moments of Inertia

Second Moments of Areas	Mass Moments of Inertia

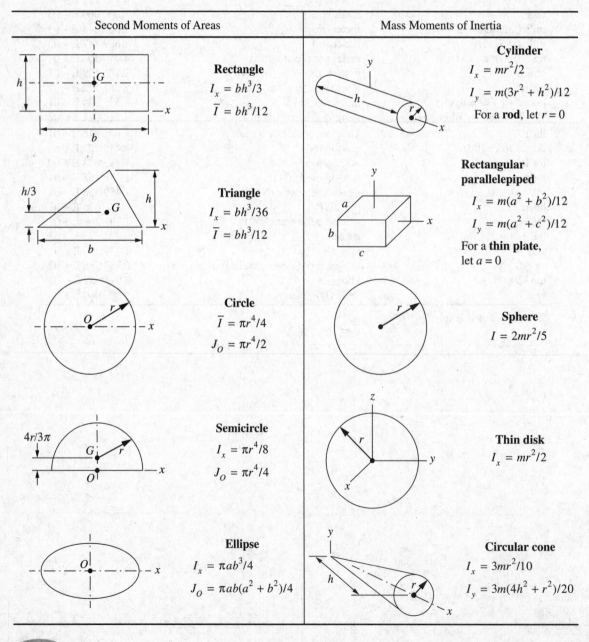

Second Moments of Areas

Rectangle
$$I_x = bh^3/3$$
$$\bar{I} = bh^3/12$$

Triangle
$$I_x = bh^3/36$$
$$\bar{I} = bh^3/12$$

Circle
$$\bar{I} = \pi r^4/4$$
$$J_O = \pi r^4/2$$

Semicircle
$$I_x = \pi r^4/8$$
$$J_O = \pi r^4/4$$

Ellipse
$$I_x = \pi ab^3/4$$
$$J_O = \pi ab(a^2 + b^2)/4$$

Mass Moments of Inertia

Cylinder
$$I_x = mr^2/2$$
$$I_y = m(3r^2 + h^2)/12$$
For a **rod**, let $r = 0$

Rectangular parallelepiped
$$I_x = m(a^2 + b^2)/12$$
$$I_y = m(a^2 + c^2)/12$$
For a **thin plate**, let $a = 0$

Sphere
$$I = 2mr^2/5$$

Thin disk
$$I_x = mr^2/2$$

Circular cone
$$I_x = 3mr^2/10$$
$$I_y = 3m(4h^2 + r^2)/20$$

Index